STARLAB

An Orbiting Space Technology Applications & Research Laboratory

FINAL REPORT
No. CR-61296

Prepared Under

CONTRACT NSR 01-003-025
UNIVERSITY AFFAIRS OFFICE HEADQUARTERS
NATIONAL AERONAUTICS AND SPACE ADMINISTRATION

with the cooperation of

THE AMERICAN SOCIETY FOR ENGINEERING EDUCATION

and

PROGRAM DEVELOPMENT
GEORGE C. MARSHALL SPACE FLIGHT CENTER
NATIONAL AERONAUTICS AND SPACE ADMINISTRATION

by

AUBURN UNIVERSITY ENGINEERING SYSTEMS DESIGN
SUMMER FACULTY FELLOWS

SEPTEMBER 1969

APPROVED:

Reginald I. Vachon
Alumni Professor, Auburn Univ.
Director

Malcolm A. Cutchins
Associate Professor, Auburn Univ.
Associate Director

J. Fred O'Brien
Associate Director, Engineering
Extension Service, Auburn Univ.
Administrative Director

Herman G. Hamby
MSFC Program Development
Co-Director

Jim E. Cox
Associate Professor, Univ. of Houston
Associate Director

William R. Payne
MSFC Program Development
Administrative Liaison

ABSTRACT

STARLAB is a 35-man orbiting Space Technology Applications and Research Laboratory preliminary design and a training exercise in the systems approach. The selected orbit is 308 n. mi. with a 69-degree inclination which enables coverage of 93 percent of the earth's surface. The onboard experiments utilize research and application laboratories with auxiliary supporting laboratories. A laser communication system with a 3000 megabits per second data rate, three synchronous communication satellites, and an earth-based administration function comprises the information management system.

Emphasis is given to earth resources, chemistry, life/behavioral sciences, physics, materials and manufacturing, communication, navigation and traffic control, manned space flight engineering, and astronomy.

The design is based on the existence of the shuttle with a 50 000-pound payload capability. Free-flying modules for selected experiments augment the 430-foot-long nuclear-powered STARLAB. Existing Saturn V capabilities are used for launch.

The abstract presents the essential elements of STARLAB. The success of the training exercise in Systems Engineering is perhaps best conveyed by the following unsolicited comment given at the close of the final presentation by one of the participants at Morris Auditorium, MSFC, 20 August 1969:

"The design of STARLAB has brought each of us closer to the space program; we now feel more a part of the space effort. In the course of this design we have learned a great deal about NASA, its organization, and its facilities. However, the most important goal for this summer was to learn about the Systems Approach. We have learned about the Systems Approach on the least two levels; personal and group.

At the beginning of the program each of us wondered how we would accomplish a good design. We worried a little. After a series of up-to-date briefings and many personal contacts with the excellent people here at MSFC, we began to feel like we could begin the design. Now at this time we discovered the real strength of the Systems Approach. In this approach a person is no longer an isolated unit trying to self-produce and stimulate himself in a solution of the problem. In a Systems Approach, a group of N persons delivers at least (N - 1) new perspectives about the problem to each other. With a certain assumed knowledge of the material, the Systems Approach enhances the productivity by multistimulation of synthetic thinking. This produces ideas!

The Systems Approach affected many of us by:

1. Broadening our horizons to interdisciplinary approaches.

2. Making us better prepared for the future.

3. Enabling us to relate more closely to students.

4. Giving our research new goals.

5. Showing us a bigger picture of science and its applications.

6. Teaching us to be better listeners.

7. Further developing our individual skills.

The Systems Approach affected us not only as individuals but as a group. All 21 of us are firmly convinced that the Systems Approach is a method of solving some of the big problems facing our society today. We felt that man-focused areas of study should be emphasized.

We have had a very instructive lesson in the Systems Approach. A lesson in design by execution of a design.

The summer Fellows of the Auburn Design Program would like at this time to express their appreciation to the Marshall Space Flight Center for helping us in this design exercise. We of course also wish to thank the Auburn-Alabama University Staff for managing an excellent program. Furthermore, we men and our families greatly appreciate the opportunity of having been at MSFC this summer during the historic flight of Apollo 11 and its mission to the moon. All of us sincerely wish that a STARLAB will someday be orbiting the earth."

PREFACE

Objective

Systems engineering, or the systems approach, is an accepted term to describe the multidisciplinary or interdisciplinary character of the "systematic design" of any large engineering system. It emphasizes and attempts to systematize, through the availability of modern techniques, the design of modern, complex, and multidisciplinary engineering systems. The term seems to have originated in the aerospace field where the complexity of modern aerospace systems demanded a systematically controlled approach to design to ensure that all factors of all subsystems, representing many disciplines, were carefully integrated into the final system. The importance of the multidisciplinary systems approach has been recognized by the National Aeronautics and Space Administration to the extent that NASA, in conjunction with the American Society for Engineering Education, conduct Systems Engineering Faculty Fellowship Programs as well as Research Faculty Fellowship Programs at NASA Centers with local universities during the summer of 1969. There were eight research-oriented programs and four Systems Engineering design programs. The Faculty Fellows were selected from applicants from throughout the nation; they either were located in a laboratory to conduct research on an individual basis, or they participated as a group to learn the system approach through a design problem. The Marshall Space Flight Center, Auburn University, and the University of Alabama conducted both a research program and a Systems Engineering design program. Other centers and universities conducting design programs were:

Ames Research Center — Stanford University

Langley Research Center — Old Dominion College

Manned Spacecraft Center — University of Houston

The purpose of the design programs is to develop a systems approach philosophy by group participation in a design training experience. The resulting design is of secondary importance to the training aspect of the program. It is hoped the faculty will use the experience to develop the systems approach in courses at their home institutions as well as use the experience to lead others in system solutions of complex multidisciplinary problems. This approach, though discussed considerably in the news media, has had little use exterior to the space program. The approach has application in solving engineering as well as socioeconomic problems.

The training exercise selected for study by the 1969 MSFC Fellows has centered on the theme of a space station laboratory. The participants were exposed to a number of MSFC and industry speakers during the first weeks of the 11-week program. A tour of the Manned Spacecraft Center, Michoud Assembly Facility, and the Kennedy Space Center was conducted at the end of the second week.

The participants established the requirements for the design-generated alternative approaches, and then through trade-offs established the present concept. The key word of the project was to avoid prejudging solutions. The result of the program, an orbiting Space Technology Applications and Research Laboratory (STARLAB), is a subpart of STARSYSTEM, an information management and application system for technology.

The enthusiasm of the participants to learn the systems approach is reflected in the accomplishment in their project. This enthusiasm should carry them through to the applications of the system approach to national, local, and world problems of a complex multidisciplinary nature. This report summarizes the study made by these 21 faculty members representing 17 colleges throughout the United States and one in Canada.

Organization

The chart on the following page depicts the basic organizational arrangement for the Systems Engineering Summer Faculty Fellowship Program. The participants elected a project leader and group leaders three times during the 11-week period. A listing of the participants, their positions, and their academic degrees also follows. Subgroups to the three groups shown were formed as needed.

The first phase of the program was to subject the participants to as many germane seminars or presentations to an orbiting space station as could practicably be scheduled and to obtain as many publications related to the problem as could be gathered and used. From this, contacts within Marshall and other areas became apparent for further perusal, and a base of knowledge necessary to design a space laboratory was established. The tour of some of the major NASA centers mentioned previously provided the "feel" for the size of the space program that could come in no other way.

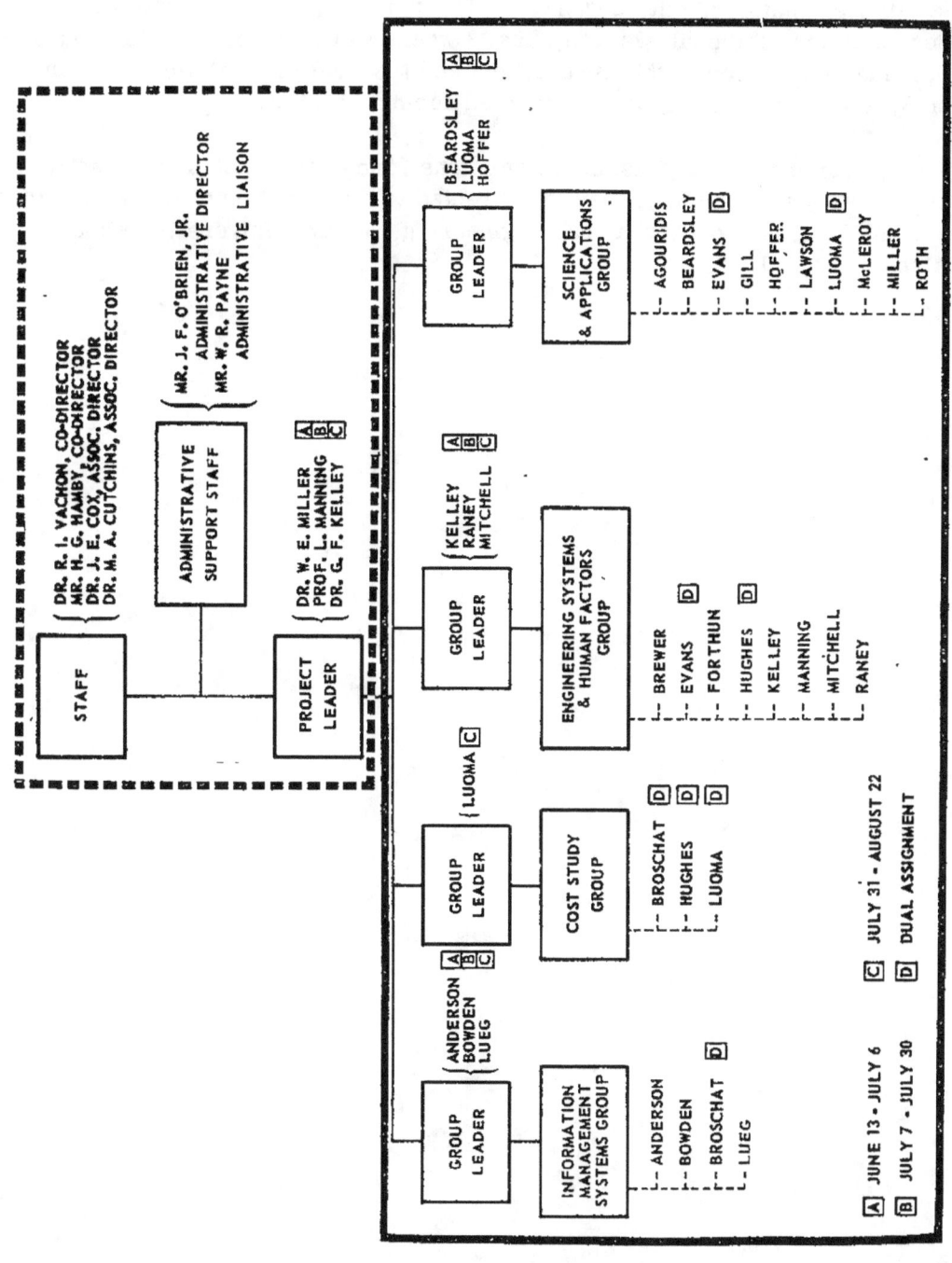

vii

After approximately 3 weeks in this first phase, the next 5 weeks were spent evaluating alternative solutions to the problems as they developed, having fewer presentations, effecting trade-offs, and accomplishing preliminary design studies and analysis. The last 3 weeks were occupied with final trade-offs and preparation of this final report and the oral presentation. Seminars on Systems Engineering were conducted during the 11-week period.

Comments, criticisms, suggestions, and questions will be answered as promptly as possible if they are directed to the attention of Dr. R. I. Vachon, Alumni Professor of Mechanical Engineering, Auburn University, Auburn, Alabama 36830.

LIST OF PARTICIPANTS

Summer Faculty Fellows

Dimitrios C. Agouridis
Assoc. Prof. of Elect. Eng.
Memphis State University

Ph.D. (EE), Univ. of Minnesota, 1966
M.S. (EE), Univ. of Minnesota, 1961
B.S. (EE), Univ. of Minnesota, 1959

Ordean S. Anderson
Assoc. Prof. of Elect. Eng.
North Dakota State University

Ph.D. (EE), Univ. of Wisconsin, 1968
M.S. (EE), Univ. of North Dakota, 1963
B.S. (EE), Univ. of North Dakota, 1961

L. Robert Beardsley
Assoc. Prof. of Mech. Eng.
University of Mississippi

M.S. (Nuclear Eng.), Univ. of Mississippi, 1963
B.S. (EE), Illinois Inst. of Tech., 1947
B.S. (ME), Illinois Inst. of Tech., 1944

Warren W. Bowden
Prof. Chemical Eng.
Rose Polytechnic Institute

Ph.D. (Chem. Eng.) Purdue Univ., 1965
M.S. (Chem. Eng.), Rose Polytechnic Inst., 1959
B.S. (Chem. Eng.), Univ. of Maine, 1949

William V. Brewer
Asst. Prof. of Mech. Eng.
University of Tulsa

Ph.D. (ME), Michigan State Univ., 1968
M.S. (ME), Michigan State Univ., 1963
B.S. (ME), Michigan State Univ., 1961

Robert A. Broschat
Instructor of Mathematics
South Dakota State University

M.S. (Physics), Univ. of Wisconsin-Milwaukee, 1966
M.S. (Math), North Dakota State Uni.., 1962
B.S. (Math), Valley City State, N. Dakota, 1960

Kenneth J. Evans
Assoc. Prof. of Biological Sc.
Chico State College

Ph.D. (Zool.), Univ. of California, Riverside, 1964
M.S. (Zool.), Univ. of Oklahoma, 1958
B.S. (Zool.), Univ. of Oklahoma, 1957

Melvin L. Forthun
Assoc. Prof. of Mech. Eng.
North Dakota State University

M.S. (ME), North Dakota State Univ., 1961
B.S. (ME), North Dakota State Univ., 1951

LIST OF PARTICIPANTS (Continued)

Joseph H. Gill
Assoc. Prof. of Mech. Eng.
Western Michigan University
    M.A. (Bus. Ad.); Michigan State Univ., 1960
    B.S. (ME), Tri-State, 1942

Jerry M. Hoffer
Assoc. Prof. of Geology
University of Texas, El Paso
    Ph.D. (Geology), Washington State Univ., 1965
    M.S. (Geology), The State Univ. of Iowa, 1958
    B.A. (Geology), The State Univ. of Iowa, 1956

Harold A. Hughes
Asst. Prof. of Agr. Eng.
University of Guelph (Canada)
    M.S. (Ag. Eng.), Univ. of Arizona, 1966
    B.S. (Ag. Eng.), Michigan State Univ., 1964

Gilbert F. Kelley
Asst. Prof. of Astro. &
Computer Science
USAF Academy
    Ph.D. (AE), Univ. of Colorado, 1968
    M.S. (AE), Univ. of Colorado, 1962
    B.S. (AE), St. Louis Univ., 1960

Edward R. Lawson
Instructor of Indust. Eng.
Purdue University
    M.S. (Bus. Ad.), Rensselaer Polytech. Inst., 1966
    B.S. (ME), Univ. of Rhode Island, 1963

Russell E. Lueg
Prof. of Elect. Eng.
University of Alabama
    Ph.D. (EE), Univ. of Texas, 1961
    M.S. (EE), Univ. of Texas, 1956
    B.S. (EE), Univ. of Arkansas, 1951

John R. Luoma
Asst. Prof. of Chemistry
North Dakota State University
    Ph.D. (Chemistry), Purdue Univ., 1966
    B.S. (Math), Ohio Univ., 1961
    B.A. (Chemistry), Ohio Univ., 1961

Lindley Manning
Asst. Prof. of Mech. Eng.
University of Nevada
    M.S. (ME), Univ. of Nevada, 1966
    B.S. (ME), Univ. of Cincinnati, 1958

Donald F. McLeroy*
Asst. Prof. of Geology
Lehigh University
    Ph.D. (Geology), Stanford Univ., 1966
    M.S. (Geology), Univ. of New Mexico, 1962
    B.S. (Geology), Univ. of New Mexico, 1960

---

*Participants in the Alabama-Auburn Summer Faculty Research Program — assigned to Program Development.

LIST OF PARTICIPANTS (Continued)

Walter E. Miller
Prof. & Chairman of Gen. Chem.
City College of New York

Ph.D. (Chemistry), New York Univ., 1941
M.S. (Chem. Eng.), City College of New York, 1936
B.S. (Chem. Eng.), City College of New York, 1935

William S. Mitchell
Assoc. Prof. of Mech. Eng.
Tennessee Technological University

Ph.D. (ME), Univ. of Houston, 1968
M.S. (ME), Univ. of Houston, 1965
B.S. (ME), Univ. of Houston, 1963

Donald C. Raney
Assoc. Prof. of Mech. Sys. Eng.
University of Alabama

Ph.D. (EM), Virginia Polytechnic Inst., 1967
M.S. (ME), Auburn Univ., 1960
B.S. (ME), Univ. of Kentucky, 1954

Thomas A. Roth
Asst. Prof. of Indust. Eng.
Kansas State University

Ph.D. (Met. E.), Univ. of Wisconsin, 1967
M.S. (Met. E.), Univ. of Wisconsin, 1961
B.S. (Met. E.), Univ. of Wisconsin, 1960

LIST OF PARTICIPANTS (Concluded)

Technical Staff

Jim E. Cox
Assoc. Prof. of Mech. Eng.
University of Houston

Ph.D. (ME), Oklahoma State, 1963
M.S. (ME), S.M.U., 1960
B.S. (ME), S.M.U., 1958

Malcolm A. Cutchins
Assoc. Prof. of Aerospace Eng.
Auburn University

Ph.D. (EM), Virginia Polytechnic Inst., 1967
M.S. (EM), Virginia Polytechnic Inst., 1964
B.S. (CE), Virginia Polytechnic Inst., 1956

Herman G. Hamby
Staff Scientist
Program Development - MSFC

M.S. (Physics), Univ. of Georgia, 1961
B.S. (Physics), Univ. of Georgia, 1950

Reginald I. Vachon
Alumni Prof. of Mech. Eng.
Auburn University

L.L.B., Jones Law School, 1969
Ph.D. (ME), Oklahoma State, 1963
M.S. (Nuclear Sc.), Auburn Univ., 1960
B.S. (ME), Auburn Univ., 1958

ACKNOWLEDGMENTS

The successful completion of Project STARLAB would have been impossible without the enthusiastic cooperation of the offices and personnel of Marshall Space Flight Center, some of their contractors, and other individuals. It is a distinct pleasure to acknowledge their efforts. It would not be possible to give recognition to each one; however, we have made an attempt to list speakers and certain personnel who have been most instrumental in the success of the project. These are listed on the following pages.

We are particularly indebted to Dr. Werhner von Braun, Director of MSFC; Dr. William Lucas, Director of Program Development; and Col. E. D. Mohlere, Director of University Affairs, and their fine staffs. For doing so much of the liaison work, making speaker arrangements, and in general doing much of the coordination between NASA and the design group. Mr. Herman Hamby of Program Development deserves our particular appreciation. Mr. William R. Payne of MSFC has been very instrumental in helping with many support areas, including much of this final report work.

Mrs. J. M. Miller and Mr. E. W. Harper of MSFC have been invaluable in their cooperation in supporting functions. The tours and arrangements made through the Protocol and Transportation Offices have been very valuable. Specifically, thanks are due to Mr. E. L. Riddick, Mr. E. S. Schorsten, and Cmdr. W. K. Martin.

The assistance of Mr. J. F. Dowdy, Chief of the Training Branch of the Manpower Office, and Mr. C. M. Hightower is certainly appreciated. The help of Mr. Byron Davis of Communication Skills, Inc., is gratefully acknowledged.

The Office of University Affairs, NASA Headquarters, Washington, D. C., funded the project, and certainly the engineering educational world is indebted for such foresighted assistance to education in this country.

Finally, the secretarial assistance of some of the MSFC administrative assistants and secretaries as well as our own has been greatly appreciated. Mrs. Bonnie Holmes, Mrs. Gertrude Conard, and Mrs. Jerry Wright of MSFC, and Mrs. Lyn Newman, Mrs. Betty Whatley, and Mrs. Sandra Page of our staff comprise this list of hard workers.

Others to be acknowledged for their assistance on this project are given in the following list.

R. E. Allen (MSFC)*

D. A. Barnes (McDonnell-Douglas)*

T. A. Barr (MSFC)*

J. A. Bethay (MSFC)*

L. W. Brantley (MSFC)*

T. Carey (MSFC)*

R. D. Crawford (MSFC)*

J. A. Downey (MSFC)*

J. B. Dozier (MSFC)*

R. Fary (U.S. Geological Survey)*

C. Hamilton (MSFC)*

J. D. Hilchey (MSFC)*

W. R. Lucas (MSFC)*

E. C. McKannan (MSFC)*

J. R. Olivier (MSFC)*

A. G. Orillion (MSFC)*

A. Siepert (KSC)*

H. Sohn (Lockheed-Georgia)*

W. G. Stroud (Goddard)*

E. Stuhlinger (MSFC)*

G. von Tiesenhausen (MSFC)*

F. L. Vinz (MSFC)*

D. Waltz (TRW)*

H. H. Watters (MSFC)*

D. K. Weidner (MSFC)*

F. L. Williams (MSFC)*

H. F. Wuenscher (MSFC)*

A. C. Young (MSFC)*

W. Weisman (MSFC)*

L. Balleau (MSFC)

J. Bensko (MSFC)

B. Benson (MSFC)

J. F. Blumrich (MSFC)

D. Cagle (MSFC)

D. Carlile (MSFC)

B. Chandler (MSFC)

J. C. Cody (MSFC)

D. Cramblit (MSFC)

J. Curry (MSFC)

K. Dannenberg (MSFC)

B. G. Davis (MSFC)

J. Dawson (MSFC)

*Denotes speakers during the program.

P. Dreher (MSFC)

J. Earl (MSFC)

E. E. Engler (MSFC)

J. Evers (MSFC)

B. Goldsby (MSFC)

J. Hethcoat (MSFC)

M. Kent (MSFC)

A. Kruprick (MSFC)

Mrs. L. Luna (MSFC)

R. McFolin (MSFC)

A. Madyda (MSFC)

B. Narrow (Goddard)

E. Nathan (MSFC)

R. J. Naumann (MSFC)

M. Nein (MSFC)

C. Nevins (MSFC)

J. R. Pierce (Bell Labs)

W. P. Prasthofer (MSFC)

T. M. Ragland (Goddard)

D. Raney (Northrop)

J. Randall (MSFC)

J. Rasquin (MSFC)

B. Reed (MSFC)

M. Rheinfurth (MSFC)

W. T. Roberts (MSFC)

B. Robertson (MSFC)

Mrs. L. M. Robertson (MSFC)

W. J. Robinson, Jr. (MSFC)

R. D. Scott (MSFC)

Mrs. L. Smith (MSFC)

G. Solomon (MSFC)

L. Spears (MSFC)

J. Stokes (MSFC)

J. Swenson (MSFC)

H. Thomason (MSFC)

H. Vaughn (MSFC)

W. Webb (MSFC)

C. Wende (Goddard)

W. Whitacre (MSFC)

J. Williams (MSFC)

C. Wood (MSFC)

W. Wood (MSFC)

H. E. Worley (MSFC)

C. Wyman (MSFC)

R. Beaudry (Capt.) (Arnold Air Force Station, Tennessee)

C. Fiquett (Wyle Research Labs, Huntsville)

B. O. Montgomery (Col.) (MSFC Resident Officer, KSC)

J. L. Stamy (MAF)

J. L. Youngblood (MSC)

TABLE OF CONTENTS

	Page
ABSTRACT	iii
PREFACE	v
LISTS OF PARTICIPANTS AND STAFF	ix
ACKNOWLEDGMENTS	xii

CHAPTER I.	INTRODUCTION	1- 1
	Mission Objective	1- 2
	Mission Constraints	1- 2
	Orbiting Space Technology Applications and Research Laboratory (STARLAB)	1- 3
	Space Technology Applications and Research Administration (STARAD)	1- 4
	Starsystem	1- 4
	Scientific Mission	1- 4
	Summary	1- 4
CHAPTER II.	SCIENTIFIC MISSION	2- 1
	DEFINITION OF SYMBOLS	2- 1
	INTRODUCTION	2- 2
	STARLAB: Its Design and Rationale	2- 2
	The Experimental Laboratory and Experimental Program	2- 4
	Design of the Laboratory	2- 5
	Integration and Management	2- 8
	Summary of Philosophy	2- 8
	RESEARCH LABORATORIES	2-10
	Purpose and Philosophy	2-10
	Areas for Research and Applications	2-11

TABLE OF CONTENTS (Continued)

		Page
	SUPPORT AREAS.	2- 94
	Purpose and Philosophy.	2- 94
	INTEGRATION.	2-103
	Time Schedule.	2-103
	Contamination Control.	2-104
	Safety.	2-106
	Automatic Checkout of Experiments.	2-107
	Reliability.	2-108
	Maintainability and Resupply	2-108
	Commonality.	2-108
	Time Line.	2-109
	Special Equipment and Instruments	2-109
	ADMINISTRATIVE NEEDS OF STARLAB.	2-109
	REFERENCES.	2-114
	BIBLIOGRAPHY.	2-121
CHAPTER III.	INFORMATION MANAGEMENT	3- 1
	INTRODUCTION.	3- 1
	Communication Center	3- 2
	Internal Communication.	3- 2
	Onboard Computer	3- 3
	STARLAB to CIMC (External Communications).	3- 4
	Ground Stations	3- 5
	Processing of Experimental Data	3- 5
	Summary	3- 8
	SYSTEM FUNCTIONAL CAPABILITIES	3- 11
	Introduction.	3- 11
	EQUIPMENT DETAILS	3- 34
	Introduction.	3- 34

TABLE OF CONTENTS (Continued)

		Page
	REFERENCES	3-43
	BIBLIOGRAPHY	3-44
CHAPTER IV.	MISSION AND ANALYSIS	4- 1
	DEFINITION OF SYMBOLS	4- 1
	MISSION DESCRIPTION	4- 2
	MISSION PROFILE	4- 4
	THE ASCENT TRAJECTORY	4- 6
	ORBITAL TRAJECTORY	4- 6
	REFERENCES	4-13
	BIBLIOGRAPHY	4-14
CHAPTER V.	STARLAB VEHICLE AND SUBSYSTEMS	5- 1
	DEFINITION OF SYMBOLS	5- 1
	PHYSICAL ARRANGEMENT	5- 4
	Introduction	5- 4
	Launch Package Modules	5- 4
	Launch Package Module Integration	5- 5
	Description of Modules	5- 5
	Docking and Resupply	5-10
	Imposed Structural Loads	5-11
	Load Bearing Structure	5-12
	Materials	5-15
	Micrometeoroid Protection	5-15

TABLE OF CONTENTS (Continued)

	Page
POWER SYSTEMS	5-18
Overview and Cost Summary	5-18
Introduction	5-18
Mission Power Requirements	5-20
Available Power Systems	5-20
Summary of Power Systems	5-29
Description of the Solar Array/Battery System	5-32
Reactor Power System	5-34
Power Distribution System	5-39
Onboard Communications	5-42
NAVIGATION, GUIDANCE, AND CONTROL	5-43
Navigation	5-43
Guidance and Control	5-44
Vehicle Orientation	5-44
Disturbance Torques	5-48
Momentum Exchange Devices	5-48
Reaction Control System (RCS)	5-55
Control Considerations	5-56
Service Propulsion System (SPS)	5-56
Proposed Vehicle Orientation	5-58
Proposed Attitude Control System	5-59
ENVIRONMENTAL CONTROL AND HABITABILITY	5-61
Introduction	5-61
Laboratory Atmosphere	5-62
Food and Water	5-64
Gravity	5-66
Extravehicular Activity (EVA)	5-70
Living Quarters	5-70
Waste Management and Personal Hygiene	5-73
Movement and Restraints	5-74
Safety	5-74
Crew	5-76
Logistics	5-77

TABLE OF CONTENTS (Concluded)

		Page
	REFERENCES	5-82
	BIBLIOGRAPHY	5-85
CHAPTER VI.	ECONOMICS AND COST OF STARLAB	6- 1
	INTRODUCTION	6- 1
	APPROACH	6- 1
	HOW MUCH WOULD STARLAB COST?	6- 2
	WHAT PART OF NASA'S BUDGET WILL STARLAB CONSUME?	6- 4
	ALTERNATIVE COSTS OF STARLAB	6-10
	RECOMMENDATIONS	6-11
	CAN WE AFFORD STARLAB?	6-15
APPENDIX A.	INVENTION DISCLOSURES	A- 1
APPENDIX B.	COMMUNICATION SUBSYSTEM AND CIRCUIT STUDIES	B- 1
APPENDIX C.	ALTERNATIVE CONSIDERATIONS	C- 1

LIST OF ILLUSTRATIONS

Figure	Title	Page
II- 1.	Schematic of Automated Experiment Package	2- 21
II- 2.	Earth Resources Use of Electromagnetic Spectrum	2- 28
II- 3.	Electromagnetic Sensors for Site Selection	2- 31
II- 4.	Earth Resources Console	2- 37
II- 5.	Linear Ground Resolution Versus Data Rate	2- 40
II- 6.	Floor Plan for Materials Science and Manufacturing Processes Laboratory	2- 52
II- 7.	Flow Diagram of Operation of Materials Science and Manufacturing Processes Laboratory	2- 53
II- 8.	Physics Laboratory Design	2- 62
II- 9.	Cosmic Ray Laboratory Module	2- 63
II-10a.	Chemistry Laboratory (Floor Plan)	2- 70
II-10b.	Chemistry Laboratory (Side View)	2- 71
II-10c.	Artist's Conception of Chemistry Laboratory	2- 72
II-11.	Handling Container	2- 75
II-12.	Pycnometer	2- 77
II-13.	Atomizer	2- 77
II-14.	Conceptual View of a Typical Support Laboratory	2- 97
II-15.	Heat Transfer and Thermodynamic Laboratories	2-102
II-16.	Interactions of STARAD	2-110

LIST OF ILLUSTRATIONS (Continued)

Figure	Title	Page
II-17.	Internal Functions of STARAD	2-112
III-1.	Research Information Flow Diagram	3-6
III-2.	The Communication System of the Information Management Network	3-9
III-3.	A 3000-Megabit per Second Laser Transmitter	3-13
III-4.	The Multiprocessor Computer	3-19
III-5(a).	Sensor-Video-Voice Input Units	3-22
III-5(b).	Channel Selection and Information Output Units	3-23
III-6.	Information Flow	3-27
III-7(a).	Laboratory Area Data Acquisition and Display	3-28
III-7(b).	Crew Quarters Data Acquisition and Display	3-29
III-7(c).	Offboard Data Acquisition and Display Manned RFM	3-30
III-7(d).	Offboard Communications STARLAB-Shuttle Link	3-31
III-8.	The Commander's Display and Control Console	3-35
III-9.	Frequency Allocation on TSP	3-37
III-10.	Frequency Allocations on Critical S-Band Channels	3-38
III-11(a).	Communications Link Parameters	3-41
III-11(b).	Laser Communication System	3-42
IV-1.	Relative Trajectory of Free-Flying Module	4-3
IV-2.	Exploded View of Two-Stage Saturn V and Payload	4-5

LIST OF ILLUSTRATIONS (Continued)

Figure	Title	Page
IV-3.	The Ascent Trajectory Ground Track	4-8
IV-4.	Star-Track, First Three Orbits	4-11
IV-5.	Star-Track, First Day's Orbits	4-12
V-1(a).	Physical Arrangement of STARLAB Vehicle	5-6
V-1(b).	Physical Arrangement of STARLAB Vehicle	5-7
V-2.	Modular Arrangement of the STARLAB Vehicle	5-8
V-3.	Typical STARLAB Module	5-13
V-4.	STARLAB Skin Structure	5-16
V-5.	Block Diagram of a Typical Spacecraft Power Supply (After Scull)	5-19
V-6.	Area Requirements for Solar Cell Arrays	5-25
V-7.	Area Requirements for Nuclear Heat Source Radiators	5-30
V-8.	Nuclear Reactor 4π Shield for a 25 kW_e Power System	5-36
V-9.	Block Diagram of SNAP-8/T.E./Organic-Rankine Power System Rated at 23 Percent Efficiency Overall	5-38
V-10.	Power Distribution System	5-40
V-11(a).	Voltage Regulator for the STARLAB Power System	5-41
V-11(b).	Booster Regulator for the STARLAB Power System	5-41
V-12.	Earth Fixed Orientation Modes	5-45

LIST OF ILLUSTRATIONS (Continued)

Figure	Title	Page
V-13.	Solar Fixed Orientation Modes	5-46
V-14.	Standard Control System Block Diagram for CMG System	5-51
V-15.	Standard Control System Block Diagram for the RCS System	5-57
V-16.	Integrated Water Management System Revised Schematic	5-65
V-17.	Preliminary Tolerable Limits for Acceptable Human Performance in Rotating Space Systems	5-68
V-18.	Floor Plan of Living Areas	5-71
V-19.	Interior View — Crew Quarters	5-72
V-20.	Laboratory Vehicle Attitude During Periods of High-Intensity Solar Radiation	5-75
V-21.	STARLAB Manpower Buildup Versus Shuttle Launch Period	5-78
VI- 1.	Probable Trillion Dollar Economy by 1971	6-10
A- 1.	Hollow Sphere	A- 4
A- 2.	Nesting Concepts "Dixie Modules"	A- 8
A- 3.	Nested Module Assembly	A- 9
A- 4.	"Drinking Barrel" Concept for Space Structure	A-12
A- 5.	Rhombic Units for Constructing a Variety of Structures	A-16

LIST OF ILLUSTRATIONS (Concluded)

Figure	Title	Page
A-6.	Sketch of Hook Location on Space Shoes, Top View and Side View	A-19
A-7.	Flexible Coupling for Docking Two Vehicles in Space	A-22
C-1.	Alternative Considerations — Deployment	C- 2
C-2.	Alternative Considerations — Launch	C- 3
C-3.	Close-Pack Parallel Installation of Shuttle Payloads	C- 5
C-4.	Deployment Scheme — "Drinking Cup" Concept	C- 7
C-5.	Details of Deployment Scheme for "Drinking Cup" Concept	C- 8
C-6.	Power Module	C-10

LIST OF TABLES

Table	Title	Page
II- 1.	Research Laboratory Requirements	2- 6
II- 2.	Support Area Requirements	2- 7
II- 3.	Module and MDA Requirements	2- 7
II- 4.	Design Requirement Summary — All Laboratories	2- 8
II- 5.	Astronomy Lab Requirements	2- 13
II- 6.	Electromagnetic Spectrum for Remote Sensing	2- 29
II- 7.	Sensors and Applications in Earth Resources Program	2- 30
II- 8.	Remote Sensor Comparison	2- 32
II- 9.	Manned Earth Resources Program	2- 35
II-10.	Natural Resource Applications Grouped by Resolution Requirements	2- 39
II-11.	Summary of Possible Events in the Earth Resources Survey Program	2- 41
II-12.	Physical Requirements for a Typical Candidate Materials Science Experiment	2- 58
II-13.	Equipment Requirements — Physics Lab	2- 61
II-14.	Manned Requirements — Physics Lab	2- 64
II-15.	Chemical Sales and the Gross National Product (GNP)	2- 68
II-16.	Molecular Beam Equipment and Power Requirements	2- 89
II-17.	Analytical Laboratory Physical Requirements	2- 96
II-18.	Estimated Requirements of Mechanical Shop	2-100

LIST OF TABLES (Continued)

Table	Title	Page
II-19.	Estimated Requirements of Electrical Support Laboratory	2-101
II-20.	Physical Requirements of Heat-Thermodynamics Laboratory	2-102
III-1.	Necessary Channel Capacity for a 50-Man STARLAB	3-14
III-2.	Classification of Channels	3-15
III-3.	Summary of Frequency Allocations	3-16
III-4.	Estimated Hardware Requirements and Specifications	3-21
IV-1.	Ground Track of Ascent Profile	4-7
V-1.	Power Allotment	5-21
V-2.	Candidate Nuclear-Electric Power Systems for the 1975 - 1985 Time Period (After Brantley)	5-27
V-3.	System Power Requirements	5-31
V-4.	Characteristics of Orientation Modes	5-47
V-5.	CMG Physical Characteristics	5-52
V-6.	CMG Performance Characteristics	5-53
V-7.	Summary of Physical Characteristics of STARLAB Estimates of Propellant Requirements and Required CMG System	5-60
V-8.	Weight Penalty for Artificial Gravity or Devices to Prevent Physiological Degradation in Zero-g	5-69
V-9.	Logistics on Life Support Resupply by Space Shuttle to 35-Man STARLAB	5-79

LIST OF TABLES (Continued)

Table	Title	Page
V-10.	Logistics for the 35-Man STARLAB -- Resupply Time.	5-81
VI- 1.	Cost of STARLAB (Optimum Budget)	6- 5
VI- 2.	Typical STARLAB (Proper) Modular Subsystem Costs	6- 6
VI- 3.	RFM's Subsystem Costs	6- 7
VI- 4.	Subdivision of Recurring and Nonrecurring Costs for the Entire STARLAB and RFM's (10 Modules Total)	6- 8
VI- 5.	Percent Program Cost of STARLAB	6- 9
VI- 6.	Alternate Deployment No. 1, Late Deployed RFM's (Low Initial Cost Budget)	6-12
VI- 7.	Alternate Deployment No. 2 Costing, No Remote Flying Modules (Disaster Budget)	6-13
VI- 8.	Estimated Effects of Delayed or Deleted FFM Deployment upon STARLAB Effectiveness.	6-14
B- 1.	Circuit Quality Chart No. 2, S-Band Critical Directly to Ground Emergency	B- 2
B- 2.	S-Band Critical Circuit Quality Chart No. 3, Satellite to CIMC	B- 3
B- 3.	Circuit Quality Chart No. 4, S-Band Critical, STARLAB to Satellite	B- 4
B- 4.	Circuit Quality Chart No. 5, S-Band Critical, Satellite to Satellite	B- 5
B- 5.	Circuit Quality Chart No. 1, VHF Voice, STARLAB to Ground (Emergency)	B- 6
B- 6.	Channel Frequency Allocations	B- 7

CHAPTER I

INTRODUCTION

CHAPTER I. INTRODUCTION

The flight of Apollo 11 has launched the space age into a new era, the era of man's exploitation of space to his own advantage. The technical difficulties of going into space and returning safely, which were not too long ago considered insurmountable, have been overcome. The launch vehicles and the technology for the exploitation of space are available today. It is no longer a question of how, but simply a question of when.

There are at least four properties of space man may exploit:

1. Earth area coverage from high altitude.

2. Lack of atmospheric scattering and absorption.

3. Weightlessness.

4. Freedom from earth interaction.

On a small scale, unmanned satellites are already exploiting the first two of these properties for man's benefit. Communication satellites are relaying audio and video data to nearly every part of the world, and meteorology satellites are providing local weather data to users all over the world via their own "backyard" receiving stations. Similar benefits can be obtained from programs in earth resources, navigation, and air traffic control if man desires to obtain them. There has been no significant effort to exploit the other two properties of space listed above. A discussion of how these two properties affect biological, chemical, and materials phenomena, and how the effects may be exploited by man, will be explained in Chapter II.

This study makes no attempt to justify the need for man's real-time, onboard participation in space applications and research experiments. Instead, it was assumed impossible to develop automatic equipment so complex that it could duplicate in an orbiting laboratory a function similar to man's role in a research laboratory on earth.

Mission Objective

The initial objective for Project STARLAB was to perform a conceptual design study of a manned orbiting research and applications laboratory. Before the study was completed, this objective had effectively broadened at least three-fold.

The first step in the design study was to establish the requirements which would guide the design. For an orbiting laboratory, the fundamental requirement is the experimental program which the laboratory supports. Long-term operation of the laboratory was desired, and a laboratory lifetime of at least 10 years was adopted. When the requirement for an experimental program was overlayed with the constraint of a 10-year lifetime, the problem of establishing an experimental program, which could by itself sustain a laboratory for 10 years, became unsolvable. An alternative was to reduce the lifetime desired and base the design on a list of experiments which could sustain the laboratory for perhaps 2 years. But it was desired that this laboratory be a follow-on to the Apollo Applications Program, and it was felt that such a limited purpose laboratory would not be appropriate for this time frame. A new approach was required. Some means had to be found to ensure the perpetuation of the experimental program and of the laboratory. The solution was a program and information management system to stimulate new research and applications experiments and to ensure effective dissemination and utilization of acquired data.

Through the above process, it was realized that our initial objective was actually a composite of three equally important objectives:

1. Define a long-term applications and research program for experimentation in near-earth orbit.

2. Define a general-purpose orbiting laboratory to support the applications and research program.

3. Define a program and information-management system to support the laboratory.

Mission Constraints

Four mission constraints were adopted to guide the laboratory design:

1. Time frame — The laboratory will be more sophisticated than any contemplated for the Apollo Applications Program, and will be deployed not later than 1980.

2. Technology — No technological advances will be required for systems design requiring more than a 5-year advance in the state-of-art.

3. Laboratory lifetime — The lifetime shall be greater than 10 years to amortize the large initial development costs over a sufficiently long period of time.

4. Logistics vehicle — All articles required for resupply shall be sized in accordance with the presently adopted NASA criteria for space shuttle performance.

Orbiting Space Technology Applications and Research Laboratory (STARLAB)

The name adopted for the orbiting laboratory is STARLAB (orbiting Space Technology Applications and Research Laboratory). STARLAB is a zero-g, 35-man station. It is composed of a main station and five free-flying modules. Two modules are devoted to astronomy, one to physics, one to life-behavioral sciences, and one to materials and manufacturing. The main station contains the crew facilities, the majority of the analytical laboratories, and the laboratory support areas. The total power requirement for STARLAB is 100 kw and is provided by a nuclear power source.

The main station is a 430-foot-long cylindrical structure with a maximum diameter of 33 feet. This structure will be transported into earth orbit in five launch packages. The launch vehicle is a two-stage Saturn V. Each launch package weighs 120 000 pounds and has a volume of 50 000 cubic feet. This size and weight is well within the performance capability of the two-stage Saturn V.

The orbit of STARLAB has an inclination of 69 degrees and an altitude of 308 n. mi. This orbit provides coverage of 93 percent of the earth's surface.

Space Technology Applications and Research Administration (STARAD)

The program and information management functions are directed by STARAD (Space Technology Applications and Research Administration). The program functions involve planning, establishment of priorities, budgeting allocations, and the overall administration of STARLAB. The information management system provides direct communication between scientists and engineers on board STARLAB and on the ground. It has the responsibility for processing, storing, and disseminating STARLAB's data.

Starsystem

Starsystem is a general purpose space system for the exploitation of space. It is comprised of the functions of STARLAB and STARAD. Included in STARAD is a synchronous communications satellite network and a ground communications network.

Scientific Mission

The scientific mission of STARLAB has a two-fold purpose; (1) to design a science and applications program for a 10-year duration, and (2) to obtain the maximum feed-back in science, technology, and applications.

The experimental program has been designed with research areas for basic science and applications. These are complimented by a series of support laboratories in which services such as analysis, standardization, repair, and modification can be performed. This flexibility in the program will allow for new and continuous experiments for the future in STARLAB.

Experimental programs with greatest emphasis in STARLAB are those that seem to offer the greatest returns in the shortest amount of time. These programs include the fields of life-behavioral science, materials and manufacturing, and especially, earth resources.

Summary

The design highlights and summary of STARLAB begins with the Applications and Science mission's design philosophy of emphasis upon immediate and man-focused benefits. Several new areas are emphasized: chemistry, life-behavioral science, materials and manufacturing, and earth resources. Primarily the design is aimed at research and support laboratories which have a great deal of flexibility. It was driven by scientific programs, not experiments.

The philosophy of the Engineering Systems and Human Factors Task Group was that of being man-centered, not hardware limited, with a goal of shirt-sleeves environment. A list of the novel design features of this group follows:

1. Modular — on-ground vehicle assembly.

2. Multipurpose remote flying modules.

3. Central docking.

4. Nondocking shuttle — container transfer.

5. Orbit chosen to maximize earth looking and pointing.

6. Zero-g environment.

7. No extravehicular activity (EVA).

8. Integrated multipurpose hull.

9. Growth and storage modes of operation.

10. Choice of two atmospheres.

11. Automated alerting, diagnosis, and self-maintenance.

12. Windows and training areas, earth-like meals.

13. Continuous crew monitoring.

14. Experiments priority subroutines.

15. Power load programming.

16. Solar lighting.

17. Power sizing (100 kw).

18. Compartmentalized structure (safety).

In achieving the design of STARLAB, the Information Management Task Group employed four unique characteristics. The most important is the integrated laser communications system with its high data rate and possible new lasing material. It was believed that the large scientific computer will be earth based with a smaller modular computer on STARLAB. Communications and information distribution on board STARLAB itself will be handled by the advanced twisted shielded pair (TSP).

In general, STARLAB was designed from model programs for a 2-year period that would self-propagate the mission for at least 10 years. Flexibility was considered to be the prime attribute for the design.

CHAPTER II

SCIENTIFIC MISSION

CHAPTER II. SCIENTIFIC MISSION

DEFINITION OF SYMBOLS

Å	Angstrom units
p	Pressure
v	Volume
n	Number of moles
R	Universal gas constant
T	Temperature
M	Mass
MDA	Multiple docking adapter
u	Velocity of diffusion
atm	Atmospheres
UV	Ultraviolet
IR	Infrared
XUV	Extreme ultraviolet
nadir	Point on the celestial sphere vertically below the observer or 180 degrees from the zenith
pH	Hydrogen-ion concentration
FFM	Free-flying module

INTRODUCTION

STARLAB: Its Design and Rationale

The design and actual erection of a 35-man earth-orbiting research facility having at least a 10 year life is a stupendous task requiring man to organize and coordinate an enormous program. The recent success of Apollo 11 has shown us that such an endeavor is not a wild dream, but is conceivable if mankind can so motivate his society and drive his technology. After STARLAB has been designed and erected in space there will come the question, "Why?"; that was raised time and time again even while Apollo 11 was in progress. If such a project warrants national and possibly international support, then the contributors are entitled to an answer. We must be prepared for the question "Why?"

This question must be answered by the productivity and relativity of the studies that are carried on in STARLAB's experimental program. STARLAB is the forerunner of a series of projects that will utilize the synergistic effects of group research; i.e., the total is much, much more than the sum of the constituents. This does not signal the end of the individual researcher but adds incentive in terms of a large-scale goal plus an emotionally up-lifting experience of a winning team effort.

STARLAB's productivity can be visualized specifically in three areas: (1) Science and exploration, (2) Technological development, and (3) Commercial applications (Ref. II-1).

1. Science and exploration — Man's exploration of our solar system and beyond is one of the most exciting and appealing frontiers for human exploration. The recognition of its importance is almost universal.

Present space technology gives us an unparalleled opportunity for scientific discovery if properly applied to the performance of experiments. In the field of astronomy, the great advantages of observation from space would include making available the entire electromagnetic spectrum and decreasing background light scattered by the atmosphere. Angular resolutions far beyond those possible from the earth could be possible. These extended conditions would contribute to further definition of (1) the status of the universe, (2) the status of the sun and the solar system, (3) the universality of physical laws and their application to new processes, and (4) the possibility of the existence of extraterrestrial life.

2. Technological development — A highly successful research laboratory in space would show the world technological and organizational leadership. A major portion of the NASA resources is spent for technology on challenging goals. These are important to the nation's military capability, industrial development, and economic health. The development of a successful STARSYSTEM requires unusual technological and organizational leadership. The overall results are exemplified by the demonstration through the Apollo Program of the capability of building a complex system that provided a first-time reliable performance while using a diverse contractor structure. In the fields of materials science and chemistry, for example, new materials and/or new methods of processing can provide better materials for living and, in addition, give the United States' economy a favorable competitive position in world markets.

The technological challenge of the program further provides motivation for the entry of young people into scientific and engineering professions, thus increasing the technological manpower of the nation.

3. Commercial application — Applications of space flight in the areas of meteorology and communications have already been demonstrated to be economical because of direct material benefit to civilian life. In the future, earth resources are expected to make a major contribution in this field. Remote sensing of the earth from orbiting spacecraft offers a bold, progressive approach and a new dimension to our efforts to advance our society. It will provide us with the necessary energy, raw materials, and food and water for the future, as well as protection from the ravages of man's environment and pollution. Hopefully, remote sensing of the earth's resources will greatly alleviate hunger and international conflict and consequently war between nations. This remote sensing could also be used as a defensive monitor.

In addition, observations of man's activity in the fields of navigation and traffic controls can result in programs of control and new direction in these fields. Therefore, our goals are the design of the experimental laboratory, the design of the experimental program, and an attempt at a method of answering the question, "Why STARLAB?"

The Experimental Laboratory and Experimental Program

The design of the vital part of STARLAB, the experimental laboratory, will be controlled by two aspects of the experimental program:

1. A candidate experimental program (which will probably be quite different at flight time) which has been used to control the design of the laboratory such that one can arrive at a reasonable estimate of the laboratory.

2. An evolving experimental program which can be used to include new and emerging studies.

The general approach to the candidate experimental program is to set forth a group of representative experiments that will require 2 years effort of approximately 35 STARLAB personnel for completion. This vast experimental exploration of our greatest new-found asset, space, will probably generate enough problems to necessitate an infinitely long program of research and technical studies. (Compare the continued programs of several research institutes and foundations.)

The candidate experimental program was obtained from various sources, but can be classified as being from NASA, or from the design group. The areas covered are:

1. Physics.

2. Chemistry.

3. Earth resources.

4. Materials and manufacturing.

5. Astronomy and astrophysics.

6. Life/behavioral studies.

7. Communication, navigation, and traffic control.

8. Manned spaceflight engineering.

It is thought that all areas of technology, behavioral, and medical science should be represented to obtain the maximum effect of the availability of the space environment upon our society.

The type of problems and experiments to be considered on STARLAB represent a wide swath of problems for study in the space environment.

Usually, each of these investigated areas has requirements with three levels of classification. These classifications or categories are shown on Page 2-47 and are the alternatives to be used in the trade-off stage.

The evolving experimental program is considered as a simple fixed proportion of the candidate experimental program. This fixed proportion can be used as a trade-off parameter, but its lower limit was set at 0.5, or 50 percent; expansion possibilities and its upper limit would be 2.0, or 200 percent

An alternative approach to the evolving experimental program is to construct STARLAB such that it can be expanded in segments.

Design of the Laboratory

Although the laboratory design is primarily influenced by the experimental program, the laboratory can serve many other functions. In general, the laboratory design provides for:

1. Work laboratory space for experimental programs in the forementioned scientific areas.

2. Support facilities for research and studies that are carried out on STARLAB.

3. Assembly space in the construction and assembly of STARLAB.

4. Back-up systems to manually maintain and repair the STARLAB.

5. Staging for future missions.

The mass, power, volume, manhour, and miscellaneous requirements for the work laboratories are given in Table II-1; the support laboratories are given in Table II-2; the modules and MDA's are given in Table II-3; and a summary for all labs is given in Table II-4. Supporting details for these appear in later sections of this chapter.

TABLE II-1. RESEARCH LABORATORY REQUIREMENTS

Identification	Power (w)	Mass (lb)	Volume (ft³)	Manned (Manhour/day)
Physics	500 (200)	1 000 (400)	2 500 (1 200)	16 (8)
Chemistry	1 000 (500)	500 (250)	5 000 (2 500)	24 (8)
Earth Resources	3 000 (2 000)	9 000 (6 000)	400 (400)	24 (16)
Materials & Manufacturing	10 000 (5 000)	15 000 (5 000)	5 000 (2 500)	56 (32)
Astronomy	500 (300)	1 500 (1 000)	12 000 (9 000)	12 (8)
Biomedical-Behavioral	7 800 (5 500)	5 000 (4 000)	5 000 (4 000)	48 (38)
Communication, Navigation, and Traffic Control	500 (300)	3 000 (2 300)	50 (35)	16 (8)
Research & Development Engineering	1 000 (500)	3 500 (2 500)	600 (400)	48 (24)
Total	24 300 (12 300)	38 500 (21 450)	30 550 (20 035)	244 (30.5 men) 142 (17.8 men)

Note: Figures in parentheses represent minimum, other figures represent optimum.

TABLE II-2. SUPPORT AREA REQUIREMENTS

Facility	Power (w)	Mass (lb)	Volume (ft^3)	Manned (Manhour)
Analytic	3000 (1500)	2000 (1000)	5 000 (2 500)	48 (24)
Dark Room	1000 (500)	100 (50)	200 (100)	2 (1)
Mechanical	1000 (500)	1500 (700)	1 500 (500)	8 (4)
Electrical	500 (300)	400 (200)	1 000 (500)	8 (4)
Heat-Thermo	1000 (500)	1000 (400)	600 (400)	8 (4)
Storage	200 -0-	200 (200)	2 500 (500)	2 (0)
Total	6700 (3300)	5200 (2550)	10 800 (4 500)	76 (33)

Note: Figures in parentheses represent minimum; other figures represent optimum.

TABLE II-3. MODULE AND MDA REQUIREMENTS

Modules & MDA's	Power (w)	Mass (lb)	Volume (ft^3)	Manhours Planned
Physics	5 000 (3 000)	40 000 (30 000)	12 000 (9 000)	32 (16)
Astronomy	1 700	40 000 (30 000)	15 000 (13 000)	
Material & Manufacturing	10 000 (8 000)	30 000 (20 000)	18 000 (9 000)	16
Biomedical Behavior	1 000 (1 000)	25 000 (24 000)	9 000 (9 000)	8
Total	17 700 (12 000)	135 000 (104 000)	54 000 (40 000)	48 (6 men) (32) (4 men)

Note: Figures in parentheses represent minimum, other figures represent optimum.

TABLE II-4. DESIGN REQUIREMENT SUMMARY — ALL LABORATORIES

Identification	Power (kw)	Mass (lb)	Volume (ft^3)	Man Requirements (Manhour/day)
Work Laboratories	24.3 (12.3)	38 500 (21 450)	30 550 (20 035)	244 (142)
Support Laboratories	6.7 (3.3)	5 200 (2 550)	10 800 (4 500)	76 (33)
Modules and MDA's [a]	36.2 (27.9)	315 000 (263 000)	111 000 (90 000)	48 (32)
Total	72.7 (49.7)	358 700 (287 000)	152 350 (114 535)	368 (207)

a. Power supplied by solar panels on module.

Note: Figures in parentheses represent minimum; other figures represent optimum.

Integration and Management

After the elements of this laboratory have been delineated, the interactions and general configurations of the laboratory must be considered. The suggestions concerning reliability, maintainability and resupply, safety and contamination, time programming, and communication are presented herein.

The final aspect that must be considered in the operation of such a vitally important and productive laboratory is the management of the experimental programs of STARLAB which is carried out by STARAD, described in a later section.

Summary of Philosophy

STARLAB's experimental program is a planned, coordinated, and systematic endeavor to extend our understanding of basic concepts (1) to permit the interplay of basic research, technical study, and production, (2) to push forward the frontiers in technical areas, and (3) to provide new and better materials, processes, and theories for our society. But, the rationale and justification for STARLAB falls squarely upon the experimental programs that are carried out by STARLAB. These must contribute to society by producing advances in technology in a very broad sense. At least three levels of study are considered for STARLAB initially:

1. Basic research — Fundamental processes in the space utilization of fundamental theories and environment.

2. Technical studies — Space environment to develop devices and processes.

3. Protoproduction — Production and study of space-made materials and processes.

STARLAB's protoproduction could be used to stimulate industrial and private involvement in the space manufacturing area. The protoproduction along with the results from the technical studies help to justify STARLAB's existence.

Technical studies are an area which could help the U. S. better use resources through surveillance, production of more efficient devices, transportation systems, and infestation and crop control.

The basic research carried out on STARLAB can be used to support society and its needs for a continuously developing technology. Basic research could also be easily coupled to an academic program and a training program, and could yield completely unforeseen advances.

The experimental programs on STARLAB must be documented thoroughly and, in fact, a detailed continuing study of each new theory, material, process, law, or idea must be made. The relation of STARLAB to the people is very important and should be developed thoroughly. STARLAB's effects should be publicized through the various media. Programs should be set up to publicize STARLAB's spin-offs in:

1. Earth resources.

2. New materials.

3. Medical research and medical care.

4. Education.

5. Tax reductions.

6. Travel and tourist trade.

7. Traffic and navigation.

8. Group social work.

The experimental program, the effects of spin-offs, and an aid to some of the national and international problems are a few of the large-scale problems that a facility of the magnitude of STARLAB can attack. The problem of data analysis, coupled with the communications problems and the so-called technical publications explosion, is an enormous auxiliary problem.

There is hardly a knowledgeable person that can deny that man must utilize and explore this newly found resource — space. The research and development that is possible from STARLAB can begin to self-stimulate much new technology, heretofore untapped.

RESEARCH LABORATORIES

Purpose and Philosophy

A working laboratory has been designed to include the necessary space and equipment for the performance of most of the actual experimental work. Support laboratories, in the form of analytical, preparation, maintenance, and storage facilities, are separate from the work laboratory.

The basic design of the working laboratory has been guided by the numerous candidate experiments that have been approved by NASA for early flight. These experiments, in general, will occupy the laboratories for a 1-year period or less. Since the space laboratory is planned for a 10-year duration, new experiments will replace the initial candidate ones. At this time it is impossible to predict all the experiments that will be included in the space laboratory for the remaining 9 years. Many of the new ones will depend upon the results of the candidate experiments, therefore, to be as flexible as possible, working space for future experiments will be available plus support facilities which include analytical, preparation, mechanical, and electrical laboratories.

These laboratories include all the basic equipment and facilities needed for the care of experimentation in the work labs and for the remaining life of the space base. Additional equipment needed for various specific future experiments can be brought up in the space shuttle.

Areas for Research and Applications

Astronomy

    a. Purpose and Objectives. "The ability to put telescopes and other instruments in orbit above the earth's atmosphere is giving astronomers an opportunity to observe the sky in a way they could only dream of a dozen years ago."

    ". . . such an (international) astronomical Space Observatory. . . , measured by its potential for fundamental discovery, . . . is by far the most rewarding scientific activity that can be envisioned for the U.S. space program in the next two decades." (Ref. I-2).

    Advances in astronomy are needed for future exploration of space.

    b. Description of Operation. Two free-flying modules will be the functional laboratories in which the advanced stellar and solar programs will be accomplished. These two modules will fly in a pattern behind STARLAB as described in Chapter IV. Each of these modules rendezvous once every 15 days and will be remotely controlled and monitored from a console in STARLAB. Docking will provide updating and resupply facilities.

    Each module, while flying its pattern, will have an attitude control system that provides the pointing required for the stellar and solar programs.

    c. Program. The astronomy program, being one of the most aggressive and well-developed programs in the space experimental domain, has an extremely well-defined and documented program (Refs. II-3 and II-4).

    The program for astronomy corresponds to a passive listening mode that covers the electromagnetic spectrum other than that which can be observed at the earth's surface.

The two aspects of the program, stellar and solar studies, will have as their main tasks the observation and understanding of stellar and solar processes.

The information management of the program will be handled through the communications system of STARLAB. In-flight experiment modification can be obtained from STARLAB's manned console. Storage and maintenance requirements are the main requirements that the astronomy modules will place upon STARLAB.

The proposed experiments for this model program were those abstracted from References II-3 and II-4. These experiments are as follows:

1. Solar experiments.

    a. Photoheliograph.

    b. High resolution X-ray spectroheliograph.

    c. XUV spectroheliograph.

    d. XUV spectrograph.

    e. UV scanning polychromator spectroheliometer.

    f. UV long wavelength spectrometer.

    g. Hydrogen-alpha telescope.

2. Stellar experiments.

    a. Manned Astronomical Space Telescope (MAST).

The astronomy lab requirements are given in Table II-5.

    d. Physical Requirements. In addition to the two astronomy modules and their scientific contents, STARLAB must provide docking, maintenance, updating, and resupply facilities for the astronomy program. Space for the assembly of large telescopes and radiation monitoring devices will be required. The mechanical and electrical support laboratories are needed for this task.

TABLE II-5. ASTRONOMY LAB REQUIREMENTS

Identification	Power (w)	Mass (lb)	Volume (ft^3)
Photoheliograph	170	650	120
XUV Spectroheliograph	99	430	39
X-Ray Spectroheliograph	350	640	87
XUV Spectrograph	490	115	43
UV Scanning Polychromator Spectroheliometer	64	335	300
UV Long Wavelength	90	508	40
H Telescope	118	220	12
MAST	266	2000	122
Total	1647	4898	763

Life/Behavioral Sciences

a. Introduction (Purpose and Objectives). The experiments to be performed in the life sciences, medical sciences, and behavioral sciences before 1975 are extensive (Refs. II-5 and II-6). However, these experiments are only preliminary in nature. Many are in the category of first experiments; others are observational and will provide the basic information on which the first experimental approaches will be based.

The life sciences are very likely to provide the most exciting and useful new knowledge to be derived from man's study of space. We expect to learn more about the biology of man. The study of other earth organisms transported into the space environment will significantly increase our understanding of life processes in general. From the new knowledge that is acquired, a new technology of biological applications may well be expected.

To obtain whatever benefits that grow from biomedical research, an aggressive program is recommended to provide a general purpose biological and medical research and technological facility on an earth orbiting laboratory (STARLAB). To provide for the greatest benefit to be derived for mankind, this program should be larger than previous programs by at least one order of magnitude.

b. Research Program. The general scientific mission of a life-behavioral science laboratory stationed in earth orbit will be (1) to study man's capabilities and limitations in this environment, (2) to use the properties of the environment to study life processes, and (3) to use these properties for the development of biomedical technology. The programs of work which are delineated in the following paragraphs will be made up of new and follow-on experiments which will be generated by the scientific community prior to 1977.

(1) Routine Medical and Behavioral Monitoring. It is anticipated that guidelines for medical and behavioral monitoring will be established as a result of space operations over the next 6 years. The goal of this program is to keep all personnel healthy.

The measurements to be made will be routine in nature and will include all physiological systems and behavioral performance, including standard psychological and psychomotion testing. This monitoring is required to measure changes in the state of well being that may be treated preventively and to assist in the evaluation of the environment habitability.

The information obtained by routine monitoring will be used to provide population statistics, but the information on any individual should be considered as confidential-personal. It should be recognized that the monitoring of behavior is not in principle much different than the monitoring of prison inmates. To try to reduce this aspect of confinement, behavioral monitoring should be carried out within work time and in work areas. There must be private areas and also public areas in which no monitoring occurs.

(2) Experimental Investigations of Mammalian Physiological and Behavioral Performance and Adaptation. In addition to routine monitoring, there must be a program of experiments to test the limits of capabilities and adaptation of mammalian physiological and behavioral measurements. These experiments will be paralleled with measurements on laboratory animals.

(a) Investigations are needed to obtain information on the causes of physiological changes that occur in the space environment. Every organ system that has been cursorily studied as yet has shown changes as a result of exposure to the environment within a space craft. It is suspected that most changes are adaptive. System functions may adjust to a new steady state which is optional for the new environmental conditions. However, there remains a possibility that some changes are degenerative.

Several good programs of study of human organ system functions have been proposed as studies for an orbiting research laboratory. (Refs. II-5, II-7, and II-8.) These proposed programs are centered about measurements on man. Experimental studies on causation and mechanisms of homeostatic adjustment require experimental animals. Primate and other mammals should be used for investigations paralleling those on man. These animals will also provide the materials needed for study of general metabolic processes at the tissue and cellular level.

(b) Psychology. The behavioral performance of man is very closely related to physiological state, and programs of study have been proposed (Refs. II-5, II-7, and II-9).

There are three important areas of research that should be continued for the life of STARLAB; these are:

1. Measurement of sensory and psychomotor performance and adaptation. These measurements are related to physiological measurements indicated above, and should be paralleled with experimental studies on other mammals.

2. Measurement of mental performance and motivation. Widely recognized and accepted testing measures of behavioral parameters of performance and well being shall be used.

3. Social dynamics — Attitudes may be affected by prolonged stays in the STARLAB environment as a result of confinement and limited interpersonal reactions. As yet, there are no specific investigations proposed for this area, but this report recommends that social interactions be studied.

(3) Investigations of Earth Organisms in the STARLAB Environment. The relationship of the biological sciences to earth-orbiting research laboratories has been studied by the National Academy of Sciences and by the American Institute of Biological Sciences (Refs. II-6 and II-10). There is considerable disagreement among biologists concerning the value of such projects in terms of providing new approaches and knowledge. While many biologists may question the usefulness of biological research in space, the following quotation should be heeded: "In any case, I wish to emphasize that our ignorance of living organization — of how living matter works — is in general so profound, and our lack of any real theory (in the sense that the physicist uses the term) so complete, that we cannot afford the luxury, or rather the scientific arrogance, of insisting that we know the space environment will bring us no surprises. In short, we cannot afford an arrogance that may cause us to lose the possibility of major discovery." (Refs. II-10 and II-11.)

To make the best use of the STARLAB environment for biological research, all levels of biotic organization, from micromolecules through populations, should be studied. Basic studies on the biological adaptations of organisms, other than man, frequently provide the first clues as to how life processes adjust to environmental demands. The STARLAB will provide facilities for programs of investigation on the effects of (1) weightlessness and (2) freedom from earth geophysical rhythms.

Studies making use of the weightlessness aspect of STARLAB should have the highest priority. While biologists have considerable knowledge of gravity sensors in some animals and the adjustment which these organisms make to gravity, nothing is known of more fundamental adjustments to gravitational forces. The program of investigation of the adaptation of life to gravity should include:

1. Cellular studies — Cytology, cytogenetics, metabolism, and biochemistry.

2. Differentiation and morphogenesis — Tissue culture, plant and animal development, and growth regeneration.

3. Studies of organ function — The adaptive function of the parts of multicellular plants and animals.

4. Adjustments of organisms — The study of whole organism responses to weightlessness and varying degrees of gravitational force.

5. Populations — The behavior of organisms and social interactions in conditions of weightlessness.

STARLAB also provides a different temporal environment than is found on earth. Provisions will be made on STARLAB for a program of study of biological rhythms in the absence of normal earth geophysical rhythms. Biological rhythms are (1) circadian rhythms and (2) rhythms of other periodicities. Of greatest immediate interest is the investigation of circadian rhythms which have a period of approximately 24 hours. Experiments can be designed to monitor activity and other parameters to determine the existence of periodicities and the environmental factors such as radiation, magnetic fields, and electrostatic fields, as well as the known synchronizing agents, light and temperature.

It is likely that there are synergetic actions on living systems in space that cannot be apparent on earth. The most important ones are, perhaps, the relationship between weightlessness and noise, or any combination of these. The effects of these various factors in combination should be studied on living systems of which the functions are fairly well understood in our earth environment.

(4) Exobiology. STARLAB will provide for continued surveys of particles from the near space environment. The exobiological survey will look for (1) organic compounds, (2) morphological evidence of life, (3) complex and intermediate biochemicals, (4) evidence of metabolism, (5) evidence of growth, and (6) evidence of reproduction. These types of experiments are presently being developed under NASA support (Ref. II-12) and they indicate the types of experiments that will be conducted in the future. The order in which the items are listed is the order of probable success of obtaining positive findings.

By the time STARLAB investigations begin, near space will have been considerably polluted by microorganisms that man has lifted into orbit. It will be of interest to determine (1) the rate of microbial pollution of the exosphere (2) the viability of microbes collected in earth orbit and (3) the rate at which microbes may be ejected into space.

The problem many people are interested in, however, is detecting life forms that may not be recognizable to us. Without knowledge of its chemistry, such life would probably be recognizable on the basis of other characteristics, principally, morphological. Sampling of microscopic life forms from an earth orbit depends upon detecting very rare events, none of which may be repeatable; that is, no two such particles collected may be the same. It would seem advantageous to combine the exobiology survey with the micrometeorite survey and provide in both surveys for the detection of organic substances; this would also require that there be provisions for microscopic examinations of particles that are collected. Exobiology survey collectors should perhaps accompany all free-flying modules.

(5) Biomedical Technology. The basic studies to be done under conditions of weightlessness are expected to generate technological material and methods to occupy the facility for its life. The facility for biomedical technology will develop applications of biochemistry, pharmacy, and medicine from the basic information obtained from the space station environment.

The facility will have the additional function of testing equipment and apparatus which has been developed on earth for use in space. For example, it is essential that a life support system be designed to remain closed for periods of 3 to 5 years if man is to explore the interplanetary universe.

    c. Research Facility. One part of the life-behavioral research facility is related to studies on human functions; a second part has, as its main function, the provision for research on organisms other than man.

    (1) Facilities for Study fo Human Functions. The medical and behavioral monitoring of man, and the experimental measurements on human subjects will be performed by a system similar to the Integrated Medical and Behavioral Laboratory Measurement System (IMBLMS) (Ref. II-5). The IMBLMS concept, developed under the direction of the Office of Space Medicine, Manned Space Flight, NASA Headquarters, proposes a capability for performing approximately 150 measurements of human behavioral and physiological functions. An emergency medical treatment facility is located adjacent to the medical-behavioral monitoring console. Medical care capabilities must be sufficient to care for routine ailments. Specific requirements will have to be determined from the experience obtained from the Apollo Applications Program. Emergencies of large magnitude must also be provided for. A large air-lock adjacent to the medical area will provide for critical patient and immobilized patient care.

    (2) Facilities for Study of Organisms. Experimental studies in mammalian physiology require a console designed to perform certain of the measurements incorporated in the IMBLMS concept on a variety of mammals other than man. However, the methods and transducers for some studies have not been perfected.

A console arrangement of instruments will be required for the study of physiological functions of nonmammalian animals and for the study of plant biology.

The laboratory area must include a work area for technicians which provides for monitoring all experiments and a work area for experiment setup and takedown.

Experiments in the life sciences are to be conducted under atmospheric conditions equivalent to sea-level ambient atmosphere. The biological materials room is within the same pressure hull. The life support system that is needed to support life other than man will be separate from the man life support system, and its capacity will be the equivalent of the support for 10 men.

The biological materials room must be divided into at least four functional areas:

1. A clean room for bacterial and protozoan cultures — constant temperature incubation required.

2. A clean room for cell and tissue culture — constant temperature incubation required.

3. Plant growth chamber, with enclosed high-intensity lighting and temperature controls, and provisions for light and temperature cycles.

4. Animal room.

    a. Primate and rodent cages, with air-flow waste removal.

    b. Enclosed moist chambers for small terrestrial animals having permeable surfaces.

    c. Enclosed aquaria for fish and aquatic invertebrates, with provisions for circulating, purifying, and aereating water.

A technician work position must be provided in the area of higher atmospheric pressure.

A large radius centrifuge must be available to provide for controls for all experiments on the effects of weightlessness on biological processes. The proposed centrifuge will have a diameter of approximately 30 feet. Spoke-like arrangements will provide for different radii and gravitational forces at the same rate of rotation.

The centrifuge should be designed for long duration running times of 2 to 3 months, for example, as its principal mode of operation. The centrifuge will be located in the pressure hull with the biological materials room and access into the centrifuge will be through the biological materials room.

A free-flying module should be available for studies which must avoid the gravitational perturbations resulting from the normal operation of STARLAB. This module will dock adjacent to the biological materials room.

d. Operational Requirements and Considerations

(1) Personnel. The personnel in the life-behavioral sciences program are identified as technicians and scientists, and astronaut selection procedures should apply to both groups. Technicians are to be trained to monitor and maintain experiments and the engineering systems of the module in which they are located. Scientists originate experiments and are onboard to monitor them. Different disciplines will be represented by the life scientists, and the optimum combination should be determined by the interaction of STARAD and the scientific community. It seems preferable to have the scientists assigned to STARLAB on a research fellowship basis so there will be continual flux of scientists through the program.

The following table indicates a desirable level of direct involvement with the experimental program of STARLAB:

	Ground	Onboard
Physicians	3	1
Life Scientists	12	2
Technicians	18	3

In the early phases of the flight program, the onboard personnel may consist of only one technician or one physician, but to achieve the most benefit from STARLAB, the optimum number of personnel should be attained rapidly.

After STARLAB has been proven as a functional and habitable system, provisions should be made for scientist passengers to visit STARLAB for limited times for the purpose of making observations on their specific experiments. While such visits cannot be routine, they should be permitted to visit whenever there is good scientific reason.

The functions of the ground scientists are to manage ground experimental controls and to manage data analysis. Ground-based technicians will be involved in continued training and class training exercises, and with the continued integration of experiments into the program of research.

(2) Shuttle. The maintenance of the life-behavioral science facility will require approximately 1400 pounds per month for the movement of experiment packages and supplies. However, the shuttle requirements cannot be accurately scheduled until the experiment schedule is programmed. Supply

kits will be designed so that new kits can be fastened into the laboratory facility structure, and old kits removed for refurbishing.

Whenever possible, experiments in the life-behavioral sciences will have a modular form; i.e., each experiment will be designed in a package that is completely automated. These packages will have (1) provisions for power input for 28 vdc, or power may be self-contained, and (2) provisions for inputs for environmental control and regulation, depending upon the requirements of the experiment. The outputs from the automated experiment package will be one videcon information output, one output with 6 to 12 channels of analog voltages from transducers, and another output unit with 10 to 20 channels of digital signals as in Figure II-1. The automated experiment package will have provisions for a technician to repair and replace some parts and to make adjustments on the environmental controls. There will also be provisions for the experimenter, if onboard, to manipulate the experimental material and alter the experiment. The onboard experimenter may manage or manipulate several other kinds of experiments in the biological area as requested by onground investigators. Automated experiment packages will be replaced by shuttle.

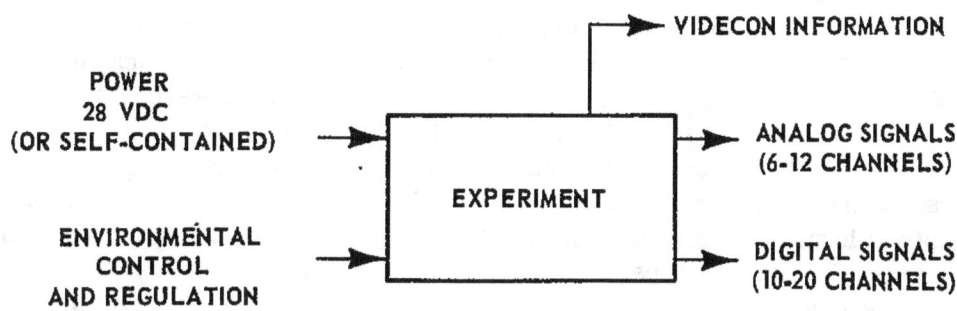

FIGURE II-1. SCHEMATIC OF AUTOMATED EXPERIMENT PACKAGE

Earth Resources

a. Purpose and Objectives. The overall objective of NASA's Earth Resources Survey Program is to develop and implement the science and technology for surveying and investigating the earth's natural and cultural resources from space. Presently, several approaches have been undertaken. The ground testing of many remote sensing devices has been accomplished. These devices have been flown by aircraft and proven capable for remote surveys. The first earth resource satellites will fly in the near future to test the usefulness of certain scientific instruments in space environments and their resolution of earth resources. Some equipment has been used on previous space flights with excellent results. Other equipment is still in the development stages, and some equipment is still conceptual. Within the next few years, highly sophisticated instruments should be developed for earth resource sensing.

As the year 2000 approaches, the need for additional resources and careful management of these resources becomes critical for our standard of living, as well as our existence. By the year 2000, the population of the United States will have doubled and demands for raw materials and energy will have increased exponentially. Pollution of the atmosphere and land will have become a problem of such proportions that it will be necessary to expend major energy for the control of this problem. Within this time span, more of our natural resources will be used than have been used in the entire history of man, and our appetite for additional resources will be just beginning.

Based on present reserves, the peak of petroleum production will occur shortly after the year 2000, as will coal production. Technological advances will prolong this production peak for some time, but certainly we are seeing the end of the petroleum and coal producing capability as we know it today. The other main energy producer is radioactive decay of certain elements. This promises to supplement power requirements for the present and supplant the organic fuel requirements sometime after the year 2000. However, at the present time, we are still far from producing the radioactive minerals necessary to see us through our needs until the year 2000. A major exploration program is now in progress for these minerals. Even if we are successful in attempts to locate new reserves of atomic energy, we are still faced with the formidable problem of disposing of the radioactive waste material. This problem may become so acute that atomic power as we know it today is not practical. The peak of energy derived from water power (rivers) has been reached, but possibly in the future it will be possible to harness tidal energy for use. We have hardly explored the use of solar energy and this may well be where the future energy source is concentrated.

With sufficient cheap, clean energy available, there would be little problem in existing with a double population, a double or triple resource requirement, and an expanding energetic standard of living. With cheap energy, it would be possible to obtain from almost anywhere on the surface of the earth the basic raw materials for our industrial requirements. With cheap, clean energy, blocks of the earth could be set aside for growing food, and produce enough food to feed our population. It is possible to cultivate beneath the sea and grow sea animals for food needs if there is cheap energy available. However, we do not have unlimited energy available under our present technology. The energy available is expensive, produces filth, and is exhaustible. The peak of energy material production could come before the year 2000 if our need for energy and that of the rest of the world expands more rapidly than anticipated. It is unlikely that we will run out of energy sources; however, they become much more expensive when we go to a different state of technology. The question is how to obtain this clean, cheap energy. It is a question that cannot be answered with present technology. As a consequence, for most of the next generation, we have to rely on those sources of energy that are presently available and utilize them for our expanding needs. It will be necessary to increase the production of the raw energy materials — coal, petroleum, radioactive elements, water, etc.

Energy sources are but one of the many resources needed to maintain our standard of living, but it is the most important one. Without cheap sources of energy, we are strapped to locating raw materials which are concentrated in the crust of the earth. Technological advances should offset this lack of cheap energy and, consequently, the cost of producing raw materials should remain about the present level for the immediate future. However, there will come a time when advances in extractive technology and easily located mineral deposits will slow down as far as producing the raw materials needed for the continued population and industrial expansion of the world. Therefore, unless it is possible to find a source of cheap, clean energy and produce raw materials from very low grade deposits, we must somehow discover new high-grade ore deposits very soon if they are to become productive by the end of the century. As previously mentioned, there is little likelihood that we will have great amounts of cheap energy; therefore, the only alternative is to begin an active exploration program for the raw materials upon which we base our economy. This has to be a worldwide effort — not only because mineral deposits are unevenly distributed over the surface of the earth, but because many nations could trade raw materials for finished U.S. goods and thereby maintain a higher standard of living. Mining and petroleum companies are undertaking a tremendous exploration program throughout the world, but unfortunately, while there have been some sensational discoveries, they are not being made at a rate which will provide the raw minerals needed in the year 2000. These exploration companies will have to redouble their efforts and use some bold new approaches for finding these necessary deposits — including the raw energy materials we need so desperately. These new efforts can benefit greatly by remote sensing from space craft.

There are many mineral exploration programs which are searching for the fertilizer minerals which are needed for productive agricultural purposes. Any new mineral exploration programs for other raw materials would certainly benefit these endeavors. It seems likely that most of the mineral fertilizers needed for productive crops and prolonged forest growth can be obtained relatively cheap, at least for the present time; however, the additional problems faced in these areas are tremendous. Plants have to have water, they must be protected from pollution of the air and water as well as additive chemical pollution, and they must be protected from disease, pests, and fire. There are many monitoring devices already developed which will greatly aid in the expansion and management of our agricultural and forested areas. Broad land-use studies can greatly aid in setting aside those land areas which could best be used for agriculture, forest, recreation, industry, and development. The land-use studies could be utilized to study soil moisture, erosion, and drainage surveys as well as offer a means of monitoring the total amount of water available in an area during any particular time so that irrigation can be more effectively instituted. All these studies will necessitate bold new approaches to land use and rapid methods of monitoring and evaluating broad areas of the earth from the standpoint of agricultural and forestry uses.

Man is literally dying in his own filth, and unfortunately he is not able to effectively police himself; therefore, there should be a policing method for pollution for the entire world, or at least large segments of the world. There is a very pressing need for sources of fresh water throughout the world. There is sufficient water in the ocean for utilization, but the cost of desalinization and transportation inland is very great. There is need for new methods of locating and utilizing our fresh water for many parts of the world. It will be necessary to greatly increase utilization of lands for living and industrial expansion with more emphasis on total environment and interface with nature. There should be consideration given to areas for industrial expansion and their relationships to agricultural, forestry, and living areas. The need for a total environmental analysis becomes highly significant when it is considered that shortly the population will be doubled. A rapid and broad-based input of data is needed now for planning for this future development. This information can be obtained rapidly and cheaply by remote earth resource surveys. Less than half the world has been mapped at a scale which can be used for general functions — geography, topography, political, vegetation, etc. With international cooperation becoming so important, there is need for some common ground for setting the boundaries of countries, the worldwide commodities, topography, and geology. This requires fairly high precision mapping.

Over two-thirds of the earth's surface is covered by the oceans. These areas still have not been completely and accurately mapped. The water bodies, their currents, depths, animal and plant populations, and subsea activities should be mapped. There is enough food within the sea to feed a hungry world, and this animal and plant population should be assessed for this purpose. A program to conserve and scientifically utilize the plant and animal populations of the oceans should be initiated now. Agricultural areas for development within the sea should be set aside within the near future. There is great need to evaluate possible areas for mineral exploration beneath the sea. The physical characteristics of these potential mining and petroleum areas should be determined to commence a utilization program with high potential gain.

Over the past few years, spacecraft have monitored weather conditions and have returned a plethora of information to ground receiving stations. There is, in the developmental and conceptual stages, additional equipment which will ultimately allow meteorologists to ascertain the causes of weather patterns on the earth. These applications should receive high priority in endeavors to plan for the year 2000 and the needs of an exploding population. Man's standard of living may be strongly influenced by his ability to predict and utilize to its fullest the water resources in the atmosphere.

While remote sensing of the earth from orbiting spacecraft is certainly not a panacea for all these problems, it offers a bold, progressive approach and a new dimension to efforts to provide man with the necessary energy, raw materials, and food and water for the future, as well as some protection from the ravages of his environment and his own pollution. This will allow him to maintain and improve his living standards.

The following is a general list of experiment areas which can readily utilize remote sensing from spacecraft. Some of the experiments can only be accomplished from space; others can be done more rapidly and cheaper from spacecraft. Under each of the broad topics, there are several experiments which could be conducted; however, the equipment needed for most of these experiments is highly versatile and can be utilized in many experiments as follows:

1. Mineral deposit exploration.
2. Geologic structures — regional tectonics.
3. Engineering geology.
4. Rock composition.

5. Stratigraphy — sedimentation.
6. Crustal-mantle studies.
7. Vegetation density.
8. Grass-brush-timberland interfaces.
9. Plant species and vigor.
10. Soil series, temperature, and moisture.
11. Irrigation water.
12. Fire detection.
13. Land use.
14. Transportation and linkages.
15. Settlement and population movements.
16. Resources utilization.
17. Climatic conditions.
18. Air pollution.
19. Topography, mapping, and geomorphology.
20. Evapotranspiration.
21. Rain distribution and infiltration.
22. Ground water discharge.
23. Water pollution.
24. Snow surveying and glaciology.
25. Effluents of major rivers.
26. Thermal conditions in ocean.
27. Sea surface roughness.
28. Shoals and coastal mapping.
29. Biological phenomena in ocean.
30. Ice surveillance in ocean.
31. Monitoring of volcanoes.
32. Detection of MASCONS.

33. Gravity variation studies.

34. Earthquake prediction.

35. Sea-fresh water interface along coasts.

36. Landslide prediction.

37. Sea current mapping.

38. Flood control.

39. Water inventory of lakes and streams and snow.

b. Experimental Program. The main equipment group will be composed of a series of sensors with recording equipment for receiving and separating wavelengths in the electromagnetic field between approximately 10^8 and 10^{18} cps (microwaves to X-rays) (Fig. II-2) (Table II-6) (Ref. II-13). There will be transmission equipment to direct microwaves to the earth's surface for reflection measurements. In addition to this sensing equipment, a magnetometer and a gravimeter should be within or attached to the spacecraft.

Table II-7 lists a few of the possible applications of these instruments. Table II-8 lists a general comparison of visible light, IR, and radar capabilities for some remote sensing requirements.

Depending on the resolution and type of data desired, it is possible to select a microwave-radio wave sensor to obtain the required information (Fig. II-3). Microwave-radio wave sensors can be used to obtain data for mapping, fault detection, vegetation type, topography, soil moisture, rock type, and stratification (Ref. II-14). At present these sensors are operational, or in the developmental stage, or are undergoing basic research into their capabilities. The best data now available for remote sensing from space are photography (Ref. II-15). However, the ultraviolet, infrared, microwave, and radio-wave parts of the electromagnetic spectrum have been only recently investigated for remote sensing of earth resources.

The usefulness of remote sensing equipment, listed below, depends upon the equipment's ability to sense certain parts of the electromagnetic spectrum which can be picked up and interpreted by present day or future state-of-the-art equipment. This equipment includes photographic systems, multispectral sensors, infrared mapping, microwave sensing, laser altimeters, and radar.

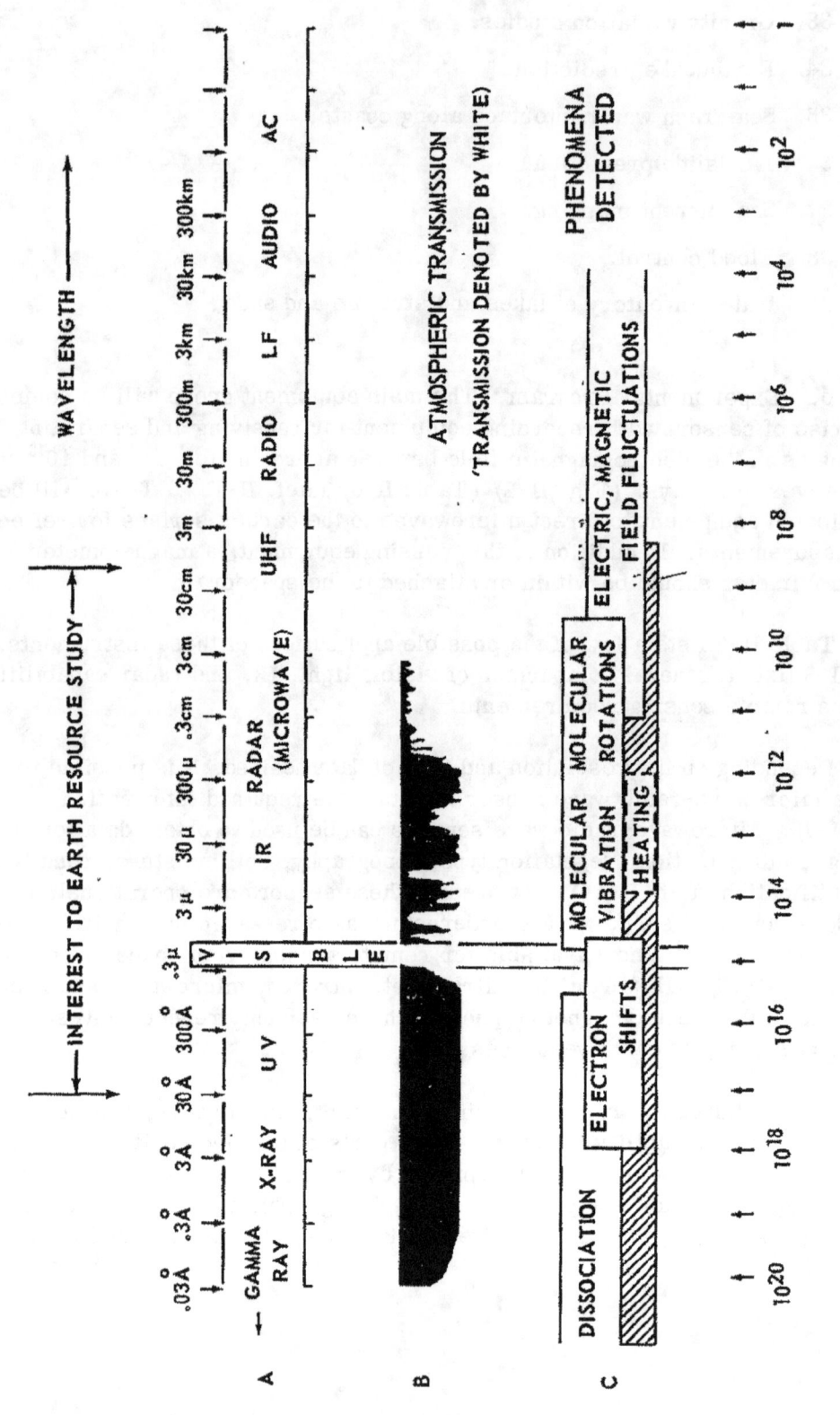

FIGURE II-2. EARTH RESOURCES USE OF ELECTROMAGNETIC SPECTRUM (Ref. II-13)

TABLE II-6. ELECTROMAGNETIC SPECTRUM FOR REMOTE SENSING
(Ref. II-13)

Kind of Waves	Wavelength (cm)	Frequency	Usual Source	Usual Method of Detection
Gamma Rays	$10^{-8} - 10^{15}$	$3 \times 10^{20} - 3 \times 10^{18}$	Atomic Explosions	Fluorescence Chemical Effect Ionization
X-Rays	10^{-7}	3×10^{17}	Cathode-Ray Impacts	Same as Above NaI Crystals
Ultraviolet	10^{-5}	3×10^{15}	Disturbances of Intermediate Electrons	Fluorescence Chemical Effect
Light	$39-78 \times 10^{-6}$	$77-38 \times 10^{13}$	Disturbance of Valence Electrons	Eye, Chemical Effect, Photo-detectors
	63.28×10^{-6}	47.3×10^{13}	Laser	
Infrared	751×10^{-4} to 1×10^{-1}	4.34×10^{13} to 300 GHz	Disturbances of Atoms and Molecules (Thermal Sources)	Thermopile Bolometer Radiometer
	$31.25 - 270 \times 10^{-6}$		IR Lasers	Photodetectors
Microwaves	$1 - 0.1$	30 - 300 GHz	Short Oscillations, High-Frequency Discharge	Diodes Bolometers
	$10 - 1$	3 - 30 GHz	Electrical Resonance in Tuned Circuits, Thermal Generated IR, Solid-State Devices	Solid-State Crystals Diodes, Tubes
UHF	$100 - 10$	300 MHz - 3 GHz	Short Oscillations Klystrons, Magnetrons, Triodes, Solid-State Devices	Electrical Resonance in Tuned Circuits, Diode Detectors
VHF	$1000 - 100$	30 - 300 MHz		
Radio	$10^6 - 10^3$	30 kHz - 30 MHz	Circuits Including Large Capacitors and Inductors	Electrical Resonance Electromagnetic Induction
VLF	$10^7 - 10^6$	3 - 30 kHz	Alternating Current Generators, Solid-State Vacuum Tubes	Oscillographs, Diodes, Heterodyne Receiver
ELF	$>10^7$	<3 kHz	Alternating Current Generators, Solid-State, Vacuum Tubes	Oscillographs, Diodes, Heterodyne Receiver
Direct Current				

TABLE II-7. SENSORS AND APPLICATIONS IN EARTH RESOURCES PROGRAM (Ref. II-17)

Sensor	Cultural Resources											Natural Resources								Earth Sciences							
	Agriculture								Urban Land Use	Transportation		Fresh Water			Forestry		Marine Life	Wildlife		Cartography	Geology	Geophysics	Geodesy	Oceanography	Natural Disasters		
	Soils		Crops		Population Distribution			Development	Control	Sources	Distribution	Pollution	Distribution	Quality		Distribution	Migration						Floods	Fires	Deformations
	Quality	Temperature	Moisture	Quality	Species																				
Metric Cameras	X			X	X	X	X	X		X	X		X		X	X	X	X	X	X		X	X	X	X
Panoramic Cameras	X			X	X	X	X	X		X	X	X	X	X	X	X	X	X	X	X		X	X	X	X
Tracking Telescope	X			X	X	X	X	X		X	X	X	X	X	X	X	X	X	X	X		X	X	X	X
Synoptic Cameras				X	X	X	X	X	X	X	X	X	X	X	X	X		X	X			X	X	X	X
Radar Imager	X									X	X	X	X		X				X	X			X	X	
Radar Altimeter/Scatterometer	X	X	X	X	X			X	X	X	X	X	X	X	X		X		X	X		X	X		X
Wide-Range Spectral Scanner (O-M)			X					X		X	X	X	X	X	X	X	X		X	X		X	X	X	X
IR Spectrometer								X		X	X	X	X		X		X			X		X	X	X	X
IR Radiometer		X						X	X	X	X	X	X	X	X	X				X		X	X	X	X
Microwave Imager (Passive)		X	X					X		X	X	X								X		X	X	X	X
Microwave Radiometer			X					X		X	X	X								X		X	X	X	X
UV Imager/Spectrometer								X				X								X		X	X		X
Laser Altimeter/Scatterometer															X				X	X	X		X		
Absorption Spectrometer										X	X														
Radio Reflectometer															X					X	X	X			
Magnetometer	X									X									X	X	X	X			
Gravity Gradiometer	X																			X	X		X		
Ground Sensors			X						X	X							X			X	X	X	X	X	X

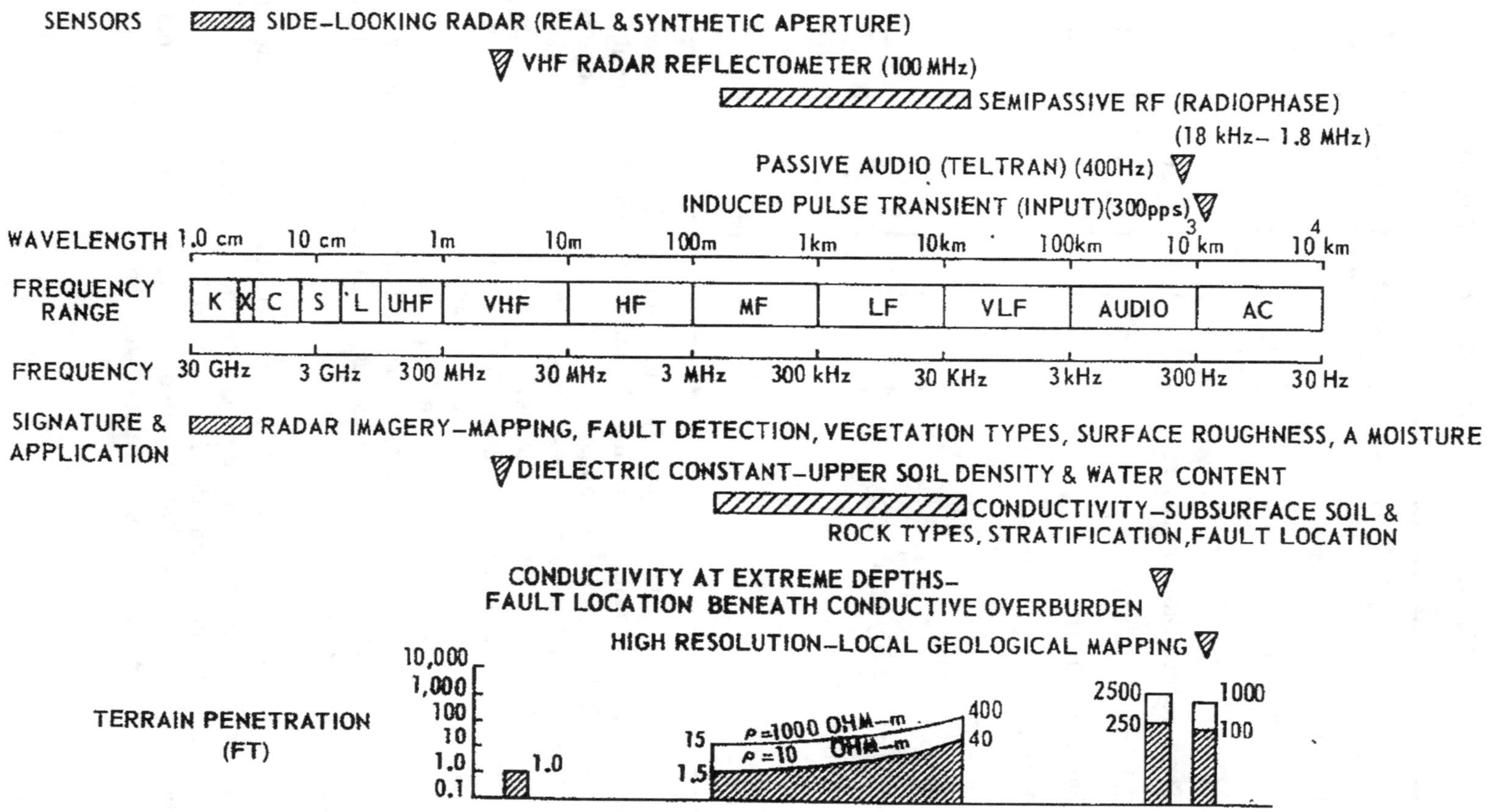

FIGURE II-3. ELECTROMAGNETIC SENSORS FOR SITE SELECTION (Ref. II-17)

2-31

TABLE II-8. REMOTE SENSOR COMPARISON

	Camera	Infrared	Radar
Day/Night	5	10	10
Haze-Fog Penetration	3	6	10
Cloud Penetration	1	2	9
Temperature Discrimination	2	10	1
Subsurface Detection	4	6	3
Stereo Capability	10	2	3
Accurate Image Representation	9	6	5
Long-Range Capability	7	4	8
Resolution	9	7	5
Interpretability of Imagery	9	6	6
Availability of Equipment	10	4	4

Poor = 0   Good = 10

(1) Photographic Systems. In the study of earth resources, the camera at present remains the most important recording mechanism. The frame camera is best suited for mapping functions but requires accurate image motion compensation to obtain the best resolution potential for the system at high-orbital velocities. The slit camera, in which the film is moved past a narrow slit in the image plane, is automatically compensated for the image motion problem found with the frame camera, but is susceptible to distortion of the images when the v/h control of the film speed is misadjusted. In the panoramic system, the use of rotating mirrors, lenses, and/or prisms permit horizon-to-horizon photography, overcoming the limitations of wide angle refraction lens systems (Ref. II-16).

(2) Multispectral Sensors. In a spectral matching sensor, the spectral reflections of each resolution element in a scene is observed by detectors which put this information in electrical form. Real-time signal processing techniques can be used to determine how well the spectrum of the resolution elements under observation correlates with the characteristic spectral

reflectance of the material for which the sensor is searching. By using an optical-mechanical scanner, the spectrum matching need not be limited to the photographic region. It need not be limited to the reflected signature and can include the thermal emission signature. Such an operational sensor would consist of an optical-mechanical line scanner with a single scanning aperture, a dispersing system, and multiple detectors, each observing at different wavelengths the radiation from a resolution element defined by the scanning aperture. The signals from the detectors are transferred into a signal processor which determines how well the spectrum correlates with the object being sought. Optical-mechanical scanners have the advantage that the output is in electrical form and can be processed and telemetered in real time (Ref. II-16).

(3) Infrared Mapping. In infrared studies, there are three types of data which can be collected: radiometry, low resolution scanning, and high resolution scanning. In the radiometry, an optical telescope orientated toward nadir collects infrared radiation and focuses it upon sensitive surfaces of the detectors. The detectors convert the radiation into electrical signals, which are then amplified and displayed or recorded. The infrared radiometer will serve two distinct purposes. The first is to provide an absolute apparent temperature profile of the region directly beneath the spacecraft. The radiometer can also provide thermal maps from space with high sensitivity and fine resolution. The use of low resolution scanning would require two multifaceted rotating prisms, and could provide TV-type scan of lightly overlapping 28- by 28-degree frames. This would provide imagery with a high degree of geometri fidelity, with overlap for redundancy and improved contrast. The low resolution scanner will produce low resolution, high sensitivity performance. The angular resolution would be 3 milliradians suitable for oceanographic work. A high resolution scanner, consisting of a nodding flat rotating prism, would provide a TV-type scan of slightly overlapping 2.3- by 2.3-degree frames. The high resolution scanner would produce high resolution, high sensitivity imagery. The angular resolution would be 0.166 milliradian suitable for observation of sea-ice and coastal geography (Ref. II-16).

(4) Microwave Sensing. The use of passive microwave techniques provide for an all-weather capability, small power requirements, small space requirements, and the obtaining of many interesting effects and useful data. Microwave radiation originates from several sources: the self-emission of materials, sky reflections, and multiple reflections from other objects. Thus, the radiation measured gives an effective temperature for the areas or objects being observed. Although the data obtained will be affected by several factors, useful information can be obtained if measurements are carefully taken and properly interpreted.

(5) Laser Altimeters. Laser altimeters can be operated only when there is no cloud cover. The cloud occlusions limit the probability for operating these altimeters to a value of 40 to 60 percent. The altimeter has a field of view (beamwidth) of 10 seconds of arc and can measure surface profiles for a very small resolution area of 17.9 meters in diameter from a 200 n.mi. satellite. The diameter of the receiving aperature for the altimeter is 1.15 meters.

(6) Radar. Radar appears to have many useful applications in space reconnaissance for peacetime applications. Radar, which operates within the microwave spectrum, has essentially an all-weather capability compared to the IR and visible spectrum. Radar returns are usually less ambiguous than passive systems since the signal return is a function primarily of the target reflectance. For passive systems, the return is a function of both reflectance and emmissivity variables. In general, the active systems can operate with higher signal-to-noise ratios thus improving the detection capabilities. Radar systems would not exhibit any dirunal variations as would most other sensors. There are also disadvantages to using radar for space reconnaissance. All active radar systems require transmitters; relatively large antennas are required for good operation. Considering all factors, it is concluded that for some applications radar sensors do provide the best means of obtaining good data and thus should be considered for a manned lab (STARLAB).

The earth resources space program will have many effects on the availability of data and the extensiveness of the data for use in our expanding economy. Many of the applications proposed for spacecraft could be used on conventional aircraft or ground observations; however, for many reasons, it is believed that a manned space earth resources program will be much more fruitful and provide the much needed basic data we require for improving our raw material outlook and evaluating our cultural resources (Table II-9). Badgley, Childs, and Vest (Ref. II-18) have summarized the reasons for doing space observations over conventional observations as follows:

1. Fulfillment of requirements for continental area coverage otherwise unobtainable by available quantity of aircraft.

2. Coverage of remote areas beyond range for practicable aircraft operations.

3. Reduced costs (considering repeated coverage).

TABLE II-9. MANNED EARTH RESOURCES PROGRAM (Ref. II-17)

			Essential	Useful	Not Needed
Cultural Resources	Agriculture	Soils { Quality	X	X	
		Temperature		X	
		Moisture		X	
	Crops { Quality			X	
	Species	X		
	Population Distribution		X	
	Urban Land Use		X	
	Transportation { Development		X	
	Control	X		
Natural Resources	Fresh Water { Sources		X	
	Distribution		X	
	Pollution	X		
	Forestry { Distribution		X	
	Quality		X	
	Marine Life	X		
	Wildlife { Distribution		X	
	Migration		X	
Earth Sciences	Cartography			X
	Geology	X		
	Geophysics			X
	Geodesy			X
	Oceanography	X		
	Natural Disasters { Floods	X		
	Fires	X		
	Deformations	X		

Man's Participation

2-35

4. Assistance to U.S. global survey operations without large on-site basing support requirements.

5. Relatively vibration-free platforms capable of acquiring high quality earth surface data.

6. Potentially better quality data of some types than from aircraft within the atmosphere.

7. Detection on a global basis of broad-scale features not apparent on high resolution, smaller area photos.

8. Repeated coverage to detect changing phenomena.

9. Reduced data acquisition time.

10. Wide area coverage to reduce the technical problems of assembly of broad-scale aircraft mosaics.

11. For any specific area, the view from space is far more orthographic (free from distortion) than from aircraft.

Physical Requirements

The operation of the earth resources laboratory will involve at least two individuals; an earth scientist and a technician. The scientist will direct the entire program based upon his own knowledge and programs established by the ground control center. The lab will be outfitted with a console which include a view screen with a zoom telescope for direct viewing of the earth's surface at varying magnifications. Control for all the sensing devices will be located on the console (Fig. II-4). During daylight hours, the scientist will be carefully scanning the earth's surface through the telescope and selecting targets of opportunity as well as predetermined targets for sensing. During the dark hours, his time would be utilized to adjust and maintain equipment, reload cameras, and plot data. For the laser altimeter, the operator (scientist or technician) will have to turn the equipment on and off, set the stabilizing platforms for proper orientation, adjust the fine pointing of the receiver for best data, take earth coordinates to be recorded with the data received, record height data from laser altimeter to be recorded with other data, and monitor the laser altimeter systems to ensure that laser altimeters are operating correctly. For the radar system, the operator's main function will be to reduce the data storage requirements or transmission requirements. The radar system

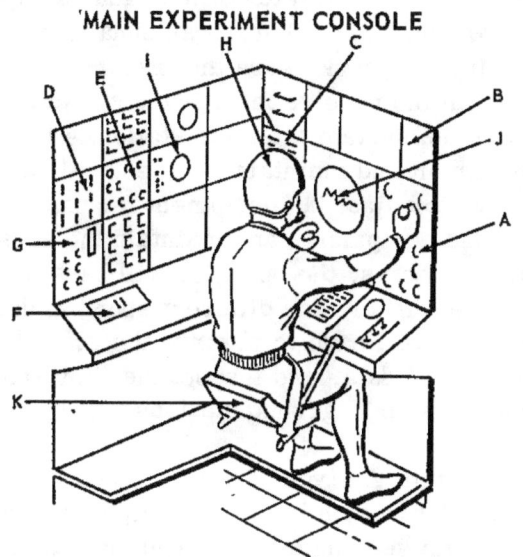
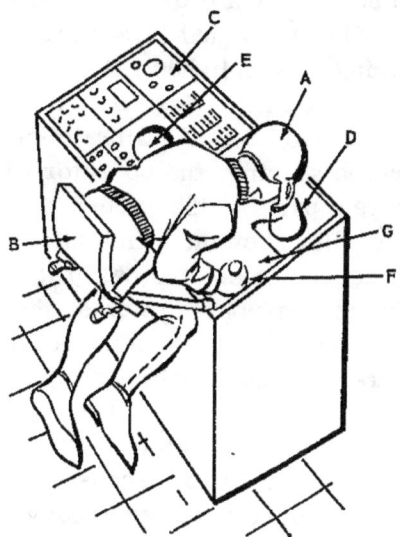

A. EXPERIMENT PLANNING & SEQUENCE
B. EXPERIMENT STATUS MONITORING
C. EXPERIMENT ACTIVATION/TERMINATION
D. EXPERIMENT SUPPORT FUNCTIONS
E. CAUTION WARNING MONITORING
F. DATA PROCESSING
G. DATA MANAGEMENT
H. COMMUNICATION
I. DATA TRANSMISSION
J. DAY/NIGHT TV MONITORING
K. PELVIC RESTRAINT

A. COMMUNICATIONS
B. PELVIC RESTRAINT
C. CAUTION WARNING
D. TELESCOPE
E. DAY-NIGHT TV MONITOR
F. TELESCOPE SLEW CONTROL
G. VOICE TAPE UNIT

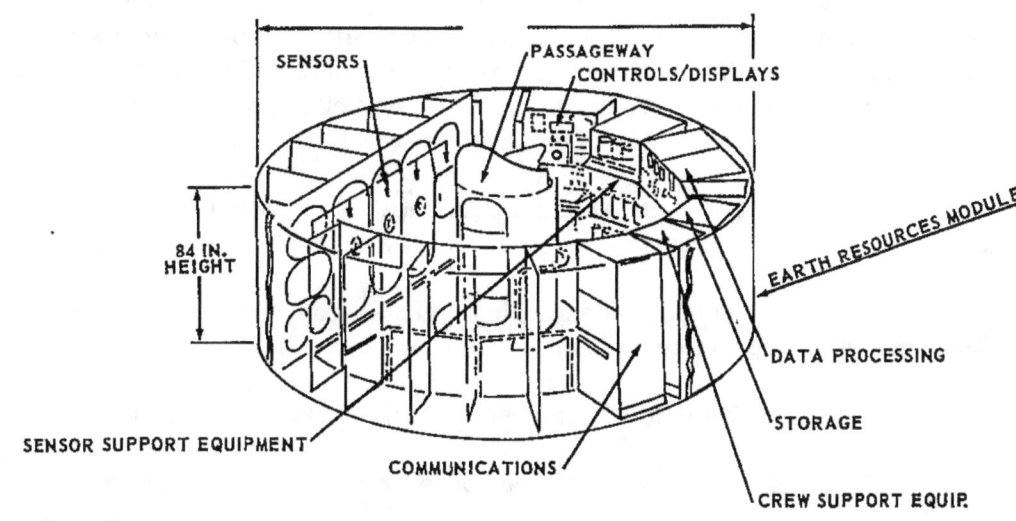

FIGURE II-4. EARTH RESOURCES CONSOLE (Ref. II-19)

will have to be maintained and calibrated. The operator of the radar equipment will have to turn it on and off, check returns to see that the antenna is positioned properly, set signal levels for recording, and make sure that recording of data is being accomplished. The operator, in the passive microwave systems, will have to assemble, install, and evaluate the system after the station is in orbit. He will also have to calibrate the equipment and maintain the electronics. In infrared sensing, the operator will have to adjust the equipment for uses over different parts of the earth, calibrate the equipment, and maintain the sensors and cooling systems. In multispectral systems, the operator will have to maintain the equipment and calibrate and select filters for different spectral data. In the photographic systems, the operator will have to observe the ground and select targets, load and unload film, record data, and inspect the film which covers perishable data for rapid transmission to ground stations.

Ground resolution varies greatly with the type of sensor used. In some applications it is not necessary to have high resolution; for example, a study of isostasy would be better accomplished with sensing over a large area at any one time; whereas, search for an ore deposit would probably require much greater resolution (Table II-10). Increasing the resolution requirements will also increase the data rate and consequently careful consideration should be given to the resolution requirements for each individual experiment. Figure II-5 shows the linear ground resolution versus the data rate. A ground swath of 100 n.mi. and a linear ground resolution of 100 feet would necessitate a data rate of approximately 10 million bits per second; a 25-foot linear ground resolution would raise the bit rate to almost 100 million bits per second. The high data rates from all the sensors will necessitate elaborate ground facilities and analyses for utilization in the field. Data handling will be one of the main problems of remote sensing from orbiting spacecraft. An entirely new concept of data handling and distribution on the ground will have to be implemented for full utilization of the data. Table II-11 is a summary of possible events in the evolution of the remote sensing of earth resources from spacecraft.

1. Power: 3500 watts when all equipment is in operation. 2000-watt average.

2. Weight: 5000 pounds excluding mounts, gimbals, and supplies. 8000 pounds with support equipment. 4000 pounds of supplies.

3. Volume: 400 feet^3 (instruments only). 2500 feet^3 total volume.

4. Manhours: Maximum 24 manhours; minimum 15 manhours.

TABLE II-10. NATURAL RESOURCE APPLICATIONS GROUPED BY RESOLUTION REQUIREMENTS (Ref. II-19)

Spatial Resolution (m)	Agriculture/Forestry	Geography	Geology	Hydrology	Oceanography
2 to 20	Timber-, Water-, and Snowline-Studies; Grass, Brush, and Timberland Interfaces; Vegetation Density; Tree Count; Tree Crown Diameter; Crop Species; Crop Acreage; Irrigation Studies; Small Fields (10 Acres or Less); Livestock Census; Infestation Surveys; Soil Texture	Population and Cultural Studies; Fishing Boat Activities; Land Use Studies; Topographic Mapping 1 250 000 and Larger Scales; Plant Cover and Soils; Forest Types; Thematic Mapping; Urban Development Survey; Classification of Facilities	Delineation of Small Folds, Small Linear Elements and Stratigraphic Sequence; Lithologic Units; Soil Compaction; Slope Stability; Permeability Studies; Ore Deposits; Local Geothermal Anomalies; Tectonic Studies; Glaciological Studies (Local)	Groundwater Discharge; Subaqueous Features of Lakes; Detection of Water Pollution, Inland Areas (Rivers, Lakes, Bays); Effluents of Major Rivers; Monitoring Lake and Reservoir Levels; Evapotranspiration; Water Surface Roughness; Rainfall; Salt Content; Drainage Basins; Water Regimens of Valley Glaciers; Snow Surveys; Reservoir Sedimentation	Ice Surveillance; Snow/Ice and Ice/Water Interface Studies; Wave Profiles; Shoals and Coastal Mapping (Bottom Topography); Currents (Long Shore); Coastal Marine Processes (Tidal Variations); Estuarine and Shoreline Morphology; Sea Level and Sea Slope; Sea Mammals Detection; Navigation Hazard Survey; Glacier Location
20 to 100	Timber- and Snowline-Studies; Fields of Larger Sizes, 10 Acres or More; Soil Temperature; Detection of Forest Fires; Farm Planning	Water Resources; Gross Cultural Studies; Geomorphology Studies; Gross Land Use Studies; Topographic Mapping, Scales Smaller Than 1 250 000; Pollution (Air, Land, Water); Thematic Mapping; Transportation Studies	Delineation of Folds and Linear Elements; Soil Compaction; Slope Stability; Gross Geothermal Studies; Geomorphic Studies; Glaciological Studies; Mineral Belts; Permafrost; Earthquake Damage Surveys	Evapotranspiration; Water Surface Roughness; Rainfall; Salt Content; Drainage Basins; Water Regimens of Valley Glaciers; Snow Surveying; Reservoir Sedimentation; Ground Water Surveys	Sea Surface Thermal Mapping; Cold Region Thermal Structure; Fresh/Salt Water Interface; Water Pollution, Large Areas, Oceanic, Harbor Areas; Ocean Waves; Currents (Offshore); Biological Studies (Fish) and Other Population; Wave Refraction Studies; Volcanic Activity
100 to 300	Timber-, Snow- and Desertline-Studies; Fields of Gross Sizes (Rangelands, etc); World Timber Inventory	Land Use Studies; Thematic Mapping; Global Population	Delineation of Large Folds and Linear Elements; Lithologic Units; Geothermal Studies; Volcanic Studies; Metallogenic Provinces; Inventory of Ice Features	Evapotranspiration; Water Surface Roughness; Rainfall; Monitoring Lake and Reservoir Levels	Currents (Offshore); Water Masses; Upwelling Areas; Fish Location; Ocean Mapping
Greater Than 300	Soil Moisture; World Gross Crop Inventory	Cloud Studies; Land Use Studies; Thematic Mapping of Regions and Continents	Delineation of Large Folds and Faults; Slope Stability; Gross and Local Geothermal Studies; Internal Magnetism; Metallogenic Provinces; Gravity Gradients; Isostasy; Continental Drift	Evapotranspiration; Rainfall; Snow Surveys	Sea State; Delineation of Pack and Cap Ice Margins; Sea Water Color Analysis

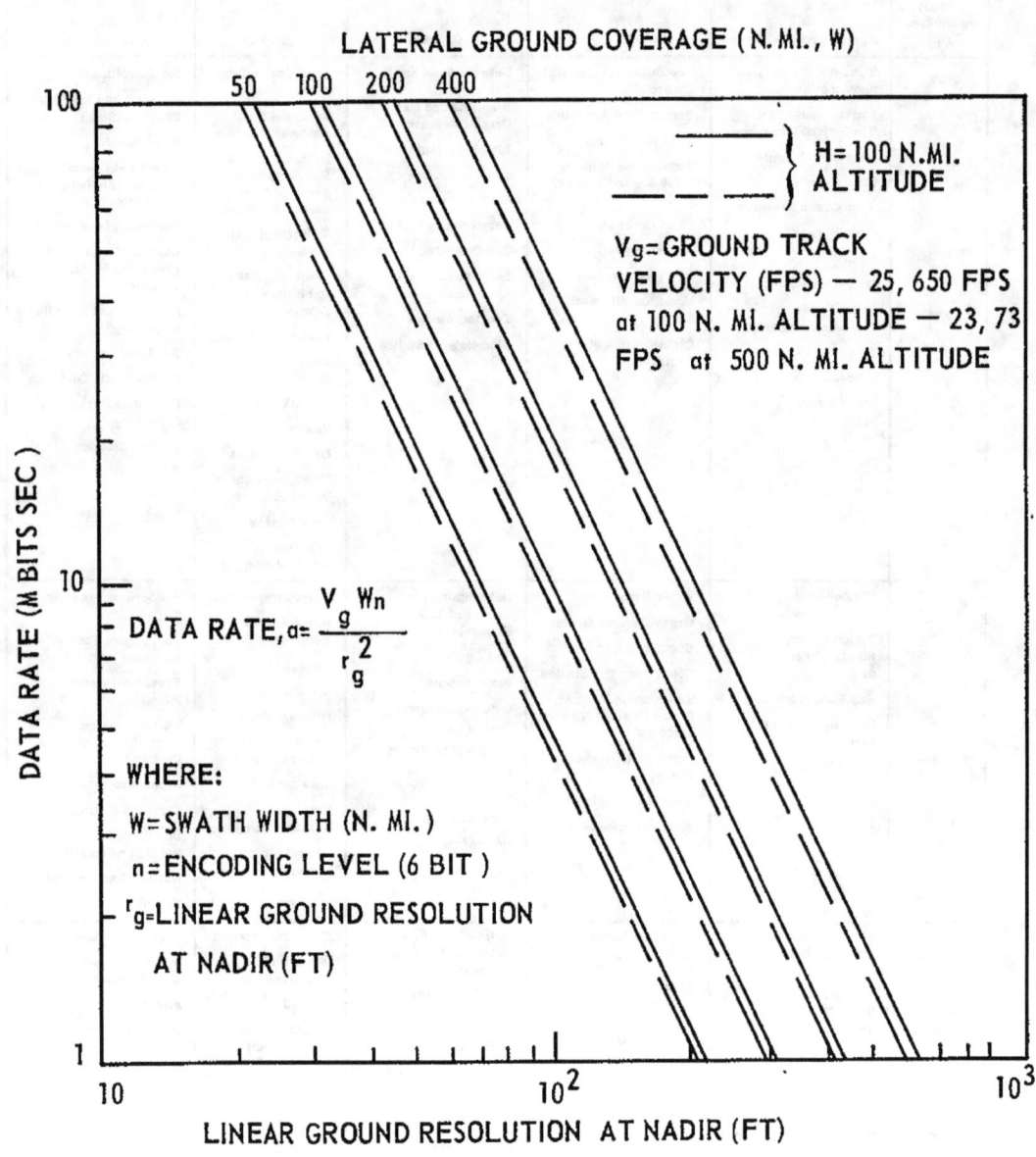

FIGURE II-5. LINEAR GROUND RESOLUTION VERSUS DATA RATE

TABLE II-11. SUMMARY OF POSSIBLE EVENTS IN THE EARTH RESOURCES SURVEY PROGRAM

Event	1968	1969	1970	1971	1972	1973	1974	1975	1976
Basic Research In the Earth Sciences (Signature Req'ts, Data User Needs)								Gov't Sponsored	
Sensor Development									
Data Management Techniques Development									
Aircraft Projects	Gov't and Commercial								
Unmanned Spacecraft R&D Flight									
ERTS A Mission					U.S. Data				Agr., Geo
ERTS B Mission						Global Data			Hydrology
ERTS C Mission							Oceanography		
ERTS D Mission								Detail Land Use Mapping	
Manned Spacecraft Flights									
Gemini Photography 1965, 1966		⟨9⟩ — 4 Camera, Multispectral Terrain Photography							
Apollo Photography									
Apollo Application Missions					Workshop Experiments				
Space Station Missions								MultiDisciplined	
Operational Earth Resources Projects									
Government		Depts. of Agric., Int., Comm., Navy							
Commercial									
Foreign Gov't/U.S. Gov't Cooperation									

2-41

5. Gravity: Zero permitted — nonrotation capability.

6. Pointing: Geocentric with 1 degree in all three axes. For those experiments needing greater accuracy, a separate platform will be used.

7. Height: 200 n.mi., preferable; 300 n.mi. acceptable.

8. Inclination: 50 to 70 degrees.

9. Constraints: Film protected from radiation. Cryogenic system for some sensors. Computer with memory for ship, aircraft, iceberg, ground control positions, etc. Many ground control stations and sensors needed. Need 4 to 5 antennas for data receiving and transmission of signals. Small sensor airlocks. Need extendable arms for magnetometer and gravity gradiometer. Automated computer calibration of instruments.

The following is a list of the major equipment needs for an earth resources laboratory (this excludes electronic, antenna, etc. equipment):

1. Spectrophotometric camera which is capable of obtaining spectral data in two selected modes covering the range from 4200 Å to 8400 Å. Data will be used in geologic mapping, pollution studies, crop studies, and oceanographic studies.

2. A microwave temperature sounder will be used for tropospheric sounding data.

3. A dual-channel scanner is used for identification of terrain features, crops, soils, shoal mapping, and water pollution. This instrument will record in graphic form the terrain radiation as observed simultaneously in the two spectral bands, 0.6 to 0.7 μ and 10.0 to 12.0 μ.

4. An electrical scanner radiometer is used to map the brightness temperature of the earth at the chosen wavelength on a global scale, and to interpret the measurements in terms of their meteorological and geophysical significance.

5. A six-channel multispectral photography is an extension of standard photographic techniques by dividing the visible spectrum into discrete spectral bands. It is to be used in studies of geology, hydrology, agriculture, forestry, oceanography, and geography.

6. A metric camera will be used for high precision metric photographs with time-correlated stellar photographs for precise location. The information will be used for geodetic and cartographic purposes.

7. An infrared interferometer spectrometer will be used for measurement of temperature profiles, vapor distribution, ozone concentrations, and minor atmospheric constituent concentrations. Infrared characteristics of surfaces in the 8- to 12-μ region will be obtained.

8. An infrared temperature profile radiometer can be used for obtaining a three-dimensional temperature field of the earth's atmosphere for meteorological observations and for spot temperatures along the orbit path.

9. Radar altimeter scatterometer.

10. Day/night camera system.

11. Coherent radar or synthetic aperture radar will provide high resolution imagery for geology, agriculture, etc.

12. Incoherent side looking radar has essentially all-weather capability with adequate resolution for gross mapping.

13. Multispectral tracking telescope.

14. Laser altimeter/scatterometer is used for accurate altitude for a very small land area. It can be used in conjunction with the metric camera for accurate topographic maps.

15. High resolution panoramic cameras.

16. Magnetometer.

17. Gravity gradiometer.

Materials Science and Manufacturing Processes

Facilities for research and experimentation embracing an interdisciplinary effort in the areas of materials science, manufacturing processes and technology, and industrial applications are to be provided on board STARLAB. Utilizing these facilities, studies into materials, materials processing, and manufacturing processes would be performed leading ultimately to product manufacturing.

Basic laboratory equipment and apparatus is provided in STARLAB's materials science and manufacturing processes facilities based on the philosophy that, although candidate experiments initiate the research program, the ultimate direction of studies cannot be predicted at this time. Results derived from preceding experimental programs, such as the Apollo Applications Program 2 flight, and from future earth-based studies will undoubtedly influence and modify any proposed program of study.

    a. Purpose and Objectives. The objective of the materials science and manufacturing processes program is (1) to investigate the behavior of materials and manufacturing processes in space, (2) to establish and evaluate materials and manufacturing processes and technology which can be used to produce products which can be made better in a space environment or cannot be made on earth, (3) meet a real and significant need of science and/or industry, and (4) have a value exceeding the cost of processing and transportation.

Many possibilities exist to utilize the unique conditions of the space environment to investigate materials and manufacturing processes which may be used to gain information of high scientific and economic value. Three effects of a zero-gravity environment form the basis for research and experimentation in materials and manufacturing processes in space (Refs. II-20, II-21, and II-22). These are:

    1. Absence of convection currents. (Thermal and mass flow.)

    2. Freedom from the need to support materials. (Levitation.)

    3. Elimination of gravity flow with atomic and molecular forces prevailing. (Container, less fluids.)

These effects can be utilized for materials research in areas involving high temperatures and transitions between the liquid or gas phase and the solid state. General areas for research suggested by these effects include studies in:

    1. The liquid state.

    2. Melting and solidification.

    3. Blending and mixing.

    4. Nucleation and supercooling.

    5. Purification.

Of the three physical states of materials, the liquid state is the least understood. Considerable enhancement of basic knowledge could be gained from observations of liquids in a zero-gravity environment. The physics of liquids in zero gravity has been suggested for study in terms of phase changes, surface tension, and dynamics. Natural extensions of the liquid-state studies would lead to investigations concerned with melting and solidification. The formation of perfect spheres, solid or hollow, caused by surface tension forces prevailing in the molten material has been proposed for study (Refs. II-23 and II-24).

The elimination of gravity forces on materials forms the basis for consideration of studies into blending and mixing materials with widely differing densities, thus producing new materials (Ref. II-25). Homogeneous multiphase materials, such as foamed materials (gas/solid) and new composite materials (solid/solid), could be produced in space.

The elimination of container requirements in a weightless environment provides the foundation for nucleation and supercooling studies. Without the restraints of some kind of container, nucleation sites for the formation of crystals from the melt are drastically reduced or eliminated. New families of glasses with less tendency to crystallize and/or with modified optical and chemical properties have been proposed as the result of the elimination of nucleation sites (Ref. II-26). Controlled nucleation with seed crystals, along with the absence of convection currents in zero gravity, has also been proposed for the production of large single crystals from a supercooled sphere of material (Refs. II-26 and II-27). The absence of convection currents in zero gravity is expected to reduce the number of dislocations in the single crystals, thus giving them greater strength and utility. Controlled nucleation in zero gravity could also be applied to study unidirectional solidified composite materials.

Since materials in a weightless environment need not be confined to a container, the pickup of impurities from the container is eliminated. The ultraclean space environment also eliminates the chance for contamination by gaseous impurities. The preparation of high-purity materials has therefore been proposed as a candidate space experiment (Refs. II-28 and II-29).

Analytical models and theoretical considerations for the above and other areas of research in materials science and manufacturing processes in space are to be found in the literature (Refs. II-28, II-30, II-31, and II-32).

The current ideas concerning materials processing techniques that take advantage of the zero-gravity environment of space have been enumerated and classified (Ref. II-33) as:

1. Levitation melting processes that are not expected to be restricted by the size or nature of the melt and would be used to produce shapes controlled by surface tension forces and for the production of cast materials without any impurity transport, nucleation, or thermal effects due to containers.

2. Floating zone refining applications in which the size of the molten zone is not restricted by the weight of the molten material that must be retained by surface tension forces.

3. Fragile structures production, such as very thin castings, long whiskers, membranes, and long, thin extrusions.

4. Production of foamed metals, dispersions, and alloys that cannot be produced on earth because of large density differences between the components.

5. Diffusion — controlled crystal growth from vapors, melts, or solutions.

The information gained from the proposed areas of study in materials science and manufacturing processes using the above techniques, would be expected to greatly enhance scientific knowledge and provide the basis for the development of technological applications unique to a zero-gravity environment.

Three mission objectives for space materials processing have been proposed (Ref. II-34). These mission objectives are:

1. Space maintenance.

2. Space-structure manufacture in space.

3. Product manufacture in space.

Initially, the first two mission objectives could be de-emphasized. But one would desire a basic level of investigation in these two areas with later emphasis coming either when (1) process development has been completed and it is desired to have long duration independent operation, or (2) commitment has been made to interplanetary travel. These two objectives will be discussed briefly below.

The area of product manufacturing in space would have the most immediate payoff and would be the primary mission of materials process development in space. Attempts would be made to utilize the various properties of the space environment, which include:

1. Vacuum.
2. Temperature extremes.
3. Radiation.
4. Low gravity.

Although all of these properties can be reproduced on earth, the combination of the properties in large amounts is difficult and offers a great possibility for space materials processing. Presently, the main purpose for including the proposed facilities in STARLAB is to investigate the behavior of materials and processes in a zero-gravity environment.

Weunscher (Ref. II-21) has divided processes into three categories:

1. Those which work also in space.
2. Those which work better in space.
3. Those which work only in space.

It is felt that processes in category 1. would only be developed for space operation as the need arose for repair or assembly. Processes in categories 2. and 3. would be prime candidates for early development. Category 2. processes offer short-term payoff which will lead to early economic inputs to STARLAB. Category 3. processes are envisioned as one of the greatest potentials of space-based research.

Crystal growth is one process which offers possibilities of significant improvement when done in space. As previously mentioned, levitation reduces possible nucleation sites and the absence of convection currents reduces the number of dislocations. Utech (Ref. II-35) has given the following six possible causes for dislocations in single crystals:

1. Introduction from the seed.
2. Externally applied stresses.
3. Stresses of thermal origin.

4. Concentration gradients.

5. Condensation of vacancies.

6. Trapping of inclusions.

It is felt that category 1. would be eliminated by levitation in zero gravity; while categories 3. and 4. would be eliminated because of the lack of convection currents. Seeds with lower dislocation concentrations are believed possible if they can be tapered to microscopic dimensions. Another possible dislocation elimination method would be seedless growth; i.e., forming the melt into a liquid stream with controlled cooling from one end. Purification of the raw materials could reduce possible inclusions, thus reducing or eliminating this cause of dislocations. Supercooling of the melt could reduce the number of vacancies present at the solidification interface, thus reducing the condensation of vacancies.

Hollow ball bearings are prime candidates for early product development. In a zero-gravity environment, a free-floating liquid would be expected to assume the minimum surface energy configuration; i.e., a sphere. By blow casting, a molten metal bubble could be formed. In the weightless environment of space, the bubble would not drain and could be solidified as a high precision hollow ball. Adhesion casting could also be used. In this technique, a thin spherical shell is formed around a substrate material wetted by the desired surface material. These two techniques require information on surface tensions and wettability and what effect, if any, the space environment may have on them.

The elimination of gravity segregation in space provides another area for study. Materials will no longer demix. This has application for the manufacture of composites from colloidal suspensions. Fibers could be arranged in a liquid matrix and thus solidified. Liquid metal foams could be formed and then solidified to become solid foams. Solutions could be made from metal systems. The properties of such composites could be studied in terms of materials as well as particle size and distribution. It has even been proposed that if the particle size approached atomic dimensions, new metallic systems with complete homogeneity could be created from previously insoluble components. Such a system would be an ultrafine composite.

The above is a brief examination of the possible experiments program in materials process development.

In the present space programs, space maintenance is not necessary. It is bypassed by redundancy, mission abort, or launching new satellites to continue operation when the original satellite fails. This maintainability may also be neglected in the development of experimental packages. One could assume that if a package fails, it is discarded or returned to earth for maintenance. However, the long-duration laboratory must be maintainable if it is to exist. This ability must also be available for long-duration interplanetary manned flights. In this mode of operation, one would soon have no access to the friendly earth with its repair facilities. There would be no rapid abort procedure. For example, a Mars mission reaches the point of no return after 1 hour out of Earth orbit. If the mission is not aborted prior to this time, it is committed for 600 days; if it is, the abort procedure still requires 14 days.

Space maintenance procedures would require the following capabilities:

1. Repair.
2. Modification.
3. Assembly.

In all three of these areas, the requirements are made for equipment development and human factors design. This is to ensure that the tasks required are within man's capabilities when operating in space. At some future date, one could envision that communications and earth resources satellites would be periodically overhauled by the space tug personnel or in a larger space garage provided at the laboratory.

Containers for satellite hardware could eventually be produced in space. However, space structures manufacture is a much larger endeavor. It goes all the way from modular design and assembly to raw materials processing. The need for space structures manufacture would be created by the desire to build enclosures for materials processing at the full-scale production level. Manufacture of a space cage for remote storage or satellite protection and EVA operation has been proposed (Ref. II-36). An atmospheric shell of glass or plastic could also enclose one area for a recreational area or for a manned repair capsule. Structures manufactured in space could facilitate outfitting of vehicles for interplanetary flight. Finally, one could ship billets of materials which would be deformed in either the liquid or the plastic state to the desired shape satellite or to reform micrometeorite protection shields.

Up to the present time, most proposals for the utilization of space have been passive in nature, using satellites as observation points to view the earth and the solar system. Perhaps the truest exploitation of space environment will come with the development of processes which will produce products of economic value on earth by utilizing the characteristics provided in the environment. With this as part of the goal of space-based research, the people of our nation could envision the space program as not only a technology driver but also as an eventual direct economic payoff in itself, over and above any spinoffs developed. As with any other operation in this extremely hostile environment, materials processing in space will become a reality only by means of a definite, positive commitment to a program of research and development. The time for that commitment is now.

b. Description of Operation. A preliminary materials research program will be undertaken in the early years of STARLAB. Initial studies should include techniques for handling and manipulating materials, if such studies have not been undertaken in preceding space programs, and exploratory experiments leading to a continuing materials science and manufacturing processes program. Ultimately, this could lead to a program in manufacturing on board a separate manufacturing satellite.

Initial investigations of candidate experiments for STARLAB will be conducted on earth to determine the critical effects of weightlessness on the materials and manufacturing processes to be studied. Criteria can be developed from such investigations for use in evaluating the feasibility of performing the candidate experiments in STARLAB. It is the intent of such earth-based studies to perform theoretical analyses and predictions of the behavior of materials and manufacturing processes in space, and to analyze each experiment with respect to potential applications, economic, and scientific value.

The materials science and manufacturing processes research laboratory of STARLAB provides the facility for an environmentally controlled volume inside the basic structure of STARLAB and will consist of the following four segments:

1. Preparation and mechanical test area.

2. Experimental area.

3. Display and control console systems area.

4. Storage area.

Each area is equipped for shirt-sleeve operations, but does not contain any facilities such as living quarters, eating and sleeping equipment, etc. A floor plan (Fig. II-6) and an operational flow diagram (Fig. II-7) for the materials science and manufacturing processes research laboratory are presented.

The preparation and mechanical test area will consist of the necessary equipment and apparatus for preparing samples of materials prior to experimentation (e.g., mixing, compressing, etc.) and after experimentation for central analytical laboratory testing (e.g., grinding and polishing, cleaving crystals, etc.) for such mechanical tests as hardness and tensile, and for preparing manufacturing processes experiments. For safety and to prevent contamination of the life-support system of STARLAB, much of the preparatory work will necessarily have to be done in a glove box or other sealed type of environment. A full-time technician skilled in the use of the equipment, its maintenance and repair, would be required for the preparation and mechanical test area.

The experimental area is the heart of the materials science and manufacturing processes laboratory. In this area, the investigations of the effect of zero gravity on materials and processes will be carried out. The main equipment housed in the experimental area will be a high temperature furnace capable of handling a 100-gram sample and equipped with zone-refining and levitation devices. Vacuum and controlled atmosphere capabilities within the furnace will be included.

Ports are provided for experiments that are very sensitive to the small gravitational disturbances that might be caused by crew movements or shuttle docking operations. Such experiments can be positioned on the outside of the STARLAB through the ports, allowed to free-fall for the duration of the experimental period, and then retrieved with a serpentuator (Ref. II-37).

The experimental area will be manned full-time by a technician trained in the operation, maintenance, and repair of the apparatus in the chamber. The supervisor of work in all areas of the research laboratory will be the principle investigator or his delegate on board the STARLAB. The principle investigator will be familiar with the apparatus of the laboratory and may wish to be his own technician on occasion. However, it is expected that he will be more concerned with the results of the experiments, observations obtained from the analytical laboratory, and preparations and modifications of future experiments. The technicians manning the preparation and experimental areas of the laboratory should be cross-trained to provide variety in their work loads and to ensure that at least one full-time technician is available in the area should a shortage in manpower occur on board the STARLAB.

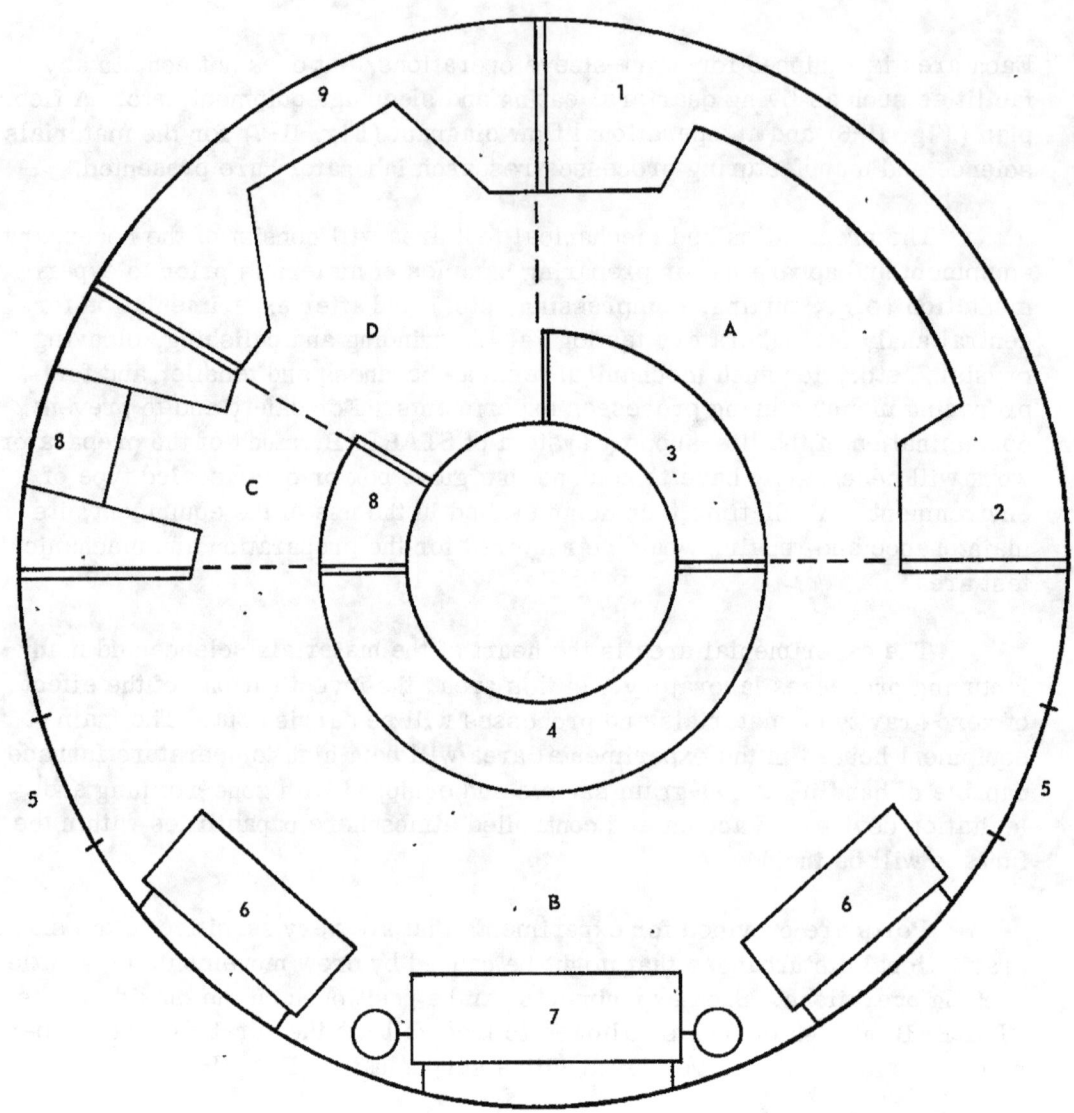

LEGEND:

A PREPARATION AND MECHANICAL TEST ROOM
B EXPERIMENTAL AREA
C STORAGE ROOM
D DISPLAY AND CONTROL CONSOLES AREA

1 SAMPLE PREPARATION APPARATUS
2 MECHANICAL TESTING EQUIPMENT
3 METALLOGRAPH AND MICROSCOPES
4 WORK BENCH
5 DOCKING PORT
6 TEST CHAMBER
7 COOLING SYSTEM
8 STORAGE CABINETS
9 TV, SENSOR, CONTROL, AN
  COMMUNICATIONS CONSOL

FIGURE II-6. FLOOR PLAN FOR MATERIALS SCIENCE
AND MANUFACTURING PROCESSES LABORATORY

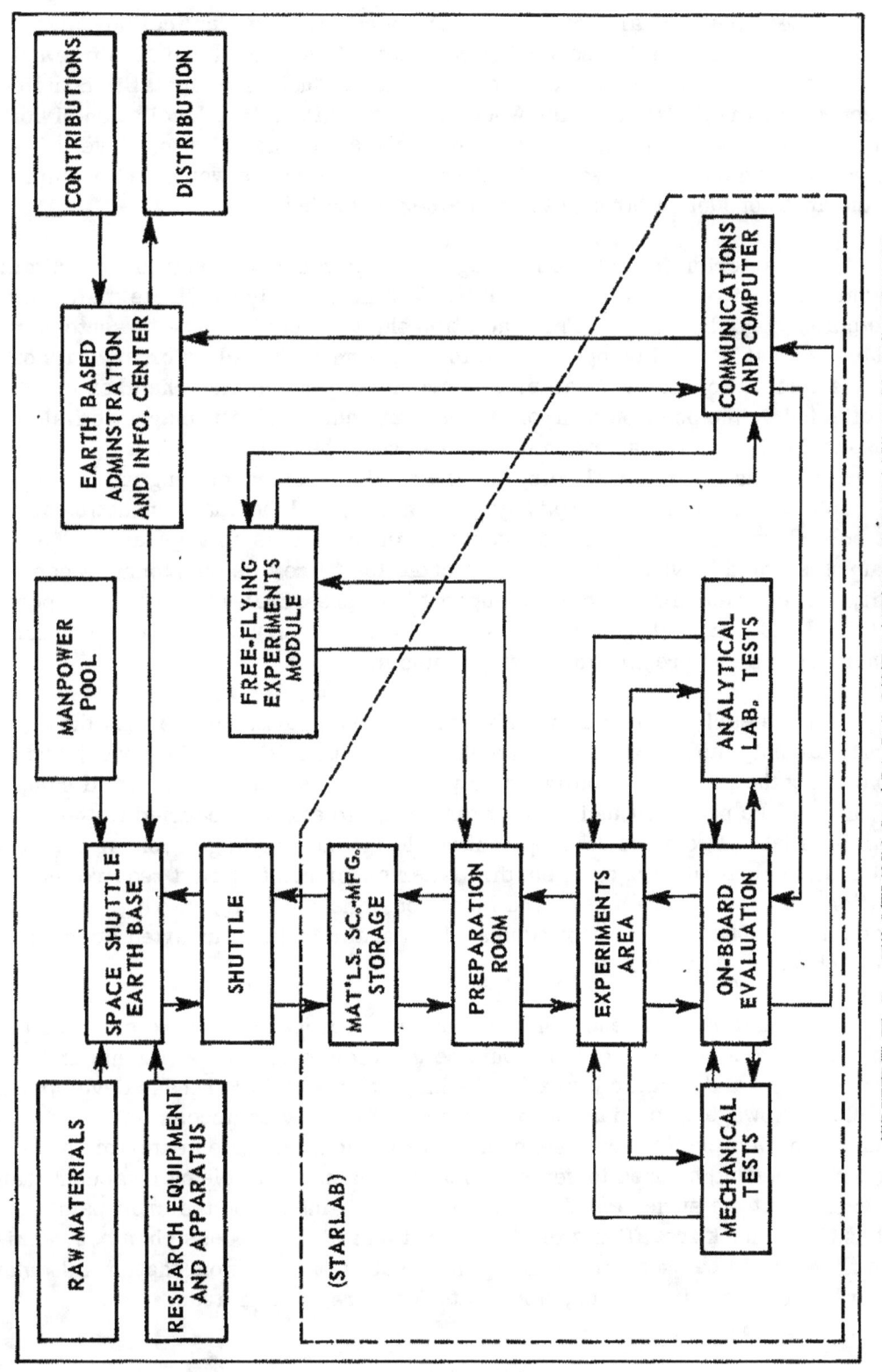

FIGURE II-7. FLOW DIAGRAM OF OPERATION OF MATERIALS SCIENCE AND MANUFACTURING PROCESSES LABORATORY

The experimental area will originally contain two identical up-dated versions of the materials processing chamber. They would contain three sets of induction coils for position control of liquid conductors or metallic containers of nonconductors. Gas jet pairs would also be available for leveitation of nonconductors. Heating would be provided by one set of the coils to a level of 2 kwh during the melting cycle. Heating of nonconductors would be accomplished by thermal conduction through the container or heated gases.

The interior wall of the melting chamber would be designed to minimize thermal radiation losses from the melt. The most likely candidate materials are polished gold or zinc. This radiation shield would be made of sectors of a sphere so that it could be opened to allow drawing of thin films or crystals or for the removal of the melt sample. Movement to the cooling chamber could be effected by the position control mechanism, pulses electromagnetic fields, or adhesion flow. The appropriate method would be used dependent upon the experiment and the material being investigated. Primary cooling devices would be a cold wall with high absorptivity, cold gases, and thermal conduction across the solid-liquid interface. These would again be used as appropriate. The chambers would be vented to a clean vacuum to utilize this environment capability. They would also be able to support high pressure work with atmospheres of varied gas composition. Electrostatic work benches would be provided for minor repairs and preparation of experiments.

In the early work, all experiments would be suitcase type with the technician only mounting them in the chamber, thus initiating their work and monitoring the results remotely. Early experiments would make small samples which would be packaged individually for return to earth or carried to the analytical lab. Prototype development would include winding of longer filaments and films for shipment back to earth. Later experiments would require modifications to control parameters and insertion of new heating, moving, and cooling devices. These would be modules which again could be inserted or mounted in the chamber with ease.

The materials science and manufacturing processes laboratory would have access to a docking port. It could be used for loading experiments and raw materials and in later years for docking of free-flying process development modules. However, its main purpose would be to provide access to the exterior of the space vehicle. Several serpentuators would be mounted exterior to the port. They would be used to remove process chambers to cleaner vacuum than is available at the surface of the ship, to place chambers on the orbit path, or to launch and dock free flights of short duration for the process chamber. This free flight would be useful in reducing the level of gravity from that of a 35-man space lab to that of free space, about 3 to 5 orders of magnitude.

Special safety measures will have to be included in the experimental area of the materials science and manufacturing process laboratory. As noted above, safety precautions for the preparation area are designed to prevent contamination of the life-support system during handling and processing of materials. Precautions against contamination of the life-support system of STARLAB by experiments in the experimental area must also be taken. Adequate exhaust venting to the outside of STARLAB is recommended. In the experimental area of the laboratory, the furnace area is of prime concern. Sample sizes should be limited in size and adequate thermal and explosion-proof shielding should enclose the furnace area. Safety considerations may require that each process chamber be enclosed in its own room when in operation. This would allow work to proceed on other chambers and has been provided for by remote monitoring of all experiments in a monitor room. Of course, any tools found to be needed during experimentation or maintenance should be carefully attended and kept from floating about the laboratory. An electrostatic zero-gravity work bench (Ref. II-38) has been proposed to help eliminate the problem of free-floating tools.

All experiments will be monitored by high-resolution television as well as having the necessary data output sensors. For this reason, a display and control console systems room is provided with a TV console, a data output console, and a control console for each process chamber. Space will be provided for the addition of more consoles in this area as they are needed.

The storage area of the materials science and manufacturing processes laboratory will operate as the warehouse for the materials brought to the laboratory by the space shuttle; i.e., raw test materials and new equipment and parts prior to assembly of experiments. Finished test materials, sample products, and any equipment to be returned to earth will also be stored in the storage area. No separate manning requirements for this area are anticipated and very little power will be required in the storage area.

The philosophy of the materials science and manufacturing process laboratory design is to provide facilities for a broad spectrum of initial experiments and maintain sufficient flexiblity for future experiments which may arise from a spin-off of the initial experiments or from original innovation. Rigid plans may facilitate design of space equipment but they can stifle innovation; therefore, an approach has been developed which provides concepts structured sufficiently for initial planning but flexible enough to permit change, individual participation, and contribution.

Original candidate experiments are now conceived as being primarily of the suitcase type. In this concept the experiment is sent to the space laboratory in a condition such that only plug-in connections are required and only minor monitoring is expected by onboard personnel.

    c. Candidate Experiments. The following is a brief listing of candidate experiments in materials science proposed for initial study on board STARLAB:

    1. Compact powders of widely differing densities, melt viscosities, and surface tensions to study infiltration, agglomeration, and dispersion rates.

    2. Grow metallic and organometallic single crystal whiskers by vapor sublimation to obtain larger whiskers.

    3. Grow complex dispersion-strengthened alloy crystals to study diffusion of vacancies and dispersed particles, dislocations, and strength.

    4. Grow Group III-IV compound crystals by solution methods to obtain larger chips for circuit bases and spherical shapes for manufacture of 4TT emitters.

    5. Grow dispersed phase nickel alloy crystals for sonar transducers.

    6. Grow ferrite crystals by solution and glassy flux methods to obtain larger ferrites.

    7. Supercool materials which normally crystallize to produce new glasses with modified optical and chemical properties.

    d. Physical Requirements. A series of factors must be considered for the design of the materials science and manufacturing processes research facilities to establish the physical requirements of the laboratory. These factors include:

    1. Heating methods and control systems.

    2. Materials-handling techniques.

    3. Thermodynamics of heat transfer.

    4. Total power requirements.

5. Vacuum venting and control systems.

6. Gas pressurization and control systems.

7. Instrumentation for monitoring experiments and collecting experimental data.

8. Packaging constraints imposed by fixed vehicle interfaces.

An indication of the physical requirements of the research laboratory is given by the information in Table II-12 which illustrates the set of apparatus that might be used in a typical materials science experiment. The total power would be required only during peak operation; i.e., when all equipment is in operation. The experiment would require the full-time use of one technician to set up, perform, and monitor the experiment.

e. Equipment List. Due to a commonality of invisioned equipment for work in the areas of materials science and manufacturing processes, a single combined research area for the two will be on board STARLAB.

The equipment and apparatus of the research facility will contain a selection of heat sources with temperature control, provisions for controlled cooling, instrumentation, and capabilities for material positioning, handling, and testing. Preparation and storage facilities will be included in the research area. Information display and control consoles will be within easy reach of the researchers working in the area.

Based upon current considerations and projected expectations, the equipment to be housed in the materials science and manufacturing processes laboratory will include:

1. Heating equipment and controls, 3000°F.

    a. Induction.

    b. Electrical resistance.

    c. Electron beam.

    d. Solar.

    e. Miscellaneous.

3. Controlled atmosphere (H_2, O_2, N_2, He, A) and vacuum chamber.

TABLE II-12. PHYSICAL REQUIREMENTS FOR A TYPICAL
CANDIDATE MATERIALS SCIENCE EXPERIMENT

	Power (w)	Weight (lb)	Volume (ft^3)	Gravity (g)	Miscellaneous
Furnace 3000°F	300	25	10	10^{-6}	1. For 100 g Sample 2. Vacuum & Atmosphere
Zone Refiner	300	25	10	10^{-6}	1. Built Into Furnace 2. Vacuum & Atmosphere
Levitation Device	100	25	10	10^{-6}	1. Built Into Furnace 2. Vacuum & Atmosphere
Quench Tank	0	100	2		Liquid Containment
Bench	0	25	5		
Specimen Preparation Desk	20	200	20		Contamination Control
Hardness Tester	0	20	1		
Tensile Tester	20	200	2		
Microscope, Optical	2	10	1		
Totals	742	630	41		

3. Levitation devices and controls.

4. Zone-refining equipment and controls.

5. Quenching facilities.

6. Work bench.

7. Liquid and gas pumping equipment.

8. Crucibles.
    a. Thermocouples.
    b. Thermistors.
    c. Potentiometers.
    d. Miscellaneous.

Equipment within the preparation area of the research laboratory would include some mechanical test equipment. The area will include:

1. Specimen preparation desk.
    a. Cleaners, slicers, and shears.
    b. Grinders and polishers.
    c. Presses.
    d. Mixers and blenders.
    e. Miscellaneous.

2. Hardness tester, diamond Tukon.

3. Tensile tester.

4. Optical microscope (transmitting, reflecting, and polarizing).

5. Metallograph.

6. Dilatometer.

7. Storage facilities for:

    a. Raw materials.

    b. Equipment and parts supply.

    c. Finished experimental test materials.

Physics

a. Purpose and Objectives. The purpose of this laboratory is to provide the means of physics research in the environment of the outer space. The properties of the space environment pertinent to the research in physics are: zero-g, clean high vacuum, low temperature, weak magnetic fields, freedom from the absorption, distortion and noise of the earth's atmosphere, and proximity to the outer cosmos. The experiments proposed for this laboratory can be classified as either experiments which can be done only in the space environment or experiments which can be done better in the space environment. The results of these experiments will yield useful information pertinent to the understanding of the physical laws of the universe.

b. Description. The physics experiments described in this section are restricted to (1) basic particles, (2) gravitation, and (3) relativity.

(1) Gravitation and Relativity Experiments. Because of precise height sensor and gradiometer afford prospects for measuring a host of significant physical parameters of the earth, these instruments should be developed; these could be incorporated in the future space station (Ref. II-39). Since the experiments must be free from any moving disturbance, a free-flying module may be advantageous. Some representative experiments are described below.

(2) Basic Particle Experiments. The objectives of these experiments would be to record, define, and determine the characteristics of particles in space and their interaction with the earth. The ultimate goal being to determine the origin of and predict the future behavior of such particles. These will include initial experiments in micrometeoroids, cosmic rays, aurora, airglow, zodiacal light, and spacecraft contamination.

Experiments in the fields of particle-earth interaction, i.e., auroras, airglow, zodiacal light, etc., will probably be completed by the time the space station is launched in 1975-1980. However, research in micrometeoroids and cosmic rays will continue in the space station. Since the instruments needed for detection and measurements of these particles need special pointing direction and orbit, a separate free-flying module is recommended for this laboratory. Provision must be made in the STARLAB for life support for an astrophysicist and storage of basic supplies and instruments. In addition, there must be a provision for periodical docking of the module adjacent to the physics laboratory for servicing. Only periodic attendance of a man is necessary. The power needed for the instruments aboard the module will be supplied by a power system aboard the module.

    c. Physical Requirements. Approximately 1200 cubic feet will be available in the general laboratory for a physics laboratory in addition to the free-flying module discussed above.

The basic physics laboratory will include two airlocks with a boom for extension and retraction of cameras and collecting devices on the outside of the space station. This will serve many future experiments requiring external access from the space laboratory. Equipment and manned requirements are given in Table II-13, manned requirements are given in Table II-14, and preliminary designs for the physics and cosmic-ray laboratories are shown in Figures II-8 and II-9.

TABLE II-13. EQUIPMENT REQUIREMENTS — PHYSICS LAB

Item	Weight (lb)	Volume (ft^3)	Power
Cosmic Ray Module (Includes Ionizatner Spectrograph, Cryogenic Refrig., 4 Experimental Bags, Film)	30 000	9000	3 kw
Physics Airlock Experiments			
Cameras	90	2	8 - 131 w
Boom (Common to All)	227	1	135 w (for 0.5 min) -10 w
Film	16	2	
Canisters	15	2	
Mass Spectrograph	9	0.3	24 w — Ave. 36 w — Mak.
Coronagraph	27	2.0	

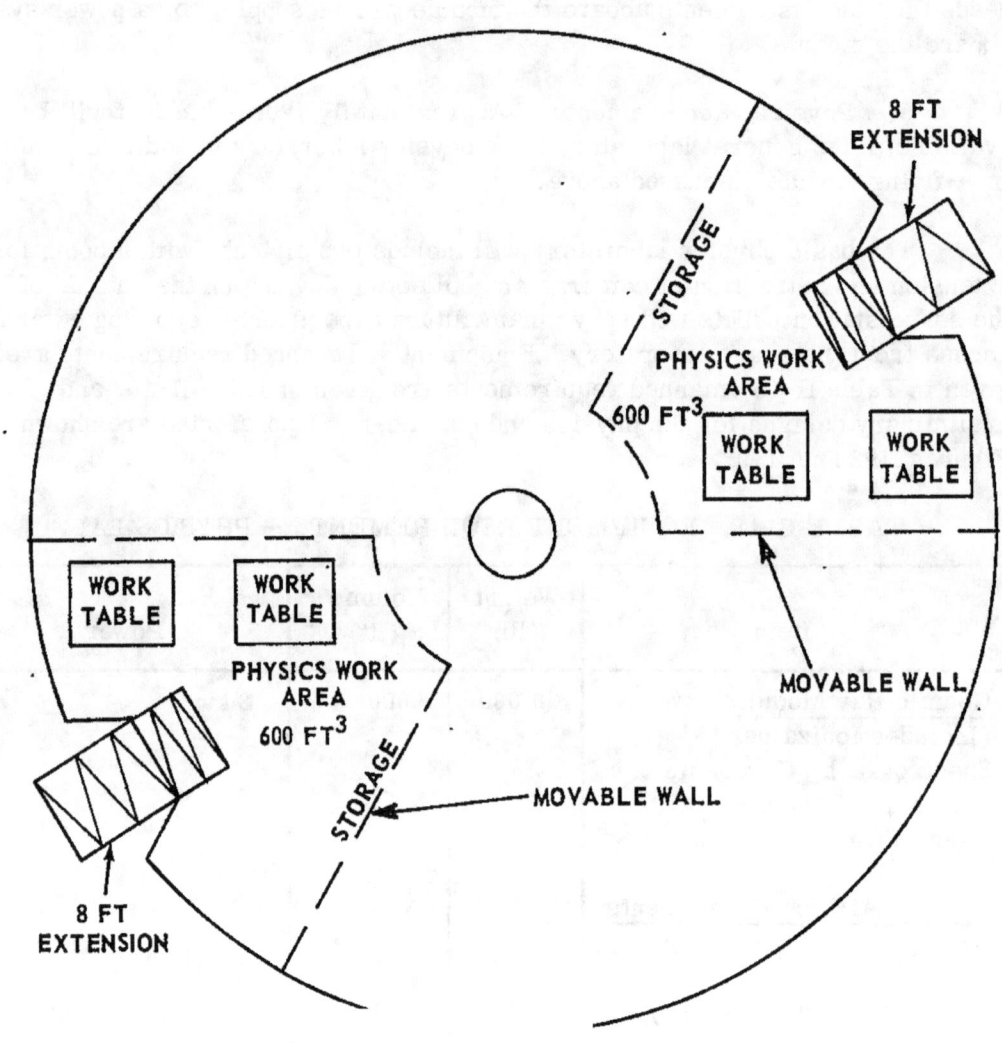

FIGURE II-8. PHYSICS LABORATORY DESIGN

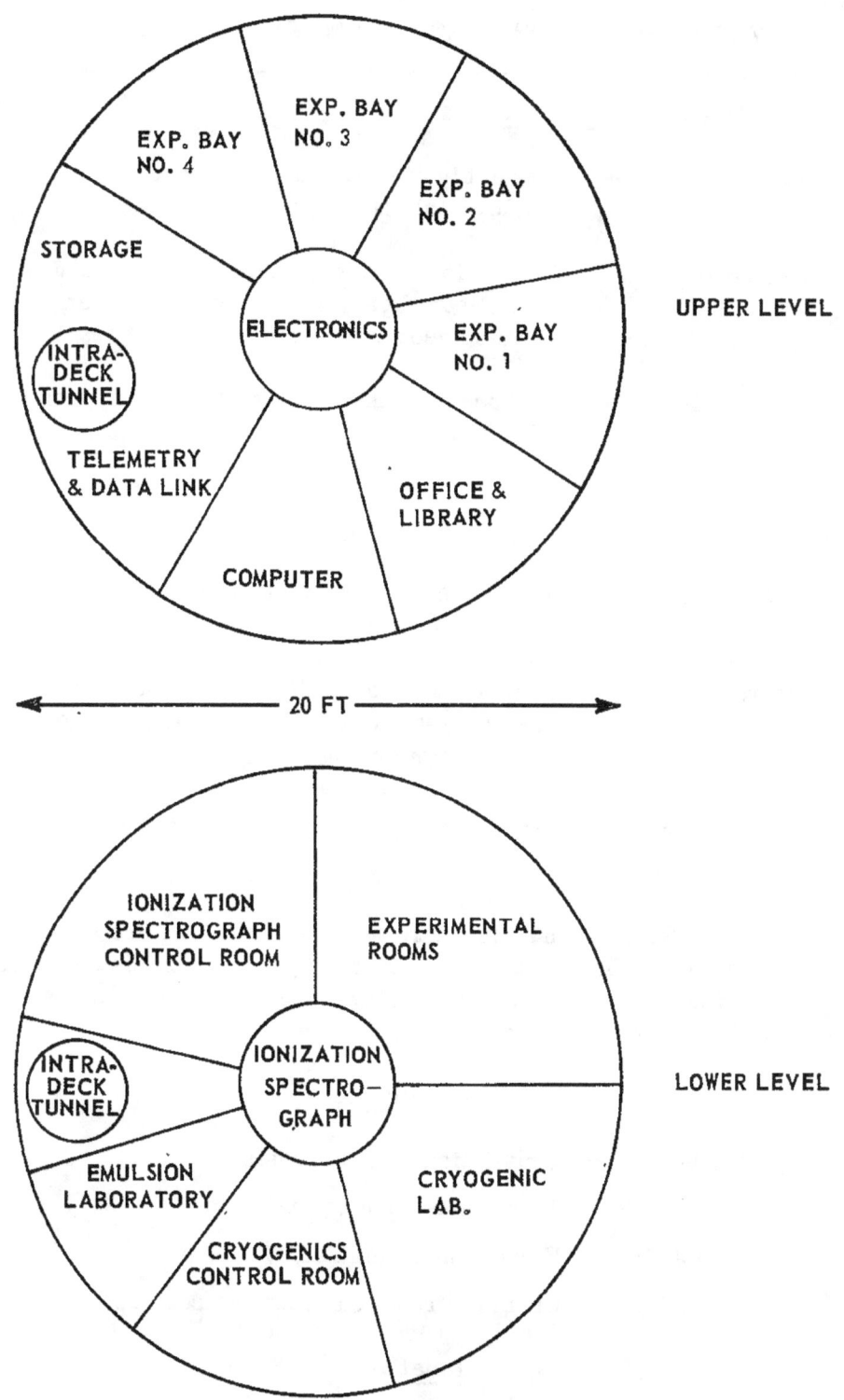

FIGURE II-9. COSMIC RAY LABORATORY MODULE

TABLE II-14. MANNED REQUIREMENTS — PHYSICS LAB

<u>Physical Airlock Experiments</u>

Preparation	—	Assembly of Components to Boom Head, Deployment and Initial Pointing of Experiments
Retrieval	—	Retraction of Boom, Removal of Components from Boom Head, Recovery of Film, and Development and Storage
Crew Skills	—	No Special Requirements; Time of Approx. 15 Min/Day

<u>Cosmic-Ray Laboratory</u>

Preparation	—	Deployment of Experiments by a Nonphysics Man — 15 Man-Days
Operation	—	Primary Responsibility for Operation Should be a Physicist; Less Highly Trained Man for Routine Monitoring of Experiments — Total 1 Hr/Day

    d. Experiments

    (1) Gravitation and Relativity. The following are some representative experiments of gravitation and relativity. The complete description of these experiments can be found in the NASA document "Manned Space Flight Experiment Catalogue" from their corresponding numbers.

Gravitation

1. EJ0471 — Gravitational mass sensing device.
2. EJ0453 — Liquid mass density under low g.
3. EJ0486 — Zero g and flammability.
4. EJ0558 — Chemical reaction in zero-g combustion.

Relativity

EJ0764 — Orbiting maser relativity test.

(2) Basic Particles. These experiments can be classified in two categories based upon their time duration.

1. Group S — Experiments that will last a short duration (up to several months).

    a. Coronograph contamination measurement.
    b. Contamination measurements.
    c. Environmental composition.
    d. Ultraviolet airglow horizon photography.
    e. Geneschein/zodiacal light.

2. Group C — Experiments that will operate continuously during most of the life of the space station.

    a. Micrometeoroid collections.
    b. Cosmic-ray investigations.

Chemistry

a. Purpose and Objectives. It is apparent that the laws of physics and chemistry apply in space as well as on earth. A rocket does fly in space, hydrogen and oxygen do combine to form water, and astronauts digest their food and give rise thereby to other physical problems.

One of the most fundamental concepts in physical science is that of equilibrium. As an example, one could cite that between N_2O_4 and NO_2 (Ref. II-40):

$$N_2O_4 = 2NO_2 \qquad \Delta H = +14.6 \text{ kcal}$$

At 25°C, the point of equilibrium, as indicated by a K_p value of 0.141, is 18.5 percent to the right. If the volume of the equilibrium system were decreased (at constant temperature), the point of equilibrium would be shifted to the left and the equilibrium mixture would contain a lesser percentage of NO_2 in accordance with Le Chatelier's principle.

Now the question is, would the same shift in equilibrium occur under conditions of zero-g? One can give the ready answer — yes — and defend such a view admirably using many earthly principles. Yet, when the principles of thermodynamics and equilibrium were worked out, no one could conceive a system under any but one g, and so the constancy of g was never in question.

In certain phenomena, such as those in which the weight of a substance is a factor, the effect of g may be evident. But in others, the effect of g may be more subtle, a second-order factor whose effect was unconceivable to a mind having always lived in the ocean of gravity.

Thus, for example, the decalcification of bone in space might be merely due to a change in solubility of calcium phosphate under the condition of zero gravity — a direct change in solubility or one induced by a change in pH of the blood (Ref. II-12). Such postulates, simple in thought and in execution, are prime experimental areas for investigation under zero gravity. Further, the major alteration in the properties of a pure substance occur with the first minute addition of a foreign material. Thus, the properties of pure water are changed markedly, by a quantum jump, by the addition of a minute amount of an electrolyte. Thereafter, additional electrolyte causes but a uniform variation in properties. Similarly, the properties of a system may be changed by the addition of only a small amount of gravity; variation of gravity thereafter may result in only an imperceptible change in properties. Therefore, it is to be expected that any major change in the properties of matter with gravity will be most evident in values of g around zero or only slightly above that. Furthermore, chemical processes are of the type that even slight improvements in the overall yield or rate of a chemical process may mean large economic returns (Ref. II-41).

The effect of gravity upon the nature and kinetics of chemical reaction is not readily apparent. One view is that the forces involved in chemical bonding are so large in comparison to the force of gravity that any effect of g is quite negligible and hence no change should be expected under conditions of zero gravity. A contrary view is that gravity is a hidden variable which has so far been ignored because of its all pervading presence on earth and its obvious constancy at any one place on a chemical bench, so that experimental results are truly a projection or section through an infinite continuum of results which depend upon the value of g. In this view, even the values of the thermodynamic variables might be subject to variation with g.

If we consider a system in equilibrium, such as the $N_2O_4 = 2NO_2$ equilibrium, and if a variation in the value of g affects both sides of the equation equally (or not at all), then no shift in the point of equilibrium is to be expected under zero gravity. But if one side of the equilibrium is affected differently from the other by a change in g, there should be an observable change in the point of equilibrium.

It is also possible that the activation energy for a chemical reaction will be affected by a change in g so that the kinetics, rate and mechanism, of a chemical reaction would change. Such change might best be discovered by an investigation of catalyzed reactions.

From a practical point of view, gaseous reactions might be easiest to follow under zero gravity. Where a gaseous reaction is concerned, homogeneous or one involving a solid catalyst, no experimental difficulty is anticipated inasmuch as a gas should fill its container completely and uniformly through kinetic energy diffusion. Such experiments can probably be set up in space with conventional equipment and glassware. Reactions with liquids will pose some difficulty inasmuch as liquid media are not readily poured in space and may have a tendency to break apart into separate aggregates, in the form of suspended droplets, especially if mixing or shaking is required. To minimize drop-forming of a liquid during a reaction, the reaction container should be filled completely, or almost completely, to leave as little void space as possible. In many cases it may be necessary to package the experiment on earth.

Chemistry is certainly a candidate for the design of an experimental program to be considered in the space program because chemistry has at least three areas that can benefit from chemistry in space.

    b. Exploration and Discovery. As in any type of science, man's natural curiosity becomes the stimulus for the searching for new ideas. New and unusual states of matter may be discovered (e.g., block holes in space, lunar glass beads, lunar mascons, etc.). Some new hints as to the distribution of the chemical elements will certainly be helpful to cosmology.

This area goes constantly forward even on earth as evidenced by the recent discovery of a previously unsuspected new state of that famous chemical, water, called polywater (Ref. II-42).

All sciences are equally entitled to opportunities for discovery and exploration; hence, chemistry should be considered in space.

c. A New Way for Solving Old Problems. In recent decades, chemical research has been governed quite strongly by advances in instrumentation. As each new instrument or technique that has a chemical aspect to it comes forth, a new wave of studies and researches occurs that essentially provides concepts, processes, or materials that will eventually be developed into consumable products.

Hence, chemistry undergoes the "fad" phenomena. The author submits that the space environment is also an immensely powerful and useful "fad" to attack anew the older outstanding and stubborn problems. Molecular beam studies of intermolecular forces (Ref. II-43) and chemical kinetics (Ref. II-44) can be quite easily adapted to the space environment. In fact, these studies can be done much better in space than on earth. Presumably they may even generate new and unique scattering studies.

The removal of convection via zero g undoubtedly will simplify the analysis of transport phenomena in the space setting. The study of compressibility near the critical state and at zero g will have a profound effect upon statistical thermodynamics of liquids (Ref. II-45).

d. Chemical Economics. The final area which supports chemistry's bid for a role in space is the economics of chemistry. Chemistry and chemical products play a very important role in our economy. Table II-15 shows that chemical sales are large and permeate our society thoroughly.

TABLE II-15. CHEMICAL SALES AND THE
GROSS NATIONAL PRODUCT (GNP) (Ref. II-46)

Year	Chemical Sales ($ B)	GNP	GNP (%)
1965	40	680	5.9
1969	56	880	6.4
1973	75	1180	6.4
$1 B = 1 \times 10^9$ $			

Most of our industries depend quite heavily upon developments in new materials. New and superior materials could be fabricated in space (Refs. II-47 and II-48).

Hence, one sees from the above arguments that chemistry is a science that has a place in the space environment. Upon first examination, one guesses that chemistry does not have the "earth-shaking" or "dazzling" type of experiments, and the days for these "dandies" are not necessarily coming to an end. As space becomes more and more available to other levels of priority, chemistry will be there. If one does not take a closed-minded point of view, chemistry may reveal some real "dazzlers" also.

This does not mean to say that chemistry is not in the space effort. On the contrary, one finds quite a bit of chemistry in the space effort:

1. Propellants.
2. Life support atmospheres.
3. Meteorite shields.
4. Space suits.
5. Seals, gaskets, and insulation.
6. Power supplies (batteries and fuel cells).
7. Adhesives.
8. Paints and coatings.

The above-given rational on the other hand is stating a case for the study of chemistry in space.

e. Description of Operation. To accomplish the chemical mission of STARLAB, the laboratory design of Figures II-10a and II-10b is proposed. An artist's sketch is shown in Figure II-10c. Instruments will be located on the "floor" and "ceiling" of one section of STARLAB. Instrument sites, on which the various instruments will be placed, will alternate with bench sites, at which utilities and working space will be available to the experimenter. On one level there will be 10 such instrument sites, 8 bench sites, 4 sites for storage of chemicals and small apparatus, and 2 openings for access to the annular space behind the instruments. The reaction handlers will be stored around the central shaft and "chairs" capable of attachment to the floor or ceiling will be available for the experimenter.

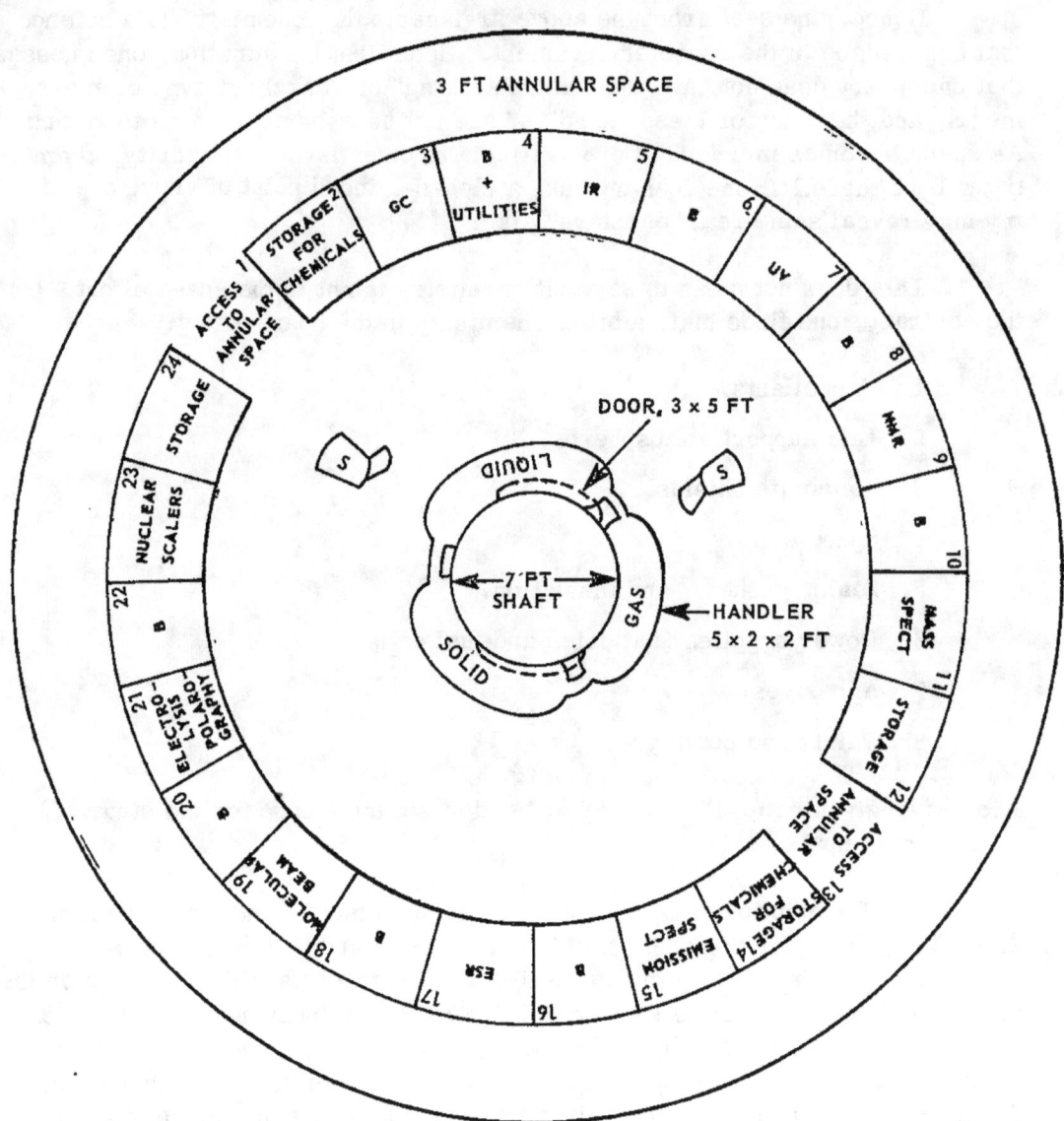

FLOOR PLAN (LOWER LEVEL)
EACH POSITION IS APPROXIMATELY 4 FT × 3 FT
POSITIONS 1 AND 13 ARE OPEN ACCESS TO ANNULAR AREA; ON UPPER LEVEL
SUCH ACCESS WILL BE IN POSITIONS 6 AND 18
B = BENCH + UTILITIES   S = SEAT (MAGNETIZED)

FIGURE II-10a. CHEMISTRY LABORATORY (Floor Plan)

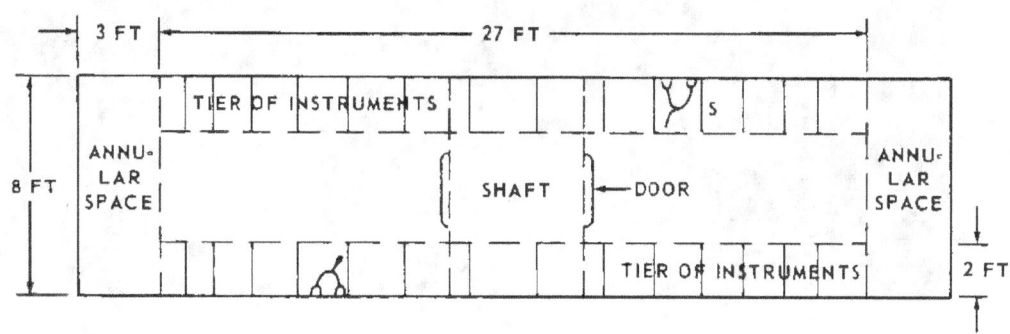

The upper level of instruments will be much the same as the lower level with alternating utility benches and instrument sites.

On this level will be six experimental sites for reaction chambers, one for combustion chemistry, two for optical apparatus such as visual microscope, polarimeter, Abbe refractometer, one for X-ray diffraction apparatus, and one for electron microscope. This totals 10 fixed instrumental sites. In addition there will be eight work bench sites, some of which will be used for necessary small laboratory apparatus; e.g., freezing point depression apparatus, electronic test equipment, melting point determination, distillation, etc. Also there will be two sites for access to the annular space at positions 8 and 16, and storage areas adjacent at positions 7 and 9, and 15 and 17. This totals 24 sites on the lower level.

FIGURE II-10b. CHEMISTRY LABORATORY (Side View)

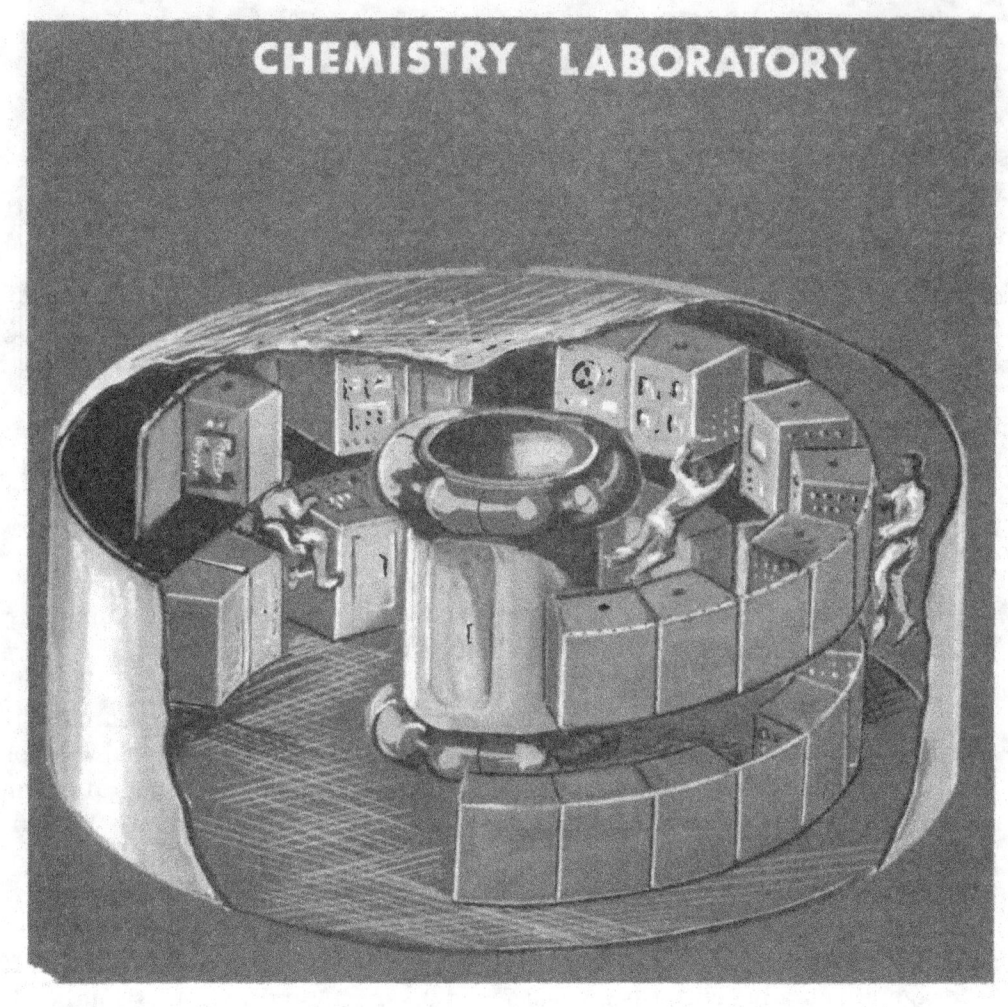

FIGURE II-10c. ARTIST'S CONCEPTION OF CHEMISTRY LABORATORY

On one level, floor or ceiling, the typical instruments to be placed could be the following:

1. Gas chromatograph (also liquid chromatograph).

2. Infrared spectrophotometer.

3. Visible and ultraviolet spectrotometer, with fluorescence attachment.

4. Nuclear magnetic resonance spectrometer.

5. Mass spectrometer.

    a. Time of flight.

    b. Quadrupole.

    c. Electric field focus.

6. Emission spectrograph.

7. Electron spin resonance apparatus.

8. Molecular beam apparatus.

9. Electrolysis and polarography equipment.

10. Radioactivity detectors and scalers; isotopic sources.

In addition there should be available, possibly on the other level, the following:

1. A weighing facility capable of weighing 0- to 20-g size samples to the nearest 0.01 mg. This will require a definite technological development by the 1975-1985 time frame.

2. A combustion chamber which would also serve as an organic hydrogenation chamber and perhaps as a calorimeter.

3. A laser station.

4. A small centrifuge.

2-73

5. An optical station, to include visual microscopes and a small electromicroscope.

6. An optical apparatus, such as light scattering equipment, nephelometer, Abbe refractometer, and a polarimeter.

7. A Mossbauer spectrometer.

8. An electrical test apparatus including a Wheatstone bridge or an A.C. conductance bridge, vacuum tube voltmeter.

9. Low energy electron diffraction apparatus.

10. Temperature bath, heating mantels, drying oven, muffle furnaces, a water till, thermistors, pressure transducers, etc.

All studies aboard STARLAB can be accomplished in three types of explosion proof handling containers or reaction vessels (Fig. II-11) which, utilizing interconnections, can effect experiments concerning:

1. Gas phase reactions.
2. Liquid phase reactions.
3. Solid phase reactions.
4. Gas-liquid reactions.
5. Gas-solid reactions.
6. Liquid-solid reactions.

The latter three categories are made possible by connections between the handlers. These can be interconnected and also individually force-flushed by the STARLAB integral vent system. A possible mode of storage for the handlers is circumferentially around the central passageway from which they would be removeable when needed. Two extra handlers should be available as backup support.

In the last analysis, these viewpoints can be resolved only by experiment. Therefore, in the belief that it is experiment which gives rise to new theory and only rarely does fundamental theory arise without root in experimentation, the following experiments were selected as prime candidates to be done aboard the space laboratory. The following four criteria were used in choosing these experiments:

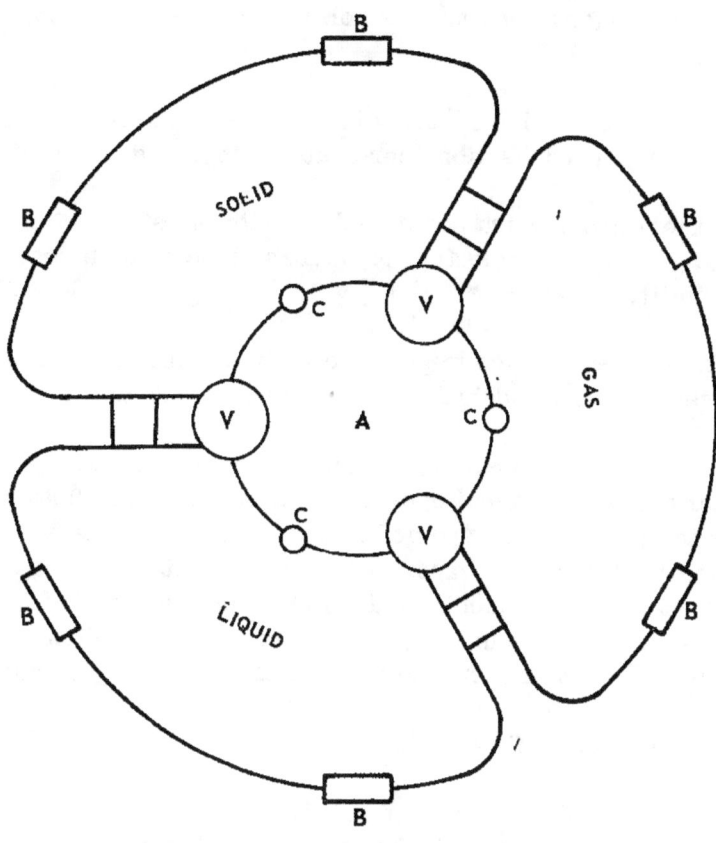

V = VENT (VACUUM OR FORCED GAS)

A = CENTRAL SHAFT (PASSAGEWAY)

B = HANDLING PORTHOLES

C = POWER AND COMMUNICATIONS OUTPUT

FIGURE II-11. HANDLING CONTAINER

1. Experiments which offered most promise in showing differences between one-g and zero-g behavior. These include surface properties; e.g., vapor pressure, surface tension, and colloidal properties; also, heat transfer and diffusion.

2. Experiments which dealt with fundamental properties of matter in bulk; e.g., boiling, condensation, chromatography, and crystallization.

3. Experiments which could be done with the simplest of equipment. For instance, if it were necessary, test tubes alone could be used for experiments in solubility.

4. Experiments which might have an immediate economic impact, such as the separation of aldehydes and ketones in essential oils.

The list of experiments recommended could be endless — if one wished to repeat every known chemical experiment under zero g. But only sufficient experiments to fulfill a 2-year period are given below. It is felt that the results obtained in these fundamental experiments would initiate, as in all research programs, a subsequent branching out which would readily fulfill a 10-year program. The program should begin with fundamental studies and essentially cover the entire domain of chemical and related physical phenomena.

f. Experimental Program.

(1) The Density of Water. Water is undoubtedly the most important chemical on earth, both in the human body and industry. Indeed, many properties of matter use water as a comparative standard; e.g., specific gravity, specific heat, various temperature scales, etc.

It appears therefore that a first effort should be made to determine whether the values for the physical properties of water are the same under conditions of zero gravity as they are under normal gravity.

A simple property to investigate and one which shows some promise of a possible change is the density of liquid water. An Ostwald-Sprengel pycnometer can be filled to mark on earth at a given temperature and sealed, and then taken into orbit under zero gravity (Ref. II-49). Since

$$\text{density} = \frac{\text{mass}}{\text{volume}}$$

and since the mass of the water is unchanged by its transport to space, and if it is assumed that any volume change of the solid glass pycnometer (Fig. II-12) will be negligible at constant temperature, any deviation from the original marks in the pycnometer can be attributed to a change in the density of water. The capillary arms of the pycnometer can be calibrated to give a quantitative value for a change in density. The experiment should be done at several temperatures.

One would expect the density of a solid to be unaffected by zero gravity but the density of a gas might so change and any such variation with temperature and pressure should be ascertained.

(2) Behavior of Aerosols and Bubbles. The following are to be investigated under conditions of zero gravity:

1. The movement of aerosol particles, both solid and liquid.

2. The coalescence of aerosol particles.

3. The formation and movement of bubbles of gas in a liquid medium and of soap bubbles in a gaseous medium.

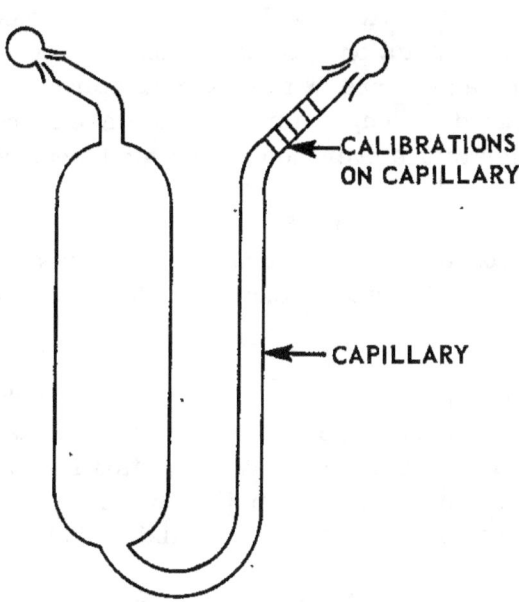

FIGURE II-12. PYCNOMETER

The movement of these particles and bubbles will be governed solely by diffusion through collision with surrounding molecules, a type of Brownian movement. Experiments are to be done at several temperatures; they are to be followed both visually and photographically.

The apparatus to be used is simple. For 1. and 2. above, an atomizer will spray droplets of various liquids and of various sizes into a container. For 3. above, an assembly of fine nozzles (much like some garden sprays or a gas burner) through which a gas can be forced, will produce a train of bubbles in a liquid (Fig. II-13).

FIGURE II-13. ATOMIZER

(3) Crystallization and Solubility. The growth of crystals is important to the physicist and the metallurgist. They generally deal with high temperature crystallization from a melt. However, the chemist has means of preparing crystals in a simple manner at room temperature; precipitation by chemical reaction and crystallization from a saturated solution are two such techniques. These methods, used under zero gravity, may yield insight to the nature of crystal formation and solubility.

(a) Supersaturation. Crystallization is to be accomplished by adding a seed crystal to a supersaturated solution, preferably one not too highly supersaturated. This experiment can be done at room temperature. The rate of crystalline growth is to be measured. Temperature measurements are to be made with time to observe any temperature overshoot in convection-force crystallization.

In the experiment above and in subsequent experiments where solids are formed, the crystals are to be subjected to microscopic examination and to X-ray diffraction analysis.

(b) Precipitation. By ordinary chemical means, the slightly soluble salts, silver chloride, zinc sulphide, and calcium carbonate are to be precipitated. The solubility of each is to be determined. Precipitation should be done with and without mixing. Without mixing, the precipitate will not "settle" and precipitation will occur only at the interface of the mixing liquids.

(c) Diamonds. Carbon is soluble in molten iron. When carbon is crystallized from molten iron, small diamonds may be produced throughout the mixture (Moissan's and General Electric's methods) (Ref. II-4 and II-5). Under zero gravity, nucleation may occur at fewer centers rather than be scattered throughout the mass. This would result in larger diamonds from a given weight of carbon. Induction heating and levitation of the iron may be feasible.

(d) Solubility in General. Both the solubility and the rate of solution as a function of temperature for inorganic salts, such as potassium nitrate and sodium chloride, are to be determined with and without mixing. Without mixing, solution will be due solely to a concentration gradient and molecular collision.

(e) Distribution Between Immiscible Solvents. The distribution coefficient of a solute between two immiscible solvents is to be determined; e.g., iodine between water and carbon tetrachloride, caffeine between water and an organic solvent, and silver between zinc and lead. These data would be of value in the extraction of organic products and in metallurgy; e.g., the extraction of silver in the Parkes Process.

(4) Boiling and Condensation. The purpose of this group of experiments is to investigate liquid-vapor equilibria under conditions of zero gravity.

The specific subjects to be considered are discussed in the paragraphs below.

(a) The Nature of Heat Transfer. In the absence of convection currents due to no gravity gradient, heat transfer should take place solely by molecular collision alone (excluding radiation which should be minimal within the body of a liquid at low temperatures). Under these circumstances it may be that a liquid acts as an insulator for heat. The experiment should be done while the liquid is static and is being mixed.

(b) The Nature of Bubble Formation. In the absence of gravity, the bubbles should not rise once they are formed but should move about solely due to molecular collisions with the liquid molecules; a kind of Brownian movement. Experiments should be done in a smooth, clean container which contains boiling chips; several external pressures should be used, up to and including the critical state.

(c) Superheating. This is related to bubble formation since superheating might be said to be the absence of bubbling above the theoretical boiling point. The degree to which a liquid can be superheated is to be determined; this may have a bearing on bubble chamber operation.

(d) Vapor Pressure. The vapor pressure of several volatile liquids should be determined as a function of temperature. Conversely, the dew point of "humid" vapor is to be measured. This will yield information which is the reverse of boiling, concerning the formation of "rain." Also the degree to which a saturated vapor can be supercooled should be investigated.

(e) Critical State. The critical data for a pure substance should be determined (Ref. II-50).

(f) Distillation of a Binary Mixture. The boiling point curve (i.e., composition versus temperature) is to be determined for a binary mixture; an organic mixture such as benzene and ethanol, benzene and carbon tetrachloride, and an inorganic solution such as sodium chloride in water. The theoretical plates required to produce changes in composition are to be calculated.

(5) Surface Tension (Ref. II-51). Surface properties, in particular the surface tension, of a liquid may be altered by conditions of zero gravity. Attempts at deriving a theory for surface tension of liquids have met with some success in the case of "normal" liquids, but little has been accomplished in the case of water.

The surface tension of a variety of liquids, e.g., water, ethanol, and benzene, should be measured by several techniques (some of which may not be applicable under zero gravity), namely, the ring method, capillary methods (vertical rise and Ferguson horizontal capillary), pendent drop method, bubble pressure method, and Wilhelmy slide method.

The surface tension as a function of temperature is to be measured at temperatures up to the critical temperature. Data obtained may give insight to the Ramsay-Shields equation.

The interfacial tension between pure liquids and between immiscible liquid metals is to be measured. The mercury-water interface is of special interest, industrially in mercury electrolysis cells and analytically in polarography.

The effect of solutes on the interfacial tension of a solvent is to be determined.

(6) Colloids and Surface Chemistry (Ref. II-52). Many of the characteristics of colloidal dispersions are surface-dependent, and since the properties of surfaces may be altered under zero gravity, colloids are a specially suitable field for investigation, particularly since much of physiological chemistry and many industrial processes involve colloidal species. The lack of settling may yield colloids of more uniform size and molecular weight, factors which may be of value in the formation of photographic silver chloride and in present-day polymeric substances such as polyurethane foams. Colloidal dispersions, normally unstable on earth, may be stable in zero gravity; normally immiscible substances such as oil and water, or a high molecular weight soap and water, properly dispersed by a blender or other technique, may be "homogenized." In particular, emulsions may show unusual behavior due to a change in the interfacial tensions of the liquids concerned.

(a) Preparation of Colloidal Dispersions. The properties of colloids are highly dependent upon their mode of preparation. Metal sols should be prepared by a chemical procedure such as the reduction of silver ion, by the physical Bredig arc method, or even by grinding. In addition, oxide, sulfide, and sulfur sols should be prepared.

The distributions of particle size should be determined.

(b) Wetting, Surfactants, and Foams. Surface tension affects the ability of a liquid to wet a solid. Fluorocarbon, silicone, and hydrocarbon surfactants normally lower the surface tension of water and hence increase the ability of water to spread. Since polyethylene has a low surface energy, it is harder to wet than more polar materials such as glass or steel.

The spreading of solutions of the above surfactants is to be measured by comparing the diameter of a droplet of pure water with that of a dilute aqueous solution of a surfactant (0.02 ml of 0.1 percent - 1.0 percent surfactant).

The thinning of the film of a foam is caused by two factors; (1) foam drainage due to gravity and (2) suction by Plateau's borders. The first of these will be nonexistent in a zero gravity environment so that the lifetime of a foam film or bubble should be increased. The absence of external convection current will also increase such lifetimes. Simple lifetime experiments can be done on soap solutions and protein solutions. The results may bear upon the separation of an ore from its impurities by the flotation method.

(c) Adsorption. Adsorption is a surface phenomenon whereas the sorption of a gas on a porous solid, such as activated carbon, involves its passage through the capillaries of the solid. Capillary condensation to the liquid state may occur and hence surface tension is a factor. Under normal gravity, the radius of curvature of the advancing liquid is greater than that of the rearward end.

Adsorption isotherms for gases such as helium, nitrogen, ammonia, methane, and benzene vapor on an adsorbent, such as carbon or alumina, shou be determined. Heats of adsorption, activation energies, and diffusion coefficients can also be measured.

(7) Diffusion. In the absence of thermal gradients under zero gravity, diffusion can occur only through molecular collision. Through diffusion experiments it is possible to check kinetic theory and the Ideal Gas Law, $PV = nRT$.

(a) Diffusion in Liquids. The rate at which diffusion occurs in a pure liquid, such as water, can be determined by using the isotopic species, H_2O^{18}. This can be added to ordinary water, H_2O^{16}, in known quantity and its concentration determined in various parts of the solution as a function of time.

Diffusion of a solute in a liquid medium is to be determined. For this purpose, radioactive isotopes will be specially useful in tagging the solute.

(b) Diffusion in Gases. This can be done, as above, using a radioactive gaseous species. The equation for the velocity of diffusion, $u = \sqrt{\dfrac{3RT}{M}}$, and the Ideal Gas Law are to be verified. The data obtained may be of value in the separation of gaseous isotopes.

The virial coefficients for a gas, such as hydrogen, are to be determined and compared with values under normal gravity.

(c) Diffusion in Solids. Two metals, placed in firm contact for an extended period of time (1 year or more) under zero gravity, are to be analyzed for the presence of the other metal. Also, the process of metallizing, developed by General Electric, offers promise under zero gravity (Ref. II-53).

(8) Preparation of Catalytic Materials. The catalytic properties of a substance depend not only on the chemical nature of the catalytic material but also on its physical structure. For example, the pore size distribution in a porous catalyst pellet can have a remarkable effect on the nature and distribution of products (Ref. II-54). Quite apart from their separation capacity, molecular sieves exhibit catalytic powers that have markedly influenced the petroleum industry.

The absence of gravity, and consequently the lack of settling, make it feasible to form catalytic materials with more uniform pores, perhaps by forced passage of fine gaseous bubbles through molten material.

(9) Osmotic Pressure and Reverse Osmosis. Osmotic pressure is a phenomenon in solution whose equation shows a striking similarity to the Ideal Gas Law. Because the process of osmosis is important in plant and animal physiology it should be investigated for any possible change under zero gravity. The following are to be determined:

1. The osmotic pressure of an aqueous sucrose solution.

2. The molecular weight of a polymer by osmotic pressure measurement; e.g., polystyrene in methyl ethyl ketone, polyvinyl chloride in various solvents.

3. The osmotic pressure of aqueous sodium chloride. This may yield information as to whether reverse osmosis is suitable for preparing pure water.

(10) Chromatography. Chromatography is intimately linked to the process of distillation and is useful in both the separation of similar chemical species and in analysis. Under zero gravity, unique separations may be feasible which may be of value for fractionating petroleum or in the separation of biochemical mixtures.

The following separations are recommended as candidates for experimentation among the almost infinite number possible:

1. The separation of the members of a homologous series of hydrocarbons such as n-hexane, n-heptane, and n-octane. In this and the following experiments, different columns are to be tested and retention times and volumes measured (Ref. II-54).

2. Separation of isomers, such as the C^7 and C^8 paraffins.

3. Separation of aromatic hydrocarbons such as benzene, m-xylene, o-xylene, and p-xylene. The meta- and para-xylenes are specially difficult to separate by ordinary means.

4. Separation of types such as aldehydes and ketones, and the C_7 and C_{11} constituents of essential oils (Ref. II-55).

5. Separation in a capillary column, using C_6 hydrocarbons. Here, surface tension may play an especially significant role (Ref. II-56).

6. Separation of gases such as the paraffin series from methane to n-pentane, and including hydrogen. Such separations may be specially affected by zero gravity.

7. Separation of hydrocarbons, using molecular sieves of various pore sizes; e.g., 4Å, 5Å, and 13Å. Also, the separation of carbon monoxide and carbon dioxide.

8. Separation of hydrogen isotopes (on an alumina column).

(11) Polarography. The migration of ionic species in aqueous solution under an electric potential is to be investigated. The conditions of concentration polarization which exist in polarographic techniques seem most promising to show differences between normal and zero gravity. A solid microelectrode will have to be used instead of the usual dropping mercury electrode.

For a solution containing nickel, zinc, cadmium, and copper ions and a solid platinum microelectrode, half-wave potentials are to be measured and quantitative analyses carried out (Ref. II-57). The experiments should be done in the presence and absence of oxygen and a maximum suppressor.

(12) Chemical Reactions. The chemical reactions listed below are comprehensive in that they cover a range of phenomena which should detect any variation with gravity, if such exists.

(a) The $N_2O_4 = 2NO_2$ Equilibrium (Ref. II-40). A quartz (or glass) tube containing a mixture of nitrogen tetroxide and nitrogen dioxide in equilibrium can be prepared on earth and its equilibrium point determined in space under zero gravity. The experiment is utterly simple and can be done spectrophotometrically.

(b) The Decomposition of Hydrogen Peroxide, H_2O_2 (Ref. II-58). The rate of decomposition of hydrogen peroxide, $2H_2O_2 = 2H_2O + O_2$, is to be determined at room temperature for the uncatalyzed reaction, and for the reaction catalyzed by an aqueous solution of potassium iodide, KI, liver catalyst, and a solid catalyst. The reaction is readily followed by measuring the pressure or the volume of oxygen gas formed. The kinetics of each reaction and activation energies are to be measured.

(c) The Oxidation of Sulfur Dioxide, SO_2 (Ref. II-59). The kinetics of the gaseous reaction, $2SO_2 + O_2 = 2SO_3$, is to be measured uncatalyzed by nitrogen dioxide, NO_2, and by a solid platinum catalyst.

(d) Fermentation (Ref. II-60). The fermentation of a sugar (e.g., sucrose, $C_{12}H_{22}O_{11}$) in the presence of an enzyme, zymase, from yeast yields ethanol, C_2H_5OH, and carbon dioxide, CO_2. Study of this reaction will also yield information concerning the behavior of yeast cells under the conditions of zero gravity. The reaction is readily followed by measuring the CO_2 produced.

(e) Decomposition of Acetaldehyde, CH_3CHO. Gaseous acetaldehyde is decomposed by iodine vapor to methane, CH_4, and carbon monoxide, CO. The kinetics of the reaction is to be measured.

(f) Inversion of Sucrose (Refs. II-61 and II-62). In aqueous solution, the inversion of sucrose is catalyzed by sucrase or by acid. This would be a homogeneous liquid reaction system. Since the optical rotation of the equilibrium system changes with time, the kinetics of the reaction can be followed with a polarimeter.

(g) Hydrolysis (Ref. II-63). The hydrolysis of an ester, such as ethyl acetate, is catalyzed in aqueous solution by acid or by base (saponification). The acid-catalyzed reaction yields an equilibrium system identical to that for the formation of the ester.

(h) Mutarotation (Ref. II-64). The kinetics of the mutarotation of glucose is to be measured in aqueous solution.

(i) Hydrogenation (Ref. II-65). The hydrogenation of an oil is carried out by bubbling hydrogen through the oil containing a finely divided nickel suspension. The hydrogenation of an oil, such as cottonseed oil, is to be investigated. The amount of hydrogen used and the consistency of the final product under zero gravity are variables to be determined.

(j) Course of an Organic Reaction and Its Different Products. Under the influence of light of 2500Å to 5000Å, chlorine and bromine react with hydrocarbons to give the monohalobenzene and further substitution products (Ref. II-66). The reaction is not selective and almost all possible substitution products are formed. Such distribution is to be investigated under zero gravity. If aromatic hydrocarbons are warmed with chlorine or benzene in the presence of iron, substitution of nuclear hydrogen occurs to give ortho-, meta-, and para-isomers. In this case, if benzene is heated with chlorine until all the benzene has reacted, the composition of the resulting mixture is 80-percent chlorobenzene, 17-percent p-dichlorobenzene, and 2-percent o-dichlorobenzene with 1-percent higher substitution products. Such distributions obtained under normal gravity are to be verified under conditions of zero gravity.

The following experiments are recommended:

1. Photochemical reaction of pentane with chlorine at 50°C.

2. Photochemical reaction of benzene with chlorine at 25°C and 100°C.

3. Thermal reaction of benzene and chlorine with iron as a catalyst at 50°C.

In each case the distribution of products is to be determined.

Other candidate experiments are:

1. Levitation polymerization; also supercured and supersized nylon fiber spinning (Ref. II-49).

2. Large, spongy metal, catalytic electrode production.

3. Convection-free combustion chemistry.

4. Gravity waves and the hydrogen-oxygen rocket engine.

5. Special relativity and the liquid oxygen-hydrogen reaction.

6. Chemical composition of Halley's comet.

7. Effect of zero gravity on DNA replication and synthesis.

8. Chemistry of finely divided propellants.

9. Orthopara hydrogen equilibrium near 0°K.

10. Determination of the atmosphere surrounding a space vehicle by infrared analysis using a technique similar to that for the determination of toxic chemical warfare agents in the field.

(13) Molecular Beam Studies (Ref. II-67).

(a) Solid-State Interactions (Low Pressure; Possibly 1/2 atm). Advanced studies of simulated meteoroid and micrometeoroids hitting various materials will be pursued.

(b) Solid-Gas Interactions (Low Pressure). A continuation of the study of accommodation coefficient, temperature regions of specularity and diffuse reflection from surfaces will be studied. Contamination, oxidation, and adsorption are the basic phenomena that are to be understood.

(c) Soil-Radiation Interactions (Low Pressure; to 1/2 atm). Surface erosion and degradation, coatings, degradation, and radiation damage are several of the technical areas that information is needed for basic knowledge and for support of material improvement. Testing of radiation shields and materials could be carried out here.

(d) Gas-Gas Interactions (Low Pressure; to 1/2 atm).

1. Interparticle Forces.

    a. Atom-atom collisions.

    b. Atom-molecule collisions.

    c. Molecule-molecule collision

    d. Ion-atom collisions.

    e. Ion-molecule collisions.

    f. Ion-ion collisions.

2. Elastic scattering.

3. Inelastic scattering.

    a. Charge exchange.

    b. Activation.

    c. Mass exchange.

    d. Angular momentum exchang

    e. Chemical reaction.

4. Imperfect gas studies (equation of state).

5. Chemical kinetics.

6. Liquid state.

7. Transport phenomena.

8. Shock tube studies.

9. Combustion studies.

10. Explosion studies.

The equipment and power requirement for the molecular beam studies are given in Table II-16 while the advantages of STARLAB for studies are as follows:

1. Low vacuum (10^{-12} atm).

2. High, essentially infinite, pumping speed.

3. Low impurities.

4. Large working volume.

5. Constant experimental conditions with no background buildup.

6. Extended experimental runs of days or weeks.

7. Absolute measurements.

8. Electron diffraction monitoring of scattering volume

Communications/Navigation and Traffic Control

a. Purpose and Objectives. The purpose of experiments in this field deal with the engineering test and evaluation of communications, navigation, and tracking techniques and related subsystems. The goals will include identification and development of efficient communications related to problems of navigation and traffic control.

TABLE II-16. MOLECULAR BEAM EQUIPMENT AND POWER REQUIREMENTS

Equipment	Weight (lb)
3 Velocity Selectors	6
3 State Selectors	6
2 Detectors	4
2 Scattering Chambers	2
2 Source Chambers	20
2 Targets	1
2 Geniometers	5
2 Ion Chambers	2
2 Source Chamber Temp. Control	4
1 Electron Diffraction Apparatus	10
Total	60

Instrumentation	Power (w)
6 Selector Controls	12
2 Detectors	20
2 Ion Chambers	40
1 Electron Diffraction Apparatus	50
2 Temperature Controls	60
Amplifier Bands	100
Enhancetron	200
Chopper	50
Phase Sensitive Detector	50
Total	600

This program is necessary to research, develop, and evaluate communications/navigation and traffic control systems. These systems are very important since they provide a data link and its capacity for STARLAB. Since STARLAB will have a tremendous data output, then the development of the communications system is extremely important.

Areas of interest included in this area are the advanced space communications systems, the space and earth navigation aids, and the close earth.

b. Description of Operation. This area of study will require the assembly of equipment such as transmitters, receivers, guidance devices, and remote sensing devices. A modification mode of operation will be necessary. This area of study will require the utilization of the standards and calibration facilities of the analytical support laboratory.

These systems will probably be made operational and checked-out manually. All data will be channeled through the communication and computer system on STARLAB.

c. Experimental Programs. A group of candidate experiments is available (Ref. II-68). This experiment program, listed below, will have a great deal of spin-off because any new developments in this program will heavily impact problems in our society.

1. Millimeter wave propagation.

2. Wideband variable high power transmitter for manned spacecraft.

3. Unified advanced laser communications.

4. Autonomous navigation experiment.

5. Noncooperative combined acquisition and tracking radar.

6. Sender antenna dynamics equipment.

7. Long-range, high-frequency communications.

8. Radar altitude sensing system for manned orbiting spacecraft.

9. Strap-down platform experiment.

d. Physical Requirements. There should be a laboratory in which the apparatus for experiments in navigation, communication, and traffic control is available. This laboratory will have pointing requirements and shields from STARLAB's stray fields. In addition, antennas and remote sensing devices must have external surface available for deployment. Mechanical and electrical lead-throughs must be available from this laboratory to STARLAB's exterior.

Due to the character of communications and navigation areas of study, this laboratory will have a larger power requirement than usual. The information management of this area of study will be handled by STARLAB's communications and computer network of analysis.

These experiments, while in progress, will be modified and will require the support of the electrical, mechanical, and the analytical support laboratories. Resupply will be required via the shuttle and the experiments will be mainly manned.

Engineering Research and Development

a. Purpose and Objectives. This area of study on STARLAB will have the tasks of evaluation of space operations, spacecraft and subsystems performance, man-machine interaction, and extravehicular activities.

These studies are quite important in the utilization of space environment and are further input into the future design of the role that man and his machines will play in space flight, space residence, and space work. A continued study and evaluation of the environment that man has built to shield himself is necessary to continue to make space an environment compatible with man.

b. Description of Operation. These studies will be carried out basically from an observational and monitoring point of view. Some of the acquired data are important for the operation of STARLAB, while other data are storable for later analysis.

Monitoring can be carried out in two modes; (1) man observed and (2) instrumentation assisted. These signals will be fed into STARLAB's computer-communication network which will sort, store, and analyze these data.

c. Experimental Program. This program consists of a set of well-documented experiments (Ref. II-69). Since most of these experiments will have been completed by the AAP, the experimental program serves merely as a pattern to control the design of STARLAB such that it will contain facilities for such studies. A descriptive list of experiments that comprise this program is as follows:

1. Jet shoes.

2. Expandable air lock technology.

3. Alternate restraint evaluation.

4. SLA panel relocation.

5. Habit ability and crew quarters.

6. Space bonding (adhesives).

7. Removal of propulsion utilization valve.

8. Test and demonstrate contamination detector.

9. Explosive metal-cutting experiment.

10. Artificial g experiments.

11. Heat pipe.

12. Compartmentalization.

13. Space suits and experiment deployment.

14. EVA program.

15. Atmospheric survey

16. Integrated maintenance experiment.

17. Gravity gradient stabilization.

18. Artificial gravity coupling system.

19. Date reclaimation unit.

20. Onboard checkout system.

21. Integrated waste collection in zero-g environment

22. Optical guidance.

23. Expendable airlock equipment.

24. Vehicular disturbances due to crew activity.

25. Centrifuge.

26. Primate transfer experiment.

27. Removal of components from stabilization platform.

28. Electrokinetic effect of a spacecraft.

29. Vibration transmission and absorption by spacecraft in orbit.

d. Physical Requirements. The main physical requirements necessary to perform the candidate experiments listed above are a monitoring and recording system, an onboard checkout system, and a laboratory facility in which certain experiments can be conducted. The monitoring system will also require instrumentation that is quite similar in many cases to the biomedical monitoring system. However, many of the experiments require no laboratory facilities since they are only data acquired from astronauts or systems participating in other experiments or duties.

The program will require resupply of new equipment that is to be space-environment tested. Tested materials must also be returned. Power requirements are quite modest (about 1 kw) and the manhour requirements serve as a cushion for some underestimation of other experimental, maintenance, and living manhours.

SUPPORT AREAS

Purpose and Philosophy

These laboratories will make routine analyses of almost any sort available to the STARLAB program. In addition, minor assemblies, standardizations, and calibrations can be performed. These labs will be manned by expert technicians that are known for their depth of knowledge, versatility, compatability, resourcefulness, and good judgment. These labs will also serve as practical consulting centers and will be capable of modifying experimental apparatus, upgrade instruments, and repair, maintain, and service a wide variety of apparatus.

Analysis facilities in orbit will provide a quarantine laboratory remote from the earth's environment. Here, the analysis of samples from interplanetary trips can start during an extended quarantine. This would alleviate many of the fears expressed during Apollo 11 concerning contamination of our atmosphere.

Analytical Laboratory. The purpose of the analytical laboratory is to provide the central analytical service section of STARLAB. Analytical equipment of general usefulness to several research groups aboard STARLAB would be housed in a shirt-sleeve environment in the laboratory and would be manned by personnel whose sole function would be to operate, maintain, and repair the equipment. Basically, the analytical laboratory would be a service facility to the experimental programs aboard STARLAB. Qualified researchers from outside the analytical laboratory staff would have access to the equipment and apparatus of the laboratory should they require use of any of it for direct experimentation. Facilities for general analyses, the calibration of experimental equipment, reference standards, and darkroom facilities would be provided through the central analytical laboratory. Samples could be reprocessed if necessary prior to shipment to earth. Orbital and terrestrial measurements could be compared to determine the effects of re-entry, its loading, and necessary handling.

The analytical laboratory of STARLAB is to be manned by a staff of specialists trained in the operations, maintenance, and minor repair of the wide variety of equipment and apparatus housed on board the laboratory. The function of the analytical facility, as noted above, is to provide researchers a central onboard test area, calibration and standardization services, and darkroom facilities. Major analytical testing equipment and apparatus would be housed in the laboratory for the common use on board STARLAB. The specialists of the analytical laboratory would generally receive samples to be tested from research laboratory workers, perform the requested tests, and report the results back to the requesting researcher. The laboratory will be linked directly to the computer facilities and to the communications center of STARLAB. Data can, therefore, be either returned to onboard investigators (via internal communications links) or to facilities on earth (via telemetry), allowing joint observation of experimental results by both. All instruments will have sensors which indicate satisfactory performance and/or trends toward unsatisfactory performance. The output from these sensors will be monitored by the executive computer and the analytical laboratory. Technicians will be assigned to correct malfunctions or perform preventive maintenance.

The analytical laboratory would be under the control of a competent supervisor who would direct the priority of work in the laboratory. Individual researchers might on occasion desire to use equipment of the analytical laboratory. The analytical laboratory supervisor's function in such cases would be to see to it that the researcher had access to the required apparatus. Storage of future schedules and past usage rates would allow researchers to plan usage prior to their arrival at STARLAB. It would also point to high usage rate or bottleneck equipment which could be duplicated.

A small darkroom facility is provided to process emulsions from experiments in earth resources, materials science, and possibly astronomy. Because film should be processed as soon as possible after exposure, it is suggested that processing take place on board STARLAB and the film not transported to earth for developing. Negatives could be fine-line scanned with high resolution TV and telemetered to earth. Through inversion, a positive could be reproduced at STARAD and at user locations around the world.

Operations to be performed in the darkroom will include development, washing, and drying of emulsions, plus facilities for storage of film and for preparation of developer, stop bath, and fixer solutions. Many of these processing chemicals could be eliminated if Polaroid-negative film could be adapted for the various cameras and sensors.

The physical requirements of the analytical laboratory of STARLAB are summarized in Table II-17.

Equipment housed in the analytical laboratory of STARLAB is intended for general usefulness to the research laboratories of STARLAB based upon current and projected expectations for analytical, calibration, and standardization

TABLE II-17. ANALYTICAL LABORATORY PHYSICAL REQUIREMENTS

Weight of Equipment	3000 lb
Volume of Equipment	300 ft^3
Power Requirements	1.5 kw (During Peak Operation)
Manned Requirements	48 Manhour/Day (2 Technicians)

requirements. An artist's concept of a typical support facility is shown in Figure II-14. The equipment and apparatus of the laboratory would include the following:

1. Abbe refractometer.
2. Analyzers for C, N, O, H.
3. Autoclaves.
4. Calibration apparatu
5. Calorimeters.
6. Centrifuge.
7. Cold-storage facilities.
8. Darkroom facilities.
9. Drying ovens.
10. Electrical devices.
    a. Multimeters.
    b. Oscilloscopes.
    c. Power supplies.
    d. Signal generators.
    e. Miscellaneous.

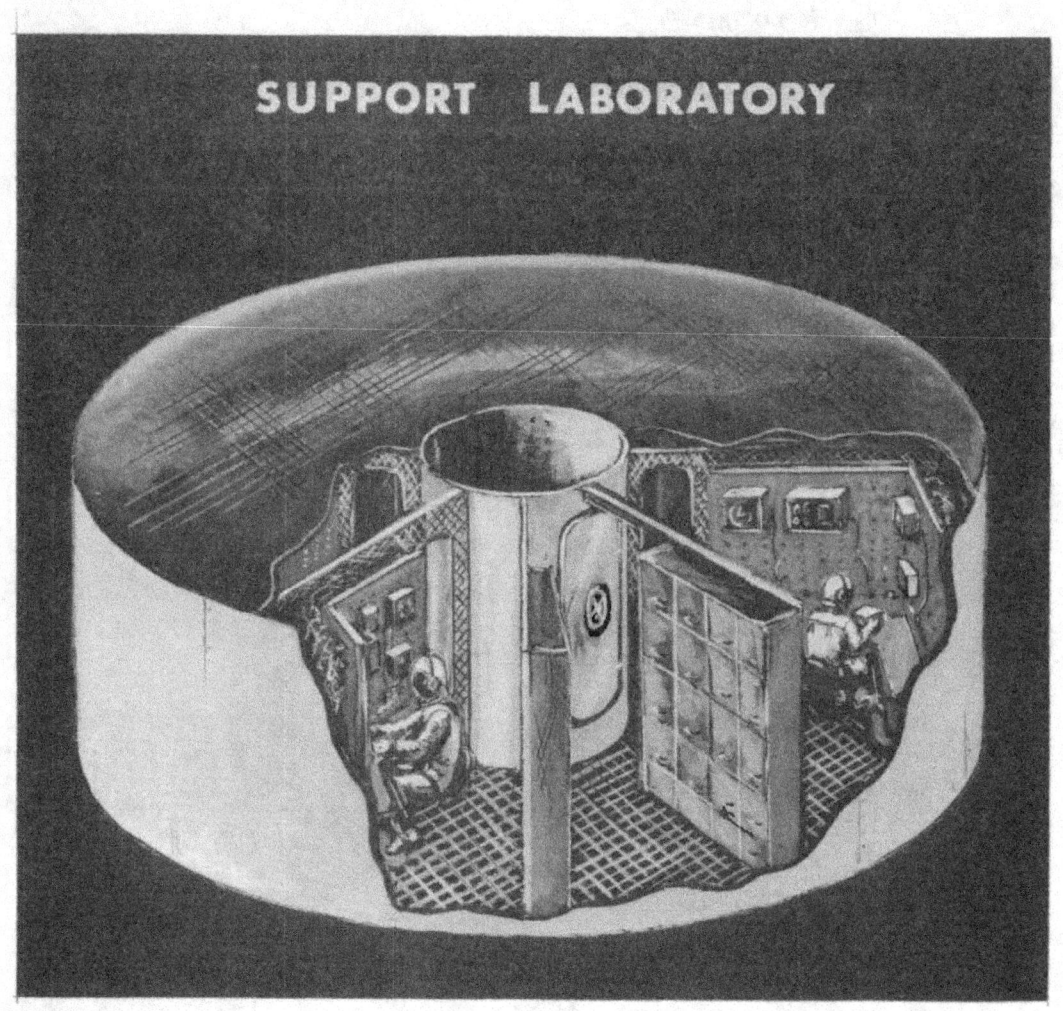

FIGURE II-14. CONCEPTUAL VIEW OF A TYPICAL SUPPORT LABORATORY

11. Electrochemical apparatus.
    a. Electrolytic apparatus.
    b. pH meters.
    c. Polarograph.
12. Electron microprobe.
13. Electron microscope.
14. Electrophoresis equipment.
15. Gas chromatograph.
16. General lab equipment.
    a. Containers.
    b. Labels.
    c. Miscellaneous.
17. Isotope tracer equipment.
18. Mass measuring devices (micro and macro).
19. Mass spectrograph.
20. Microscopes.
    a. Polarizing.
    b. Reflecting.
    c. Transmitting.
21. Optical pyrometer.
22. Osmometers.
23. Radiation detectors.
24. Reference standards.

25. Spectrophotometers (UV and IR).

26. Storage facilities.

27. Tensiometers.

28. Turbidimeter.

29. Ultrasonic cleaner.

30. X-ray diffractometer.

<u>Mechanical Area.</u> The purpose of having a mechanical laboratory aboard STARLAB is to provide a facility to make emergency repairs to STARLAB systems or make changes in the experiment packages. The fabrication of new parts and the modification of existing parts within the experiments will save considerable time which would be lost if the experimenter waits for the logistic vehicle. It may be necessary to update the changes in the experiments based on responses for previous data taken on these or other experiments or the development of new concepts. Modification of the experiments will also provide a means of overcoming unexpected defects in the design or the implementation of new ideas in a STARLAB system.

The shop should be equipped with machines necessary to do a limited amount of work and the instruments necessary to do the measurements required in the measurement or testing of mechanical devices or parts. The equipment may be added, deleted, or exchanged as the requirements aboard STARLAB change or as new developments are made on earth. The work to be done in the mechanical shop would be that which could be accomplished on the following machines: a lathe, milling machine, grinder, drill press, and a heat-treating furnace. It seems that by the time STARLAB is launched, many of these machine functions will be integrated into a single machine which will have adequate capacity to do the various jobs on material sizes necessary to make repairs and parts as required. In addition to the machining operations, welding will be necessary which can be done by a Heli-Arc welder with the various modes of current usually provided. Cutting of material for the machining operations or special contour cutting will be done by either a band saw or a laser beam cutter.

Sheet metal working equipment, such as a brake, shear, punch, and the supporting tooling, will be necessary when sheet metal work is required. Hand tools normally used on a bench will be necessary in space, except that

that these tools will probably need to be somewhat specialized because of the zero-g condition. In addition to the standard measurement tools used in the machining operations, inspection tools such as guage blocks, surface plates, optical flats, and dial indicators of various sorts will be provided to check physical dimensions of parts or assemblies. A limited amount of machinable stock in rod, tube, sheet, and plate form should be provided to make the parts necessary for the modification of experiments on STARLAB systems.

The physical requirements of the mechanical shop are given in Table II-18. The technician should be qualified in some other field to make most efficient use of his time aboard the laboratory.

TABLE II-18. ESTIMATED REQUIREMENTS OF MECHANICAL SHOP

Space	700 ft^3
Power	1000 w
Weight	1000 lb
Manned Requirements	8 Manhour/Day (Technician)

Electrical Area. The purpose of this laboratory is to provide electrical instruments to the laboratories of the space station such as power supplies, signal generators, and measuring and signal displaying instruments. Electrical standards may be included among these instruments if needed. These instruments may be used in the instruments laboratory or may be removed for short time intervals. Instruments intended for continuous use must be provided by each laboratory.

The communication system of the space station (see Chapter III) is planned to include an input-output terminal in each experimental station which is connected to a central computer whose function is to make electrical measurements and display the readings automatically on the output display. Thus, most of the electrical measurements will be made using these terminals. The instruments in the electrical laboratory will be general-purpose instruments for measurements of voltage, current, magnetic flux, etc., of wide magnitude and frequency range. As such, the instruments laboratory should include several voltmeters, ammeters, frequency meters, power meters, flux meters, oscilloscopes, signal generators, power supplies, and other common-use equipment.

In addition, there should be a storage area where some electronic components can be kept, such as resistors, transistors, etc. The physical requirements of the laboratory are given in Table II-19.

TABLE II-19. ESTIMATED REQUIREMENTS OF ELECTRICAL SUPPORT LABORATORY

Space	700 ft^3
Power	About 300 w
Weight	About 300 lb
Manned Requirements	8 Manhours/Day

Heat Transfer and Thermodynamic Area. The purpose of the heat transfer and thermodynamic support laboratories is to provide a facility for servicing and monitoring the thermal problems that will exist throughout the STARLAB facilities and the experimental programs. Facilities will exist for the storage and handling of cryogenics for the experimental programs, servicing of heat transfer and thermal equipment, and for the calibration and measurement of thermal data. In addition, the mechanisms for the control and monitoring of temperatures throughout the STARLAB will be located in the heat transfer support laboratory.

The operation of this laboratory will require qualified personnel who can assist in the complex problems that are involved in handling the liquid or slush form of cryogenics that will be required for specialized experimental programs. Also, these expert technicians will be trained in disciplines enabling them to service onboard heat transfer equipment for either STARLAB or the experimental programs and conduct either thermodynamic or heat transfer measurements of calibrations.

A major function to be performed by this laboratory will be the system regulation and monitoring of the equipment necessary to regulate the internal temperature of STARLAB. The onboard equipment will be the principle source of heat, and radiation to space will be the major method of dissipating heat. Intelligent, constant monitoring and performance of this equipment is absolutely essential for the life and comfort of onboard personnel.

The laboratory will be equipped with sufficient thermal equipment and test instrumentation so that scientists can perform fundamental experiments on thermal and radiation properties or materials in this support laboratory. Equipment will also be sufficient for development and engineering tests of thermal systems, such as heat exchangers and radiators, under space environment conditions. A preliminary design of the heat transfer-thermodynamics laboratory is shown in Figure II-15. The physical requirements are given in Table II-20.

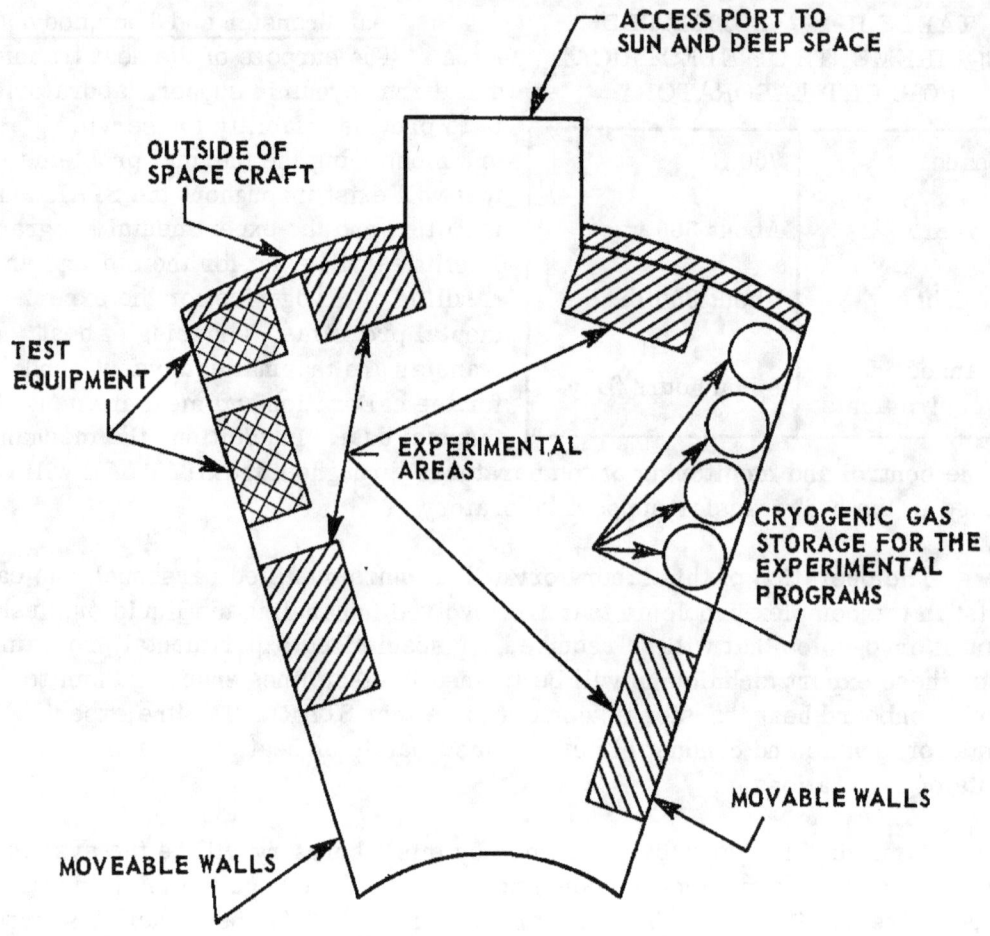

FIGURE II-15. HEAT TRANSFER AND THERMODYNAMIC LABORATORIES

TABLE II-20. PHYSICAL REQUIREMENTS OF HEAT-THERMODYNAMICS LABORATORY

Volume	600 ft^3
Weight of Equipment	1000 lb
Power	1000 w
Manned Requirements	16 Manhour/Day

Storage Facilities. In the operation of a large integrated laboratory and operations complex such as STARLAB, a programmed storage volume is required. This volume will perform many functions such as spare parts repository, supplies storage, future and completed experiments storage, equipment storage, and storage of waste. In addition, this volume could be used for an emergency shelter area, aging, curing or exposure chamber, or an auxiliary assembly area.

This storage area must be well-lighted and have ventilation and atmospheric flushing equipment. This storage area should be equipped with an emergency control panel and auxiliary life support units.

There should be docking facilities for the shuttle and FFM's, which includes parts that open wide for bulky cargos. This storage area must have fire protection, sterilization facilities, and meteorite insulation characteristics.

    a. Volume. It is estimated that a volume of 2500 cubic feet will be required to fulfill the functions of the storage area.

    b. Manpower. Approximately 3 manhours per day will be necessary to keep the storage area organized and efficient.

INTEGRATION

Time Schedule

The proposed 10-year life of STARLAB necessitates an early buildup of maximum power, men, and equipment to make the most efficient utilization of the space laboratory. It is hoped that the research and support facilities can be completely operative at 6 months or at a maximum of 1 year after initial launch.

The critical factors controlling the buildup are available power for life support and experiments and rate of launch. Assuming that STARLAB can be completely assembled within 6 months of initial launch (5 launches) and that 100 kw of power is available, then the scheduled experimental program can start at full scale. Twenty-four of the total 100 kw are alloted for the life support of 35 men and 37 kw are alloted for the experimental program.

If sufficient power for the total experimental program is not available at the time STARLAB is assembled, then one of two alternative approaches will have to be adopted. One, the experiments will have to be programmed, through STARAD on the ground, to utilize the available power. Priorities will be established and experiments run according to a specific schedule. The second alternative is to simply allow a longer period of time for experiments. Programs that are planned for a 2- to 3-month duration would be extended from 4 to 6 months. The first alternative seems to be the most attractive.

Contamination Control

Minimum standards must be adhered to provide adequate safety and to provide control of experiments to ensure against contamination of the atmosphere in the laboratory. Some of these are discussed in the following paragraphs.

<u>Nonexplosive Atmosphere; Combustible Gases and Vapor Control</u>. No combustible gases should be allowed aboard the laboratory. This would be the easiest way of protecting the vehicle from the hazard of catastrophic explosion caused by gases and vapors. This prohibition would include all fuel gases and liquids and a few others which might become hazardous because of the low total pressure of the atmosphere or the high oxygen concentration.

Experiment packages which require the use of combustible gases, vapors or liquids for producing flames and those which might produce these combustible gases, and vapors or liquids in a concentration which could burn shall be in explosion-proof containers rated for the mixtures which could be formed. Experiments which might produce the hazardous gases, vapors, or liquids as by-products should be charged with a fire suppressant such as nitrogen to ensure that a combustible mixture will not occur. The experiments should be connected to a vent.

The vent system is to be a pipe or tube system with a relief valve to prevent back-flow and control the pressure in the package. The valve may be part of the package. Flexible vent piping should be designed to prevent serious stressing of the piping due to minor variations in the structure of the vehicle during launch, maneuvering, or docking when structural loading may be severe.

The vent pressure should be below the normal pressure of the laboratory but sufficiently high to permit positive operation of the relief valves. The vent system would terminate in a nozzle to provide a thrust on the vehicle in the direction of its velocity vector and through the center of mass of the entire vehicle. This thrust would assist in keeping the laboratory from slowing in its orbit due to ballistic effects of the earth's atmosphere. The low mass rate of gas passing through the vent system probably will not create detectable acceleration of the total mass. However, the small forces could easily affect the attitude of the laboratory if the exhaust was allowed in a direction whose line of action was distant from the center of mass. Therefore, for reasons of attitude stability it may be necessary to exhaust the vent systems in a direction whose line of action is through the mean center of mass, even though the exhaust may be from any point on the surface of the vehicle.

The reason for the nozzle and the exhaust at a minimum number of points is to remove the contaminating exhaust gases as far as possible from the surface of the vehicle so that the experiments which require a hard or clean vacuum may be vented through the vehicle at other points.

A second vent system with pressure below 1 torr may be necessary to "evacuate" the experiment packages which need to operate in vacuum, but not in a hard or clean vacuum. This system would produce a rough vacuum for the purpose of preventing dangerous concentrations of combustible and other gaseous materials. This system should exhaust in the same manner as the pressurized vent system previously described to remove the contamination from the vicinity of the vehicle.

Experiments requiring the use of hazardous materials which will be removed by a vacuum can have their enclosures evacuated through the low pressure of the vent system. If the experimental procedure precludes the evacuated enclosure, the pressurized vent will provide the required safety if the purge gas is used to prevent a combustible mixture.

If a combustible mixture must be obtained in the operation of the experiment, and no explosion-proof enclosure is provided, it should be flown aboard an unmanned free-flying module. Thus, if an accident occurred, no person would be in jeopardy. The module (FFM) can be made safe by purging or evacuating before docking to the laboratory.

Contamination Protection From Noxious Gases and Vapors. Experiments using or producing noxious gases or vapors should not be brought into the laboratory.

The use of remote manipulators would keep the contamination out of the life support system. Several materials such as arsenic, beryllium, cadmium, chromium, copper, mercury, phosphorus, carbon monoxide, zinc, chlorine, cyanide, fluorine, and ozone have threshold limits. Freon contamination should be completely avoided (Refs. II-70, II-71, and II-72).

The vent and purge system or the vacuum vent system for the packages which may have contamination will prevent this contamination from being introduced into the interior atmosphere of the laboratory. The safe limits prescribed by health authorities must not be exceeded (Ref. II-72).

Particles and Dust. Particles and dust must be removed from the atmosphere in the laboratory. Since there is no tendency for particles to settle out of the air in zero gravity, a recirculating collector system would have local hoods and diverters at the locations where the particles are released. These hoods will be used on all machines which produce chips, particles, or dust.

Where practical, the machinery which produces particle contamination should be mounted inside glove boxes or similar bare-hand enclosures which operate below the atmospheric pressure in the laboratory. The exhaust air is cleaned by filters before it is returned to the atmosphere of the laboratory. It may have to be returned to the life support system to obtain temperature control before it is returned to the atmosphere.

Radiation Safety. Radiation safety is the adequate shielding of personnel from electromagnetic and nuclear radiation to reduce the levels to the limits specified by the Atomic Energy Commission (AEC). Suitable containers must be used for shielding radiative material.

If the weight of shielding is too great or other requirements make it impractical to reduce the radiation level to the specified limits by shielding alone, the use of a remote operator or a free-flying module may be used to reduce the radiation to a safe level (Ref. II-42, II-73, and II-74).

Experiments which require radiation levels below those required for personnel should be located in the laboratory at points where the level is adequately low considering the radiation from nuclear power sources and that from other experiments.

Safety

A review of conditions and uncertainties related to experiments, especially chemistry experiments, may present some surprises for the experimenters. One example is that liquids in gravity tend to stratify, and mixing begins from the surface between the layers. In zero gravity, this stratification is improbable and the two materials will mix on a volume basis, thereby increasing the rate at which the chemical action occurs by a large factor, perhaps to cause a violent reaction.

Safety precautions to be considered in each experiment should include the following:

1. Remote location such as aboard a free-flying module or remote manipulator.

2. Vent the package to vent system which will prevent contamination of the atmosphere of the laboratory.

3. Identify all by-products which might contaminate the life support system, especially the gaseous products.

4. Use samples as small as practical to prevent accidental damage to the package.

5. Make allowance for damage to the experiment by a meteor to limit further damage to the laboratory or other experiments.

Automatic Checkout of Experiments

Due to the large amount of data which become available from the experiments in STARLAB, the sensors will have to be checked periodically to ensure they are functioning correctly. This checkout procedure will be done automatically by the onboard computer. Any defects in the experiment equipment will be brought to the attention of the researchers or the commander for appropriate action. This automatic check-out would include the inputs to the experiments in terms of environmental variables such as package atmospheric pressure, temperature, oxygen content and position, volume time, etc. of events controlling the experiments. With the automatic check-out system, the validity or dependability of the results is established within limits indicated by the type and frequency of the errors or defects.

Reliability

The reliability of the experiment should be sufficient against a failure which would stop the experiment, or make the experiment a hazard, or render the data questionable during the expected lifetime of the experiment. Defects which would cause obvious errors or small components should be limited to not more than a predetermined number in the life of the experiment program. This number of defects should never be more than the number of spare devices of each kind per lifetime of the experiment or shuttle period. Calibration errors or deviations can be allowed in the data processing.

In experiments which must be operated on a schedule (such as earth resources) and replacement of parts during the regular schedule of use cannot be accomplished in the time available between scheduled activity, therefore the reliability must be better than one failure in the life of the experiment. The alternative would be to have another system available in a stand-by mode.

Maintainability and Resupply

Each experiment package should be capable of being maintained in good working order by the replacement of parts or modules which can be changed in zero gravity by one man in the time available between uses. Some types of equipment, such as cameras, might require a complete system or unit to operate instead of the defective one. This would give time to remove the defective unit and return it to earth via the shuttle for repair and return. Routine changes of parts, such as film packs, should be replaceable in zero gravity by one man in a few minutes and require as little effort as possible from the man, especially effort such as loosening or tightening of rotating locking devices.

Stocks of films and supplies for the experiments, which are necessary to keep them functioning, shall be available in the laboratory. Storage space for these supplies as well as the used items is provided in an adequately protective environment such as temperature, humidity, shielding, and darkness.

Commonality

To minimize the number and types of spare replaceable (either on earth or in orbit) parts and components, those which are similar in function and size should be made standard, interchangeable parts or components. This standardization is necessary to bring the cost of experiments down to a practical level. This cost item is more important as the other costs of space such as launch cost, vehicle cost, and time between trips is reduced. Experiment enclosures may not lend themselves to commonality, but the sensors, manipulators, servos, heaters, controls, and connectors are a few of the items within the experiment package which can be standardized. The additional weight created in some experiments by standardized parts is not serious since by the time the laboratory is launched, it is expected that the cost of launching these extra weights will not be large.

Time Line

Man, machine, power, and laboratory space must be used most effectively to return the greatest value for the country's investment in STARLAB. For this to be accomplished, these resources must be related with each other in a time-related schedule. Time is a resource which is finite and limited and must be used most effectively. This schedule could be programmed into a computer with the program being up-dated as new equipment becomes available, as well as the time when each job or operation is finished, thereby making facilities available again. The schedule would account for the available man-hours and the time each man was aboard STARLAB and his specialty field. In this manner, the scheduling of the men, equipment, power, and available space can be made nearly optimal.

Special Equipment and Instruments

The equipment of the Analytical Laboratory and the Standards Laboratory will be mounted before launch instead of bringing these items to STARLAB as parts of experiments as required. The launching of fully equipped laboratories is a better choice since the laboratory is not weight-limited, even with these equipments installed.

ADMINISTRATIVE NEEDS OF STARLAB

The huge investment in this program, and the large volume of data which will result from its operation, make it necessary to have an office which takes dollars from Government, industry, private, and the scientific communities and return information available only from activities in space. STARLAB and its administrative function, STARAD, will play a role in space very similar to that of AEC in stimulating researching and developing the nuclear reactor as a power source.

Figure II-16 shows the interactions of the starsystem components and the outside interests. STARAD provides two functions, information management and program administration. Information management will be discussed in a later chapter, therefore this section will concentrate on program administration.

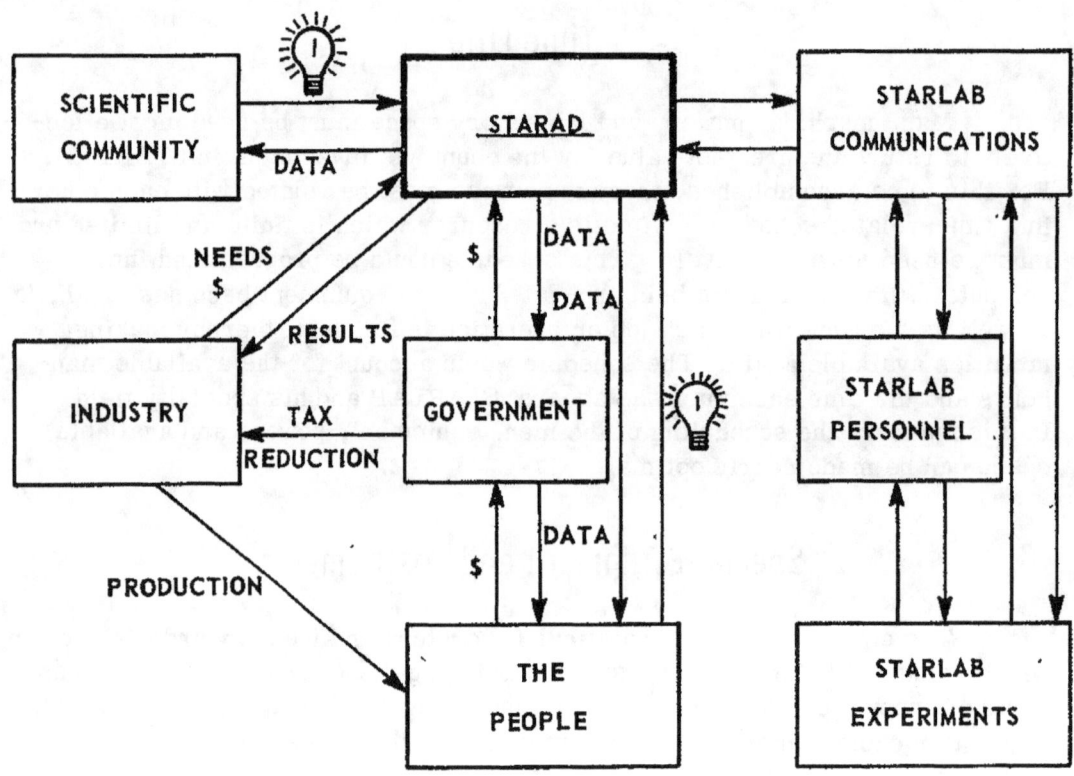

FIGURE II-16. INTERACTIONS OF STARAD

The scientific community will submit ideas to STARAD and work with their program people to develop the necessary hardware to make the experiment fly. The data will be reduced and compressed and the analysis will be done by the investigators.

Industry can be considered a supporter as well as a consumer of STARAD. It is not expected that the individual companies can support basic research in space, but the benefits of the unique environment of STARAD should be an incentive for the small and medium size companies to pool their research resources and thereby be capable of affording the heavy expenditures involved in long-term research in space. The funds from the pooled research program should be used efficiently because of:

1. Long-term amortization of research facilities.

2. Centralized program direction.

STARAD's output to the government will be in the form of condensed scientific and technical data. Also, it will have available a management summary which would include:

1. Survey of the areas considered for STARLAB.

2. Critical topics and burning questions.

3. Suggested answers or studies.

4. Spin-off possibilities and information.

5. Funding plans.

Government can provide incentives for industrial research by such methods as tax credits. These government incentives would make basic research more interesting to industry and thereby improve the product for the consumer. Thus, these incentives by the government would benefit industry and the public.

STARLAB also has a link with the people. When the results of the work in space is of general interest to the people, these results will be communicated to the public.

Figure II-17 shows the internal functions of STARAD. The information management group will acquire, store, process, relay, reduce, and disseminate data in either machine data form or hard copy. The program administration group will solicit and develop research programs in science, applications, or industrial processing. It will set priorities on programs and furnish management. Its Public Affairs section will make the general public as well as the

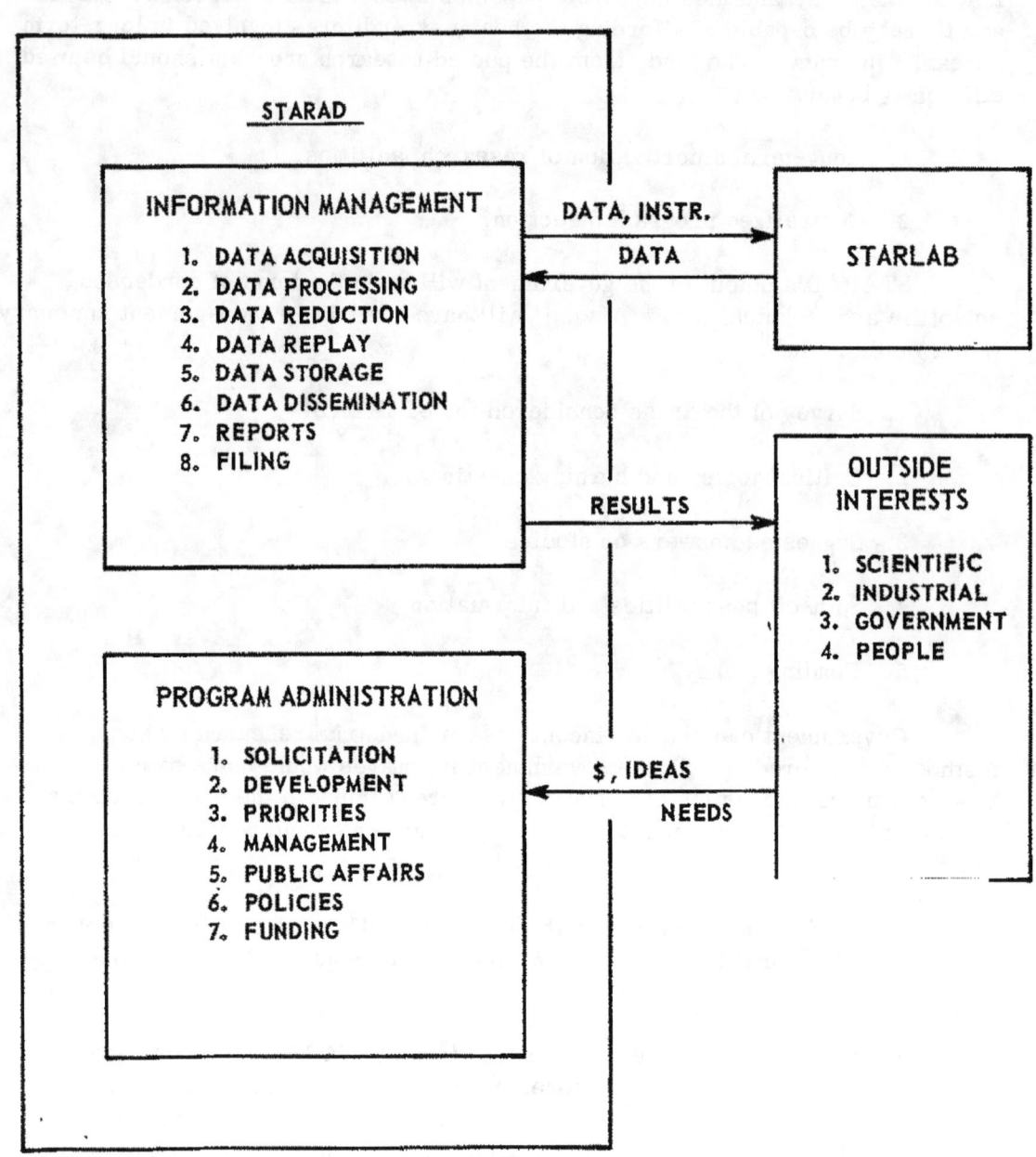

FIGURE II-17. INTERNAL FUNCTIONS OF STARAD

prospective researcher and sponsor cognizant of the benefits of working in space. It will set a policy for STARLAB as well as the rest of starsystem and provide funding programs to keep STARLAB operating efficiently.

REFERENCES

II-1. NASA Science and Technology Advisory Committee for Manned Space Flight. Vol. I Proceedings, NASA SP-196, 1969.

II-2. Goldberg, Leo: Ultraviolet Astronomy. Scientific American, Vol. 220, No. 6, 1969, p. 92.

II-3. Advanced Astronomy Mission Concepts. ATM Follow-On Study, ED 2002-797.

II-4. Manned Space Flight Experimental Catalogue. Vol. 1, Astronomy.

II-5. A Biomedical Program for Extended Space Missions. Manned Spacecraft Center, May 1969.

II-6. Space Research — Directions for the Future. National Academy of Sciences, National Research Council, Publ. 1403, 1966.

II-7. Kail, L. T.: Requirements Study for a Biotechnology Laboratory for Manned Earth Orbiting Missions. Final Report, McDonnel-Douglas Astronautics Co., February 1969.

II-8. Vinograd, S. P.: Medical Aspects of an Orbiting Research Laboratory. NASA SP-86, 1966.

II-9. Hoffman, D. B., and Kontaratos, A. N.: Medical Requirements in Support of Long Duration Manned Spaceflight. Bellcomm, Inc., TR-67-710-1, 20 November 1967.

II-10. Olive, J. R., and Beem, D. R.: Final Progress Report to National Aeronautics and Space Administration. Bioscience Research During Earth-Orbiting Missions, Manned Orbital Space Station. American Institute of Biological Sciences, 1967.

II-11. Pittendirgh, Colin S.: Scientists' Testimony on Space Goals. Hearings Before the Committee on Aeronautical and Space Sciences. U.S.S. 88th Congress, June 10-11, 1963, U. S. Government Printing Office, p. 75.

REFERENCES (Continued)

II-12. Significant Achievements in Space Science. NASA SP-155, 1966, p. 83.

II-13. Reeves, R. G. (ed.): Introduction to Electromagnetic Remote Sensing. American Geological Institute, Short Course Notes, 1968.

II-14. Kennedy, J. M.: Microwave Sensors for Water Management and Hydrology From Space. AIAA Paper No. 68-1076, 1968.

II-15. Lowman, P. D.: Geologic Applications of Orbital Photography. NASA Technical Note, TND-4155, 1967.

II-16. University of Michigan: Peaceful Uses of Earth Observation Spacecraft. Vol. II and III, N66 37029 and N66 36030, 1966.

II-17. Anon: The Needs and Requirements for a Manned Space Station. Vol. 3, Earth Resources, NASA Publication S 0000038, 1966.

II-18. Badgley, P. C., Childs, Leo, and Vest, W. L.: The Application of Remote Sensing Instruments in Earth Resource Surveys. Geophysics, Vol. 32, 1967, pp. 583-601.

II-19. Waltz, D. M.: Technological Base for Planning of Space Flight Missions to Obtain Data on the Earth's Resources. AIAA Paper (TRW DOC. 8800.8.10-32), 1968.

II-20. Candidate Experimental Program for Manned Space Stations. NASA, OMSF, 1 May 1969, p. 16-01.

II-21. Wuenscher, Hans F.: Manufacturing in Space. Manufacturing Technology Unique to Zero Gravity Environment, NASA, MSFC Form 454, 1 November 1968, p. 20.

II-22. Space Station Phase A Report. U.S. Government Memorandum, MTD/Deputy Director, Adv. Manned Mission Program, 21 November 1968, pp. 8-9.

REFERENCES (Continued)

II-23. Wahl, B. W.: Analysis of Selected Opportunities for Manufacturing in Space. Douglas Paper 10119, February 1969, pp. 8-11.

II-24. Buzzard, W. C.: Hollow Ball Bearing Technology. Manufacturing Technology Unique to Zero Gravity Environment, NASA, MSFC Form 454, 1 November 1968, pp. 170-180.

II-25. Wahl, B. W.: Analysis of Selected Opportunities for Manufacturing in Space. Douglas Paper 10119, February 1969, pp. 11-13.

II-26. Olsen, R. P.: Containerless Production of New Glasses and Other Materials. Manufacturing Technology Unique to Zero Gravity Environment, NASA, MSFC Form 454, 1 November 1968, p. 60.

II-27. Wahl, B. W.: Analysis of Selected Opportunities for Manufacturing in Space. Douglas Paper 10119, February 1969, pp. 3-8.

II-28. Frost, R. T.: Crucibleless Melting, Purification, and Solidification of Materials in Zero-G Environment. Manufacturing Technology Unique to Zero Gravity Environment, NASA, MSFC Form 454, 1 November 1960, pp. 82-116.

II-29. Wahl, B. W.: Analysis of Selected Opportunities for Manufacturing in Space. Douglas Paper 10119, February 1969, pp. 13-14.

II-30. Steurer, Wolfgang H.: Zero-G Manufacturing of Unique Structural Materials and Components. Manufacturing Technology Unique to Zero Gravity Environment, NASA, MSFC Form 454, 1 November 1968, pp. 139-169.

II-31. Reinfurth, Mario H.: Low Gravity Gradient Mechanics. Manufacturing Technology Unique to Zero Gravity Environment, NASA, MSFC Form 454, 1 November 1968, pp. 181-196.

II-32. Wahl, B. W.: Analysis of Selected Opportunities for Manufacturing in Space. Douglas Paper 10119, February 1969, pp. 1-17.

REFERENCES (Continued)

II-33. Candidate Experimental Program for Manned Space Stations, NASA, OMSF, 1 May 1969, p. 21-01.

II-34. Wuenscher, Hans F.: Manufacturing in Space. Manufacturing Technology Unique to Zero Gravity Environment, NASA, MSFC Form 454, 1 November 1968, p. 29.

II-35. Utech, Harvey P.: Growing Crystals in Space. Manufacturing Technology Unique to Zero Gravity Environment, NASA, MSFC Form 454, 1 November 1968, pp. 197-214.

II-36. Wuenscher, Hans F.: Manufacturing in Space. Manufacturing Technology Unique to Zero Gravity Environment, NASA, MSFC Form 454, 1 November 1968, pp. 32-36.

II-37. Candidate Experimental Program for Manned Space Stations. NASA, OMSF, 1 May 1969, p. 16-03.

II-38. Study of an Electrostatic Zero-Gravity Work Bench Prototype. Final Report, Phase I, NASA Contract No. NAS8-21385, Chrysler Corp., Space Division, New Orleans, La., 23 April 1969.

II-39. Space Research — Directions for the Future. National Academy of Sciences, National Research Council, Publ. 1403, 1966.

II-40. Getman, F. H., and Daniels, F.: Outlines of Physical Chemistry. Wiley, 1943, pp. 278-297.

II-41. Maslan, F.: Process for Thermal Fixation of Atmospheric Nitrogen. Proceedings of Sixth Space Congress, Space, Technology, and Society, Vol. II, 1969, p. 1-1.

II-42. Lippincott, E. R., Stromberg, R. R., Grant, W. H., and Cessac, G. L.: Polywater. Science, Vol. 164, June 1969, p. 1482.

II-43. Hirschfelder, J. A. (ed.): Intermolecular Forces. Advances in Chemical Physics, Vol. XII, 1967, p. 389.

REFERENCES (Continued)

II-44. Ross, J. (ed.): Molecular Beams. Advances in Chemical Physics, Vol. X, 1966, p. 319.

II-45. Study of Liquid Drop Dynamics in Zero Gravity. Electro-Optical Systems, Inc., EOS Report 7170-Final, p. 5.

II-46. Fedor, W. S.: Trillion-Dollar Economy. Chemical and Engineering News, 4 November 1968, p. 90.

II-47. Olson, Frost, Steurer, and Utech: Manufacturing Technology Unique to Zero-Gravity Environment. Papers, MSFC Form 454, pp. 60, 82, 139, 197.

II-48. Scientific American. Entire Issue Devoted to Materials. Vol. 217, September 1967.

II-49. Daniels, F., et al.: Experimental Physical Chemistry. McGraw-Hill, 1962, p. 90.

II-50. Study of Liquid Drop Dynamics in Zero Gravity. EOS Report 7170, 2 November 1967, p. 5.

II-51. Chemistry and Physics of Interfaces. American Chemical Society, 1965, pp. 15-41.

II-52. Ibid, pp. 73-109.

II-53. Cook, N. C.: Metalidizing. Scientific American, 1969, p. 38.

II-54. Satterfield, C. N.: Future Trends in Catalysis. Industrial and Engineering Chemistry, June 1969, p. 4.

II-55. Burchfield, H. P., and Storrs, E. E.: Biochemical Applications of Gas Chromatography. Academic Press, 1962, p. 371.

II-56. Ibid, p. 5.

REFERENCES (Continued)

II-57. Kolthoff, I. M., and Lingane, J. J.: Polarography. Interscience, Vol. II, 1952, pp. 493, 503.

II-58. Moelwyn-Hughes, E. A.: Physical Chemistry. Macmillan, 1964, p. 1195.

II-59. Glasstone, S.: Textbook of Physical Chemistry. Van Nostrand, 1946, p. 838.

II-60. Karrer, P.: Organic Chemistry. Elsevier, 1946, p. 84.

II-61. Noller, C.: Chemistry of Organic Compounds. Saunders, 1960, p. 393.

II-62. Morrison, R. T., and Boyd, R. N.: Organic Chemistry. Allyn and Bacon, 1963, p. 789.

II-63. Ibid, p. 170.

II-64. Ibid, p. 766.

II-65. Ibid, pp. 128, 498.

II-66. Noller, C.: Chemistry of Organic Compounds. Saunders, 1960, pp. 422, 433.

II-67. Ross, J.: Advances in Chemical Physics. Vol. X, Interscience, 1966, p. 140.

II-68. Manned Space Flight Experiment Catalogue. Vol. V.

II-69. Manned Space Flight Experiment Catalogue. Vol. 10, Manned Space Flight Engineering.

II-70. Duke-Elder, Sir Stewart: Textbook of Ophthalmology. The C. V. Mosby Co., St. Louis, Mo., Vol. VI, 1954, pp. 6443-6579.

REFERENCES (Concluded)

II-71. Official FDA Tolerances. Nat. Agric. Chem. Assoc. News and and Pesticide Rev., Vol. 20, No. 3, February 1962.

II-72. Threshold Limit Values for 1963. J. Occup. Med., Vol., 5, 1963, p. 491.

II-73. Cronkite, E. P., Bond, V. P., and Dunham, C. L. (ed.): Some Effects of Ionizing Radiation on Human Beings. U. S. Atomic Energy Commission Document TID-5358, 1956.

II-74. Saenger, E. L. (ed.): Medical Aspects of Radiation Accidents. U. S. Atomic Energy Commission, U. S. Government Printing Office, Washington, D. C., 1963

BIBLIOGRAPHY

A Biomedical Program for Extended Space Missions: Manned Spacecraft Center, May 1969.

Adelberg, M., and Forster, K.: The Effect of Gravity Upon Nucleate Boiling Heat Transfer. Plenum Press, 1961.

Advanced Astronomy Mission Concepts, ATM Follow-On Study. ED-2002-795, Vols. I, II, III, IV, NAS8-2400, Martin-Marietta, Denver Division.

Ambrose, D., and Ambrose, B. A.: Gas Chromatography. G. Newnes, London, 1961

Anon: Remote Sensing of Natural Resources, List of References 1960-1965. Prepared for NASA by the U. S. Army Corps of Engineers, Ft. Belvoir, Virginia, 1966.

ASP: Manual of Color Aerial Photography. American Society of Photogrammetry, Falls Church, Virginia, 1968.

Barringer, A. R.: Research Directed to the Determination of Sub-Surface Terrain Properties and Ice Thickness by Pulsed VHF Propagation Methods. Contract No. AF19(628)2998, Air Force Cambridge Labs., March 1965.

Basic Matter and Energy Relationships Involved in Remote Reconnaissance. American Society of Photogrammetry, Photogrammetric Engineering, Vol. 29, No. 5, 1963, pp. 761-799.

Bauer, George: Measurement of Optical Radiation. The Focal Press, London and New York, 1965.

Bauer, R. J., Cohn, M., Cotton, J. M., and Packard, R. F.: Millimeter Wave Semiconductor Diode Detectors, Mixers, and Frequency Multipliers. Proc. IEEE, Vol. 54, April 1965.

Beal, C. H.: Reconnaissance of the Geology and Oil Possibilities of Baja California, Mexico. Geol. Soc. America Mem. 31, 1948.

BIBLIOGRAPHY (Continued)

Beckman, W. A., Jr., and Whitten, E. H.: Statistical Problems Involved in Remote Sensing of the Geology of the Lithosphere-Atmosphere Interface. Journal of Geophysical Research, Vol. 71, No. 24, 1966, pp. 5873-5890.

Beckmann, Peter, and Spazzichho, Andre: The Scattering of Electromagnetic Waves from Rough Surfaces. New York, Macmillan, 1967.

Benedikt, Elliot T.: General Behavior of a Liquid in a Zero or Near-Zero Gravity Environment. Plenum Press, 1961.

Benedikt, E. T., and Lepper, R.: Experimental Production of a Zero or Near-Zero Gravity Environment. Plenum Press, 1961.

Burchfield, H. P., and Storrs, E. E.: Biochemical Applications of Gas Chromatography. Academic Press, 1962.

Cameron, H. L.: Radar as Surveying Instrument in Hydrology and Geology in Proceedings of the Third Symposium on Remote Sensing of Environment. University of Michigan Institute of Science and Technology Infrared Physics Laboratory, 1965, pp. 441-452.

Candidate Experiment Program for Manned Space Stations. NASA, 1 May 1969

Clark, H.: Solid State Physics. St. Martin's Press, 1968.

Cole, R. H., and Coles, J. S.: Physical Principles of Chemistry. Freeman, 1964.

Conel, J. E.: Infrared Thermal Emission from Silicates. Jet Propulsion Laboratory Technical Memorandum, No. 33-243, 1965.

Cooper, J. T.: Geologic Evaluation, Radar Imagery of the Twin Buttes Area, Arizona. U. S. Geological Survey, Unpublished Report, 1966.

Cummings, R. L., and Grevstad, P.: Orbital Force-Field Boiling and Condensing Experiment. Plenum Press, 1961.

BIBLIOGRAPHY (Continued)

Daniels, F., et al.: Experimental Physical Chemistry. McGraw-Hill, 1962.

Dellwig, L. F., Kirk, J. N., and Walters, R. L.: The Potential of Low-Resolution Radar Imagery in Regional Geologic Studies. Jour. Geophysical Research, Vol. 71, No. 20, 1966, pp. 4995-4998.

Fraser, T. M.: Review of Physiological Measurement Techniques for Applicability to Space Flight Conditions. NASA CR-1277, 1969.

Gillerman, Elliot: Investigation of Cross Polarized Radar on Volcanic Rocks. CRES Report 61-25, 1967.

Glasstone, S.: Sourcebook on the Space Sciences. D. van Nostrand Co., Inc., 1965.

Gunn, A., Gould, T. C., and Anderson, W. A. D.: The Effect of Microwave Radiation on Morphology and Function of Rat Testis. Laboratory Investigation, Vol. 10, No. 2, 1961, pp. 301-314.

Gushee, David, E.: Chemistry and Physics of Interfaces. American Chemical Society, 1965.

Harvey, D. I., and Myskowski, E. P.: Physics of High Altitude Photography in Photographic Considerations for Aerospace, Hall, H. J., and Howell, H. K. (Eds.), Lexington, Mass., Itek Corp., 1965.

Hawkins, G. S.: The Physics and Astronomy of Meteors, Comets, and Meteorites. McGraw-Hill Book Co., 1964.

Hemphill, W. R.: Application of Ultraviolet Reflectance and Stimulated Luminesence to the Remote Detection of Natural Materials. Presented at the Annual Convention of the America Society of Photogrammetry, 1968.

Hess, W. N. (Ed.): Introduction to Space Science. Gordon and Breach Science Publishers, 1965.

Hoffman, D. B., and Kontaratos, A. N.: Medical Requirements in Support of Long Duration Manned Space Flight, Bellcomm, Inc., TR-67-71D-1, 20 November 1967.

BIBLIOGRAPHY (Continued)

Hunt, G. R., Salisbury, J. W., and Reed, J. W.: Rapid Remote Sensing by Spectrum Matching Technique. J. Geophys. Res., Vol. 72, 1967, pp. 705-719.

Janza, F. J.: The Analysis of Pulsed Radar Acquisition System and a Comparison of Analytical Models for Describing Land and Water Radar Return. Phenomena, Sandia Corp., Monograph SCK 533, 1963.

Janza, F. J.: A Comparison of the Microwave Scatterometer and Radiometer for Sea-State Measurements, 49th Annual Meeting of AGU, 1968.

Kail, L. T.: Requirements Study for a Biotechnology Laboratory for Manned Earth Orbiting Missions. Final Report, McDonnel-Douglas Astronautics Co., February 1969.

Kennedy, J. M., and Edgerton, A. T.: Microwave Radiometric Sensing of Soil Moisture Content. Publication No. 78, General Assembly of Berne, International Union of Geodesy and Geophysics, Berne, Switzerland, 1967.

Kennedy, J. M., Edgerton, A. T., Sakamoto, R. T., and Mandl, R. M.: Passive Microwave Measurements of Snow and Soil — Instrumentation and Measurements. Tech. Report No. 2, Geography Branch, Earth Sciences Div., Office of Naval Research, Vol. 1, 1966, p. 1-217.

Kinsman, Frank E.: Some Fundamentals in Non-Contact Electromagnetic Sensing for Geoscience Purposes in Proceedings of the Third Symposium on the Remote Sensing of Environment. 1965, pp. 495-515.

Le Galley, D. P., and Rosen, A. (Eds.): Space Physics. John Wiley and Sons, Inc., 1964.

Lenz, R. W.: Organic Chemistry of Synthetic High Polymers. Interscience, 1967, p. 6.

Levy, S. L.: The Integrated Circuit and Its Adaptability to Manufacture in Space. AIAA, Fourth Annual Meeting and Technical Display, Paper No. 67-813, Anaheim, California, 23-27 October 1967.

BIBLIOGRAPHY (Continued)

Lowman, P. D.: Geological Applications of Orbital Photography. NASA Tech. Note D 4155, 1967.

Lyon, R. J. P.: Evaluation of Infrared Spectrophotometry for Compositional Analysis of Lunar and Planetary Soils. Stanford Research Inst., Proj. No. PHU 3943, Final Report I, Under Contract No. NASr-49(04), Sept., 1962, NASA Tech. Note TN-D1871, April 1963, p. 118.

Manned Space Flight Experiment Catalogue. I-S/AA, Vols. I-XII, 1964.

Mansfield, Joseph: Heat Transfer Hazards of Liquid Rocket Propellant Explosions. AFRPL TR-69-89, URD 706-5, URS Research Co., 1969.

Marton, L.: Methods of Experimental Physics. Academic Press, 1959.

Merifield, P. M.: Photo Interpretation and Photogrammetry of Hyper-Altitude Photography, Contract No. Nas 5-3390, LR 17666, Lockheed Calif. Co., 1964.

Miller, W. E., and Babor, J. A.: General Chemistry. Wm. C. Crown Co., 1965.

Moore, R. K.: Radar Scatterometry — An Active Remote Sensing Tool. CRES Report No. 61-11, University of Kansas, Lawrence, Kansas, 1966.

Morrison, R. T., and Boyd, R. N.: Organic Chemistry. Allyn and Bacon, 1963.

Moxham, R. M., Green, G. W., Friedman, J. D., and Gawarecki, S. J.: Infrared Imagery and Radiometry. Summary Report, U. S. Geol. Survey Interagency Report NASA-105, 1967, p. 51.

Nebergall, W. H., et al.: General Chemistry. Health, 1963, p. 428.

Noller, C. R.: Chemistry of Organic Compounds. Saunders, 1966.

BIBLIOGRAPHY (Continued)

Olive, J. R., and Beem, D. R.: Final Progress Report to National Aeronautics and Space Administration. Bioscience Research During Earth-Orbiting Missions, Manned Orbital Space Station, American Institute of Biol. Sci., 1967.

Parnley, A. G.: Evaluation of Airborne Gamma Ray Spectrometry in the Bancroft and Elliot Lake Areas of Ontario, Canada. Fifty Symposium on Remote Sensing of Environment, University of Michigan, 1968.

Partington, J. R.: An Advanced Treatise on Physical Chemistry. Wiley, 1962.

Proceedings of Tri-Service Conference on Biological Hazards of Microwave Radiation, 1957. Rome Air Development Center, 1960.

Rasquin, John R.: Heat Sources for Space Manufacturing Processes. MDM-3-69, Marshall Space Flight Center, 3 March 1969.

Raytheon/Autometric: Geoscience Potentials of Side-Looking Radar. Contract No. DA-44-009-AMC-1040(x), U. S. Army Crops of Engineers for NASA, Office of Manned Space Sciences, 1965.

Reeves, R. G.: Radar Geology in McGraw-Hill Yearbook of Science and Technology. McGraw-Hill, New York, 1968, pp. 322-328.

Reynolds, William C.: Behavior of Liquids in Free Fall. Journal of the Aero/Space Sciences, Vol. 26, No. 12, December 1959.

Sears, F. W., and Zemansky, M. W.: University Physics. Addison-Wesley, 1963.

Simons, J. H.: Some Applications of Side-Looking Airborne Radar. Proc. Third Symposium on Remote Sensing of Environment, University of Michigan, 1964.

Space Research — Directions for the Future. National Academy of Sciences, National Research Council, Publ. 1403, 1966.

Space Station Phase A Report. U. S. Government Memorandum from MTD/Deputy Director, Advanced Manned Mission Program, 21 November 1968.

BIBLIOGRAPHY (Concluded)

Steiner, Dieter, and Guterman, Thomas: Russian Data on Spectral Reflectance of Vegetation, Soil, and Rock Types. United States Army, European Research Office, 1966.

Strangway, D. W., and Holmer, R. C.: The Search for Ore Deposits Using Thermal Radiation. Geophysics, Vol. 31, No. 1, 1965, pp. 225-242.

Usiskin, C. M., and Siegel, R.: An Experimental Study of Boiling in Reduced and Zero-Gravity Fields. Publication No. 60 HT-10, American Society of Engineers, 1961.

Vinograd, S. P.: Medical Aspects of an Orbiting Research Laboratory. NASA SP-86, 1966.

Williams, James R.: Space Manufacturing Modules. Proceedings Sixth Space Congress, Space Technology, and Society, Vol. II, Cocoa Beach, Florida, 17-19 March 1969, pp. II-31 - II-48.

Wuenscher, Hans F.: Low and Zero-Gravity Manufacturing in Orbit. Paper No. 67-842, AIAA Fourth Annual Meeting and Technical Display, Anaheim, California, 23-27 October 1967.

Wuenscher, Hans F.: Space Manufacturing Unique to Zero-Gravity Environment. S&E-ME-IN-69-4, Marshall Space Flight Center, 15 May 1969.

CHAPTER III

INFORMATION MANAGEMENT

CHAPTER III. INFORMATION MANAGEMENT

INTRODUCTION

The information management system of STARAD is subdivided into two categories; the communication subsystem and the user participation subsystem. The communication subsystem is comprised of the hardware that distributes and controls the information flow while the user participation subsystem is concerned with software; that is, the programming, the decisions by man, and the feedback necessary to make optimum utilization of the gathered information.

The basic underlying assumption used in the design of the communication network is that many of the principal investigators (PI) and research groups will remain on earth while their experiments are performed in STARLAB. The STARLAB astronauts will, in many cases, provide close assistance to the PI by means of high quality voice-video channels. To improve the astronaut-PI rapport where detailed observations are mandatory, a high resolution TV display of 5000 to 6000 lines per raster is deemed necessary. High resolution TV is also of considerable assistance to the ground-based astronomer using an orbiting observatory.

It is estimated that about 6 channels of high resolution TV and 32 channels of regular TV (525 lines per raster) are necessary to serve the needs of STARLAB, and this number of television channels require a rather large bandwidth of several hundred megahertz in the transmission link.

In addition to television channels, thousands of channels to provide voice, data, and automatic control links around STARLAB to the ancillary modules and to earth are needed. If all of the communication channels are used simultaneously, a data rate of about 3000 megabits per second will be required. To handle this large data transmission rate, it is almost mandatory that lasers be used in the primary transmission link from STARLAB to earth.

The essential elements of the communications and information management system of the Starsystem project are:

1. Functions of the communications center on board STARLAB.

2. Internal STARLAB communications.

3. Functions carried out by the onboard computer.

4. Communication links between STARLAB and FFM's, shuttle, and earth.

5. Management and distribution of information on earth.

Communication Center

The control of all communications for STARLAB is through a single communication center located on board. The location of the communications center on STARLAB should be close to the antenna system so that the loss of signal power due to long wire leads, waveguide, or optical fibres is minimized. To provide the STARLAB commander with rapid communications, the communications center should be in the same vicinity as the commander's area. All communications and information channels are routed and controlled by this single center. Provision is also made in the STARLAB emergency shelter area for control of the entire communications and information system in case of any major emergency on board STARLAB. The control panels in the STARLAB communication center are automated to the extent that by pushing a button for a given area of the STARLAB, all of the environmental information for that area is displayed on a TV screen. For example, by pushing another button to interrogate the sensors on an FFM, the environmental information in the FFM is displayed on another TV screen. Thus, the communications control center has video information readily available without the necessity of reading meters or gauges. The area and the specific cause of any emergency would be automatically displayed on the TV screen along with a warning tone and a light or appropriate visual display.

Internal Communication

A twisted shielded pair (TSP) of wires with triple redundancy is used on board STARLAB to carry all communications and data signals from the various areas on board to the onboard computer for distribution. The TSP keeps the amount of wire and the weight to a minimum. Communication and

and data signals from the FFM's, EVA's, and the shuttle, as well as the areas on board STARLAB, all come through the TSP to the onboard computer to be routed or processed and distributed. Jacks for the experimental sensors, body sensors, and voice and video communications are provided at each area to get the signal onto the TSP.

To minimize size, weight, and logistics it is highly desirable to provide for commonality of equipment since each area on board can in many cases use the same type of transmitter, receiver, and even sensors. The commanlity concept for instrumentation design and usage is expected to improve the reliability and logistics problems significantly as well as reduce the cost of the instrumentation phase of STARLAB.

The emergency and warning system on board STARLAB will override all other channels. This emergency system is tied into the complete communications system including links to the ancillary modules and the Central Information Management Center (CIMC) on ground as well as on the STARLAB. An automatic warning tone would be heard throughout this communication system and an overriding voice channel is available to control the emergency action. Safety of the human beings aboard the STARLAB is of utmost importance, and with an automated emergency system, this goal is most quickly achieved. Emergency control of STARLAB functions is to be from the communication center with provisions for control in the emergency shelter area on board STARLAB.

Onboard Computer

The onboard computer is the heart of the STARLAB communications center. This computer is used primarily for station keeping, housekeeping, and maintenance with most experimental or laboratory data being processed on the ground. The computer is used to switch the signal to its proper place (RFM, EVA, shuttle, STARLAB, or CIMC) when it is received by the computer from the voice-video-date sensors. The signal has the proper signature so that the computer knows where to distribute it. Navigation and guidance of the STARLAB is controlled by the computer. Maintaining location of the FFM's shuttle, the satellites, the STARLAB, and the control of the FFM's is an important part of the work of the computer. The docking maneuvers for the shuttle and the FFM's, both manned and unmanned, are controlled by the onboard computer. This includes radar and any external TV that is used. Manual control could be exercised where necessary.

The emergency warning system is also controlled by the computer. Any emergency situation relayed to the computer from the sensors automatically sounds the warning tone, takes preprogrammed corrective action, and makes a channel available for voice communication throughout the system. The onboard computer acts as a monitor of the life support and environmental control systems. Environmental data such as atmosphere, pressure, temperature, radiation, meteoroid detection, and human body sensors are monitored at appropriate time intervals each minute by the onboard computer. The computer is given the format to monitor all systems on the control panels in the communication center. Most of the experimental data processing is done on the ground by a much larger computer; however, it is possible at times to use the onboard computer for limited experimental data processing. Storage in the onboard computer is primarily for station keeping, including navigational information or environmental data, and as a buffer for information storage.

Experimental data or reference data and similar types of information are stored in the computer on the ground. High resolution TV is available to transmit pages of a book, microfilm, magazines, special photographs, and like material to the STARLAB. The amount of paper on the STARLAB is kept to a minimum. Light-pen note pads are to be provided as a scratch pad or for temporary storage of written information. Data which must be converted into permanent, hard-cover written matter are sent to CIMC for processing and printing. High quality TV along with voice links to the ground allows well-trained technicians to conduct experiments in space while the PI directs, observes, and advises from the ground.

STARLAB to CIMC (External Communications)

Communications from the FFM's and the shuttle to the STARLAB are transmitted by C-band. Control, guidance, tracking, voice and video links, and data sensors are relayed to STARLAB by means of the C-band channels.

A system of three synchronous satellites surrounding the earth are used to communicate from STARLAB to the ground. Because of the anticipated wide bandwidth requirements, the primary communications link to the ground through the satellite system is the laser. The S-band is the backup for the laser in the STARLAB-to-ground communications link. The transmission of high quality, high resolution TV, regular, commercial TV, voice links, and the high telemetry rate between STARLAB and the ground make the laser

particularly attractive. The laser has advantages over the S-band in weight
and power even though its efficiency is low. Because of its narrow beamwidth
the laser beam has a high degree of security from undesired reception by
others.

Ground Stations

One of three ground stations located in areas of low cloud cover, such
as the Southwestern U.S., receive the signal from the synchronous satellite.
One specific, synchronous satellite is always in contact with the ground
station network, hence, the STARLAB is in continuous contact with the ground
stations through the synchronous satellite system. Maintaining three stations
in areas of low cloud cover virtually assures continuous contact with the
STARLAB. The CIMC contains one of the receiver-transmitter sites while
the other two stations are alternate receiver-transmitter sites. A microwave
or laser communications system connects all three ground stations so that
the STARLAB is in continuous communication with CIMC.

The computer on the ground is a much larger computer and would
have much more storage capacity than the onboard computer. This computer
would also be capable of repeating everything which the onboard STARLAB
computer does. The bulk of the experimental data is processed and stored
as necessary on the ground. The perishable data received from the STARLAB
are sent through the existing channels directly to the user; e.g., earth
resources data, such as forest fires, crop disease, storms, and air and
water pollution. These perishable data are transmitted directly to the user
through existing channels, such as the branches of the Forestry Service,
Agriculture Department, and the U.S. Weather Bureau. It is important that
perishable information obtained by STARLAB is sent to the user and not
stored at ground stations. Nonperishable data can be stored, processed, and
distributed as deemed appropriate.

Processing of Experimental Data

The research information flow diagram (Fig. III-1) presents some
concept of the problems involved in information management. Basically,
various inputs to appropriate sensors are translated into electrical or other
signals which are then either fed into a computer or physically sent back to
earth (film). The computer translates the electrical signals into digital form

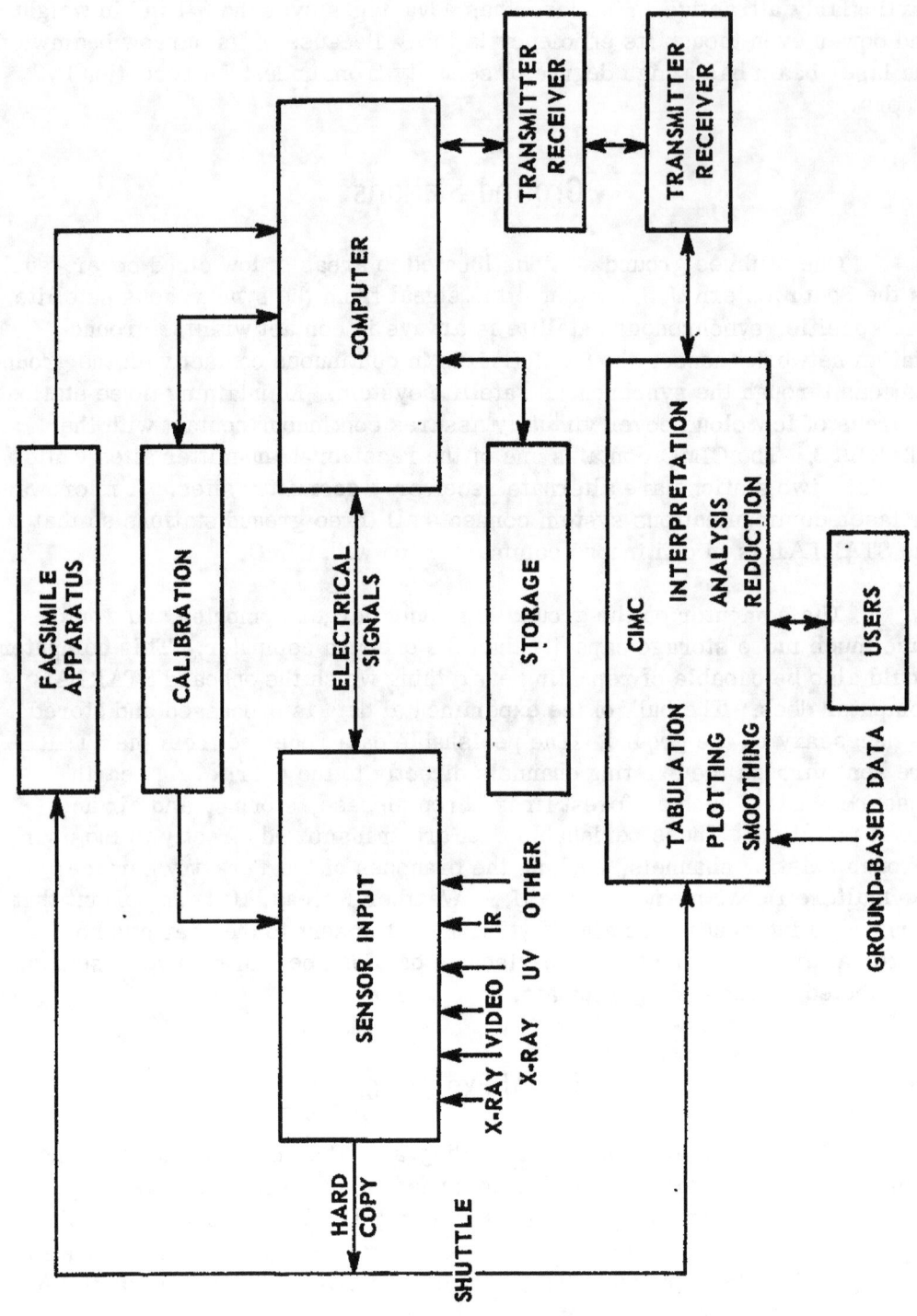

FIGURE III-1. RESEARCH INFORMATION FLOW DIAGRAM

3-6

to be sent to earth or stored for delayed transmission. The CIMC then converts the raw data into a form meaningful to various users. The users, in turn, give the CIMC personnel indications as to the usefulness of the data. Transmissions from CIMC to STARLAB explain how STARLAB can be more selective in its data acquisition. The onboard computer is programmed to assign priorities to various types of data to determine whether it should be sent to earth immediately, stored for later transmission, or processed on board.

Perhaps the most glamorous part of the earth-orbiting laboratory is the data collection subsystem. A large amount of money will be spent on developing this subsystem. However, if a commensurate amount is not spent on techniques, hardware, and people to process, analyze, duplicate, and disseminate the information gathered, then much of the money spent on data collection will have been wasted.

For example, multisensor imagery and nonimagery data (for earth sciences use, meteorology, oceanography, etc.) will be collected, filed, plotted, duplicated, enlarged, reduced, combined, mosaics, may be made, maps produced, true and false color prints made, and specialized atlases and encyclopedias published. At the CIMC, imagery interpreters who are expert in plots, radar, and infrared interpretation will process the incoming data both in film form and facsimile form. The following requirements should be available:

1. Specialized viewers and projection apparatus.

2. Measuring instruments.

3. Automatic isodensiometers and recording microdensitometers.

4. Darkrooms.

5. Photo labs.

6. Mass printing for black and white and color film end products.

7. Library and filming facilities.

8. Computers and their peripheral equipment (operators, systems personnel, and programmers).

The reader will note that all loops are closed in form, thus contributing to the objective of obtaining useful, valuable data and minimizing the common practice of gathering large quantities of data whose significance is questionable.

Summary

The points discussed in the introductory remarks are best summarized by considering the distribution of information within the continental United States (Fig. III-2). The synchronous satellite maintains continuous contact between the STARLAB and the earth. All information from STARLAB is transmitted to the CIMC, which is a part of STARAD. If the main receiver site is unable to receive the signal, two alternate receiver-transmitter sites are available to receive the STARLAB transmissions and relay them to CMIC. Almost all of the data sent to CIMC from STARLAB are in the unprocessed or raw form. At CIMC, the data are processed into the form required by the ultimate user. The two general categories for all data are (1) perishable and (2) nonperishable. The perishable data are sent by an automatically encoded arrangement directly to the user concerned while the nonperishable data are processed in a less immediate priority fashion.

The CIMC portion of STARAD is responsible for sending properly processed data to the user through already existing ground-based communication channels such as the postal service and the telephone networks. As an example of how perishable information might be handled, consider the situation of a forest fire or crop blight that is detected by the earth resource sensors of STARLAB. If it is to be useful, this information must be sent immediately to the user; namely, the forest rangers or the farmer. The information is sent to CIMC and encoded as perishable data. The CIMC computers send out a preprogrammed call to the proper district information center (DIC). The district centers are comprised of the already existing and structured county agencies such as the Agriculture Stabilization Committee, the Civil Defense Office, the Forest Ranger's Office, the Game Warden's Office, the Sheriff's Office, etc. Of course, city, state, and federal agencies such as the Weather Bureau might also be involved as deemed necessary and appropriate. In the case of the forest fire, a telephone in the proper forest ranger's office would ring and if crop blight is detected, a phone in the Agriculture Stabilization Office would ring. The proper farmer or farmers concerned about the crop blight would then be notified. In case the primary telephone did not respond to the call, secondary and other numbers would be dialed by the CIMC computers. If the user wished further information other

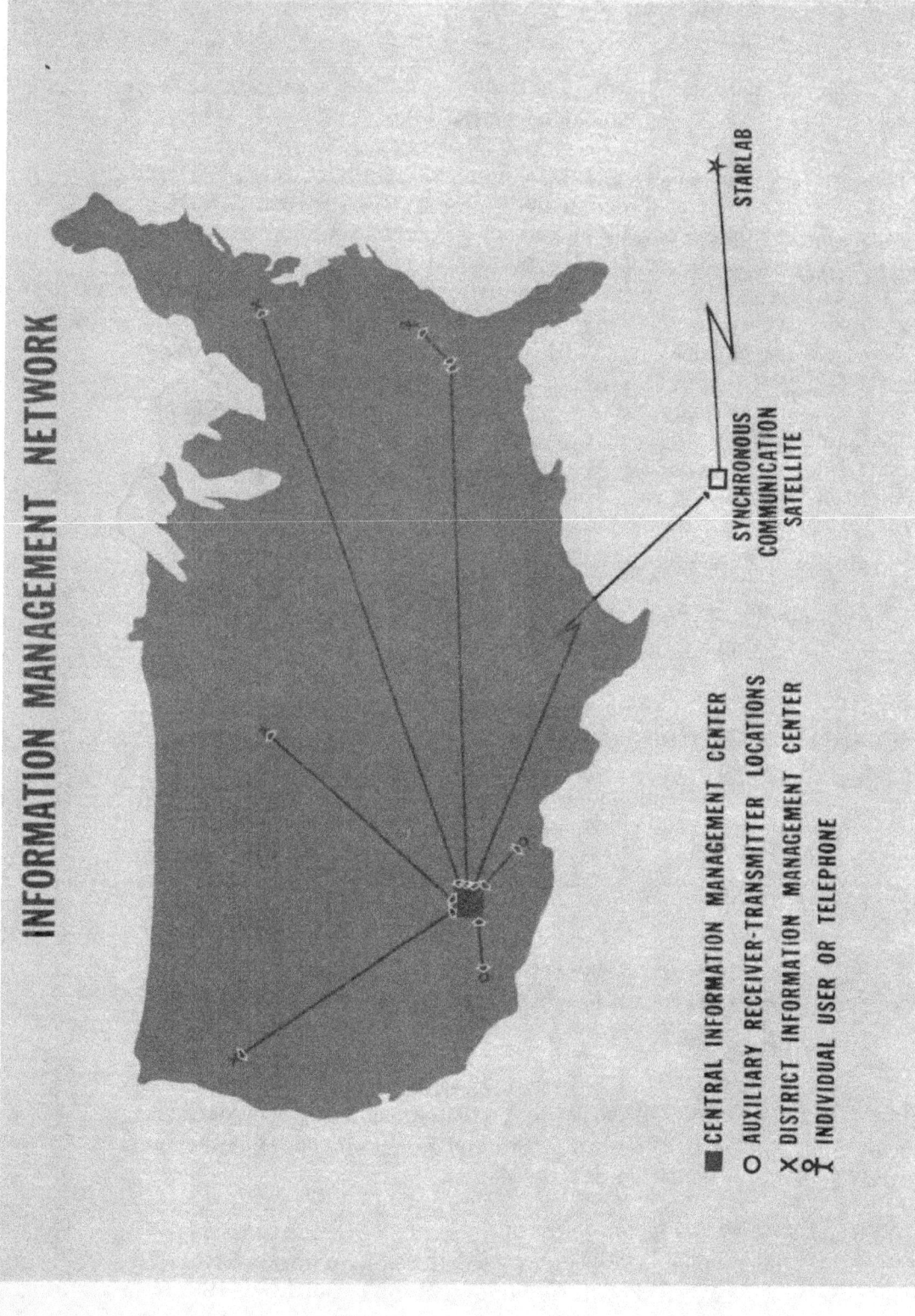

FIGURE III-2. THE COMMUNICATION SYSTEM OF THE INFORMATION MANAGEMENT NETWORK

than the "canned" message, he could call CIMC or if the emergency is severe he could converse with the commander on STARLAB.

Initially, it is anticipated that all communication links to STARLAB would pass through CIMC, which is located in the United States. As the program develops, users in countries around the world would be invited to participate first by coming to CIMC in this country to direct their work, and second by establishing their own communication networks and receiving STARLAB data directly. A worldwide advisory service could also be established after experience is gained from the United States operation.

An important aspect in any information management scheme is how to utilize intelligently the data that are accumulated. The unnecessary hoarding or storing of unneeded data is not acceptable from a cost-storage standpoint. It is earnestly recommended that an Information Management Control Committee (IMCC) be established to administer a reasonable set of policy guidelines. Some of the guidelines might be as follows:

1. A principal investigator may obtain his data in any form he desires provided funding is allocated.

2. Data of national or international interest will be stored or archived with three degrees of priority:

    a. Class A is of permanent and lasting importance. This information or data is stored permanently in STARAD files. Detailed reasons for classifying data with this priority rating will be filed with the data. A review of the priority ratings of Class A data must be made every 10 years by IMCC.

    b. Class B is of temporary importance. This information or data is normally stored for a maximum of 2 years in the STARAD files. The data may be stored for another 2 years by making petition to the IMCC.

    c. Class C is of immediate use only. This classification includes real time television, slow motion replay of video information and data that, in general, can be discarded almost immediately. Upon request by the user, these data could be stored for up to 1 month.

Normally these data will be given to the PI in raw or reduced form on magnetic tape. These tapes can be purchased, loaned, or leased to the PI or his sponsoring agency. If the PI has his own storage facilities, then of course, he may keep the data as long as he desires, but if STARAD is involved in the data storage, the above-mentioned classes of priorities must be adhered to.

The communication and telemetry system for the STARLAB will be largely automated and capable of handling a large amount of information and communication. Since the STARLAB is a laboratory in space, provisions must be made to handle a large flow of information and communication and to distribute it properly. The technology and hardware to develop this communication system is either presently available or at least should be in the time frame designated for STARLAB.

SYSTEM FUNCTIONAL CAPABILITIES

Introduction

The general assumption made in this report concerning communications is that there is a positive need for a sophisticated and comprehensive communications system or network both on board and on the ground to support properly and effectively the STARLAB functions. Crew effectiveness, efficiency, morale, and general well-being will be greatly enhanced if the STARLAB is equipped with the many conveniences as well as necessities afforded by a well-designed communications system. Since the cost of putting a man on board STARLAB is considerable, it is almost mandatory that his time be used as effeciently and effectively as possible. Although the sophisticated communications network described in this report is quite expensive, the added efficiency it affords should more than offset the additional cost. The crew, including the researchers, will be expected to perform a variety of functions including station housekeeping, experiments, manufacturing, earth resource investigations, traffic control, and perhaps function as a space university.

The effective management of information on STARLAB, between earth and STARLAB, and on earth (STARAD) is expected to be formidable, yet extremely important, task if the STARLAB project is to realize its full potential. There are three general areas into which the information flow can be divided:

1. The establishment of criteria and priorities to separate the critical and emergency functions from the routine video, voice, and data functions.

2. The allocation of channel capacity for the service functions which include assistance to the researcher and experimenter, to earth resource observations and data gathering, to the manufacturing aspects of STARLAB, to the academic community and to the ship, airline, and surface traffic controllers.

3. The provisions for crew recreation and personal business such as commercial TV, AM and FM stations, reading matter, and private correspondence and telephone calls. Because space and weight will be at a premium on STARLAB, almost all of the above-listed functions will be performed electronically. The primary readout display will be a video picture which means that almost all of the data will be displayed in digital form. Of course, real-time and slow or fast time-domain responses as well as other types of displays could also be provided using the video technique.

Communication System. The onboard computer performs both switching and computational functions and is the heart of the communication and instrumentation systems. To provide the desired communication functions a large bandwidth of over 500 MHz is required. It is felt that a laser link between the STARLAB and the earth is highly desirable for two important reasons. First, the wide bandwidth capability of lasers means that a great deal of information can be placed on one laser beam. Present laser technology indicates that a laser can be made to lase at several colors simultaneously. Each color can be separated by optics and modulated separately at up to 500 megabits per second (Fig. III-3). A six-color beam could result in a very wide bandwidth of 3000 megabits per second for a laser beam. This is equivalent to a 500-MHz bandwidth of analog signal.

The need for such a large channel capacity is obvious if one considers the expected video, voice, and data requirements for STARLAB communication purposes (Table III-1). It is also desirable to summarize the classification of channels and define certain abbreviations (Table III-2). The allocations given in Table III-1 are summarized in Table III-3, where up and down datalink needs are shown. If needed, extra channels to serve the auxillary modules can be added. Considering only the STARLAB-CIMC communication links, a total of three transmitting and three receiving high resolution TV (HRTV) of 5000 to 6000 lines per raster will require approximately 300 megabits per second per channel or 1800 megabits per second total if all six channels are operating simultaneously. The 32 channels of regular TV would need 30 megabits per second per channel or a total of 460 megabits per second. The 1000 channels of 5 KHz each also take a total of 150 to 180 megabits per second; thus, a grand total of about 3000 megabits per second is required to

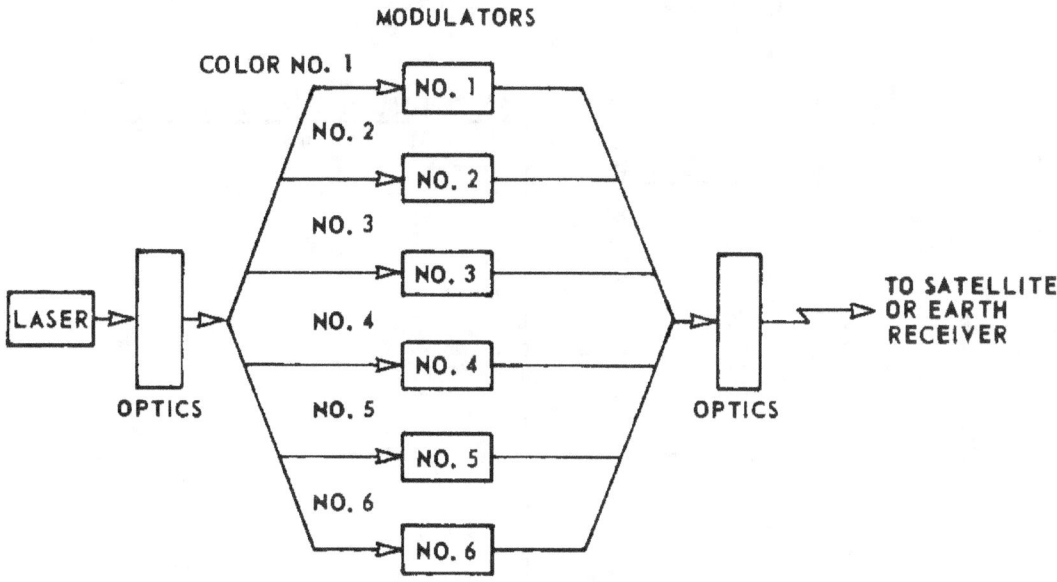

FIGURE III-3. A 3000-MEGABIT PER SECOND LASER TRANSMITTER

handle all of the video, voice, and data needs if all the equipment is used simultaneously. Since it is rather unlikely that all video, voice, and data channels will be used simultaneously, the voice and data channels could be increased greatly if several of the video channels were used for voice and data transmission. If desired, a reasonable amount of data could be stored in the STARLAB computer and dumped to earth at a faster-than-real-time rate. The onboard STARLAB computer is able to allocate channels as desired or according to a predetermined priority.

Computer Subsystem

    a. General Philosophy. It is assumed that the onboard computer performs all computational, logical, and storage functions necessary to maintain STARLAB in a safe and productive state for an extended period. All other functions, except for a relatively small amount of data processing and storage, are done by the CIMC computer.

    (1) Processing of Experimental Data. We take the position that processing of major amounts of experimental data should be accomplished at a ground station. Our reasons for adopting this policy are:

TABLE III-1. NECESSARY CHANNEL CAPACITY FOR
A 50-MAN STARLAB

A. Television Needs	Number of Channels	
	High Resolution Television	Regular Television
1. STARLAB to Earth		
(a) Astronomy	2	
(b) Biology	1	2
(c) Manufacturing	1	2
(d) Experiments	1	4
(e) Shuttle		1
(f) EVA		1
(g) Outside Windows		4
2. Earth to STARLAB		
(a) Entertainment (Commercial TV)		6[a]
(b) Other Needs	1	12

B. Audio Needs

2000 Channels of 5 kHz for Each Channel for Voice and Low Data Rates (Includes Commercial Radio).

4-10 Channels of FM at 50 kHz Each.

C. Data Needs

200 Channels of 50 kHz Each for Medium Data Rate.

60 Channels of 500 kHz Each for High Data Rate.

a. 4 UHF and 2 VHF and Educational TV.

TABLE III-2. CLASSIFICATION OF CHANNELS

Channel	Bandwidth	Definition
LDR	0 to 5 kHz	Low Data Rate Channel: Capable of Handling Voice and Analog Signals up to 5 kHz. Also Capable of Handling Time Multiplexed Signals, with a Maximum Sampling Rate of 1.5 kHz.
MDR	0 to 50 kHz	Medium Data Rate Channel: Capable of Handling FM Voice and Analog Signals up to 50 kHz. (Wide Band FM with a 10 kHz and Modulation Index of 5 Yields 50 kHz as the Highest Frequency Sideband.)
HDR	0 to 500 kHz	High Data Rate Channel: Capable of Handling Data Rates up to 200 kbits/sec.
TV	0 to 6 MHz	Capable of Handling Good Quality Commercial TV and FM Voice.
HRTV	0 to 50 MHz	Capable of Handling a High Resolution TV Channel With Real Time.

TABLE III-3. SUMMARY OF FREQUENCY ALLOCATIONS

Up Data Links	
Maximum	Channels
1000	LDR
100	MDR
30	HDR
16	TV
3	HRTV
Down Data Links	
Maximum	Channels
1000	LDR
100	MDR
30	HDR
16	TV
3	HRTV
From Other Sources (RFM's, Shuttle, etc.)	
Receive	Channels
1000	LDR
100	MDR
30	HDR
8	TV
1	HRTV

Note: All Emergency Functions, Automatic Signals That Can Be Done in 20 MHz

3-16

1. Principal investigators may not know what calculations they wish to carry out, or, if they do have preconceived ideas, they may be forced to change their minds after examining the data. At any rate, the PI's would quite often need to examine raw data over an extended period, make extensive program changes, etc. We maintain that this type of activity can best be done on the ground.

2. Some experiments generate data at enormous rates and thus require the services of a very large computer for processing and reduction. Cursory cost calculations indicate that ground computation is less expensive than computation in space. Furthermore, we do not feel that serious degradation in the quality of data will occur while being transmitted to ground via a laser communications system. Thus, we are led to the conclusion that processing of large amounts of experimental data should be done on ground.

(2) Navigation and Guidance Calculations. The computer program required for navigation and guidance (N/G) calculations is rather large and the amount of computation is extensive. We take the position that these calculations should be done on the ground except in emergency situations. The amount of input data for N/G calculations is small and hence could be sent to ground either by the laser or the S-band backup system. Thus, there is a very small probability that the ground station could not provide results of N/G calculations. In an unusual situation, the onboard computer system can use a stored program to carry out essential navigational and guidance calculations.

(3) Services Provided by Onboard Computer. The onboard computer is to provide for such services as:

1. Attitude control.

2. Power management.

3. Environmental and life support (EC/LS) monitoring and control.

4. Propulsion and fuel management.

5. Instrument and system checkout monitoring.

6. Computer operating system (software expansion and change).

7. A moderate level of scientific computation.

8. Navigation and guidance calculations in emergency situations.

9. Signal switching.

(4) Comments. The onboard computer, as visualized, has hardware which is redundant, modularly-expandable, and fails gracefully. The software is conceived as being flexible, man- and problem-oriented. The multiprocessor computer configuration simplifies the problem of providing the investigator with direct access to computer facilities (Fig. III-4). Each investigator can use the computer as his area in any way he finds advantageous to his work without the complex software package which would be needed if all investigators used a single large computer.

b. Hardware Characteristics. The system hardware consists of a fast central computer, two or more peripherial computers, and a mass storage unit (Fig. III-4). We visualize the hardware to have at least the capability of that of the late 1960's (Refs. III-1 and III-2).

c. Software Characteristics

(1) Executive Computer. The software package of the executive computer is constructed so that the former:

1. Controls I/O to and from central computer and storage units.

2. Monitors EC/LS, attitude control, fuel, and instrument and system checkout data.

3. Carries out attitude control and fuel and power management calculations.

4. Switches all signals.

5. Carries out automatic emergency sequence operations and controls formatting of information to be displayed on TV consoles.

It is expected that time-sharing techniques can be utilized to an advantage by the executive computer.

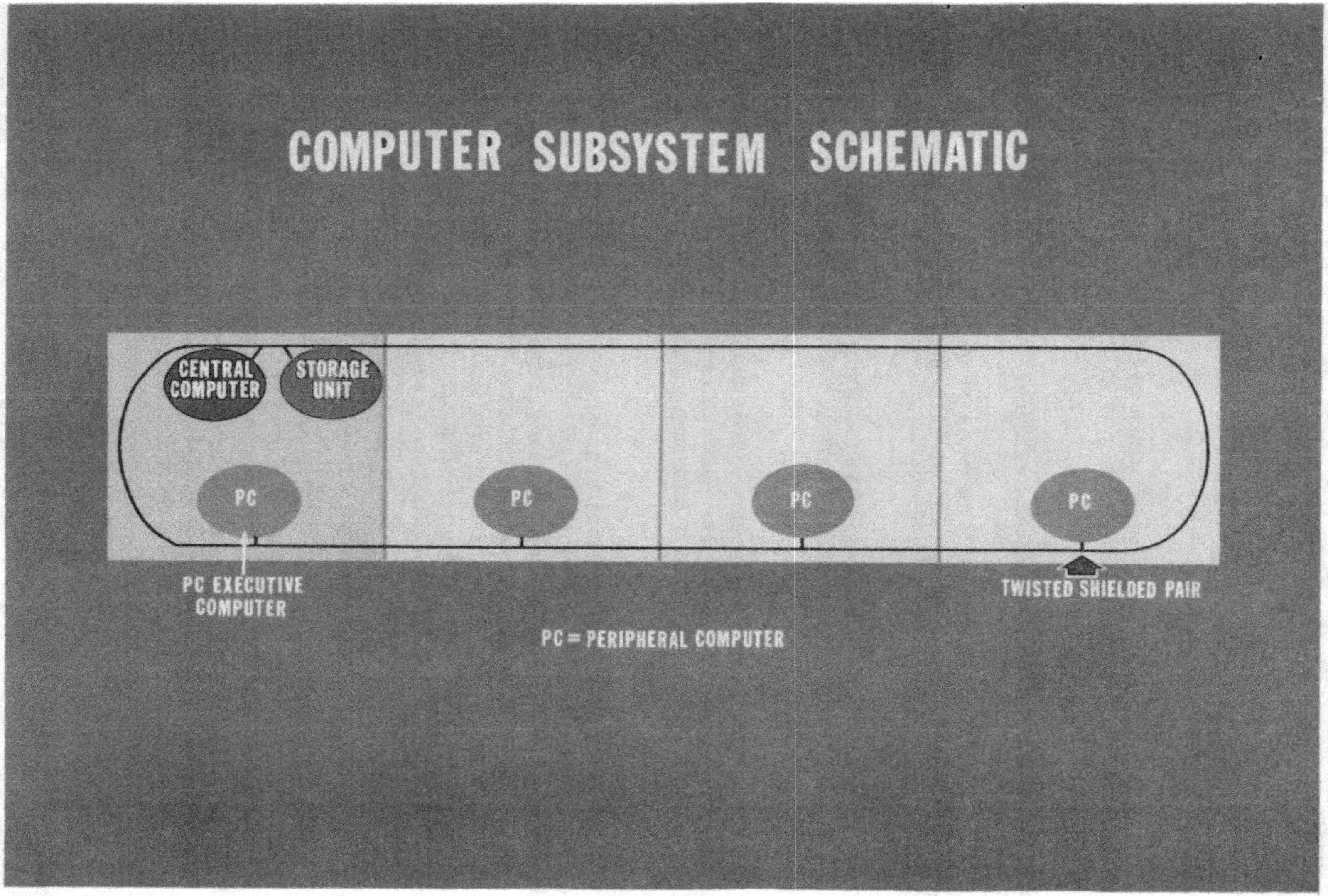

FIGURE III-4. THE MULTIPROCESSOR COMPUTER

(2) Area Computer. Normally under direct control of an investigator, the area computer will:

    1. Carry out relatively minor calculations and data processing tasks.

    2. Give the investigator access to the central computer and mass storage unit through the executive computer.

    3. Act temporarily as the executive computer when the regular executive computer is being maintained, checked out, modified, or is out of order.

Communication with the computers can be facilitated by the use of such software packages as DIALOG and SYNTAX (3. above) which have recently been made available.

(3) The Central Computer. Considerably larger and faster than the peripheral computers, the central computer processes moderate amounts of experimental data, and is capable of doing navigational and guidance calculations should be need arise.

Assuming the memory requirements given in Table III-4, the peripheral computers require a working storage of about 36K words. With time sharing, the 36K-word peripheral computer can serve as the executive computer. Initially, the central computer can be 32K to 64K words. At least two peripheral computers would be required initially. The total working storage is about 138K and mass storage about 100K. The power requirements are 750 to 1000 watts, 300 to 400 pounds, and 5 to 7 cubic feet.

The Instrumentation System

Before proceeding to a general discussion of the information management system, it is appropriate to give the general description of instrumentation which is on board STARLAB. Because of the anticipated difficulty in repairing equipment on board, it is very desirable to have a great deal of commonality among the equipment used in the experimentation, manufacturing, and other phases of the activities requiring communication equipment. Although it is desirable to have only one type of receiver and one type of transmitter for the onboard and offboard communication purposes, a more practical solution is to have three types of receivers and three types of transmitters, divided as follows: (1) a receiver-transmitter combination

TABLE III-4. ESTIMATED HARDWARE REQUIREMENTS
AND SPECIFICATIONS

Memory Requirement Words (Per Unit)	Function	Computer
32K	Navigation and Guidance	Central Computer
16K	Attitude Control	Executive Computer (EC)
8K	Power Management	EC
4K	EC/LS	EC
4K	Propulsion and Fuel Management	EC
24K	Checkout Monitoring	EC
8K	Operating System (Software)	EC
4K	Switching	EC
32K	Scientific Operations	Peripheral Computer

to handle all audio and data needs as tabulated under Items B and C in Table III-1, (2) another receiver-transmitter unit that tunes to the regular TV and FM channels, and (3) a receiver-transmitter unit to handle the high resolution TV channels. A central stockroom containing a reasonable number of these units would dispense them as required by the experimenters.

The sensor, or sensor-transducer unit, also incorporates the concept of commonality. Let us consider how the various subassemblies are packaged to provide maximum flexibility and a reasonable degree of commonality [Figs. III-5(a) and III-5(b)]. The experimenter or researcher goes to the central stockroom and to check out the necessary pieces or items of equipment to assist him in carrying out his duties. Before being checked out to the user, each piece of equipment is plugged into a test assembly to determine whether or not the unit is operational and in proper calibration.

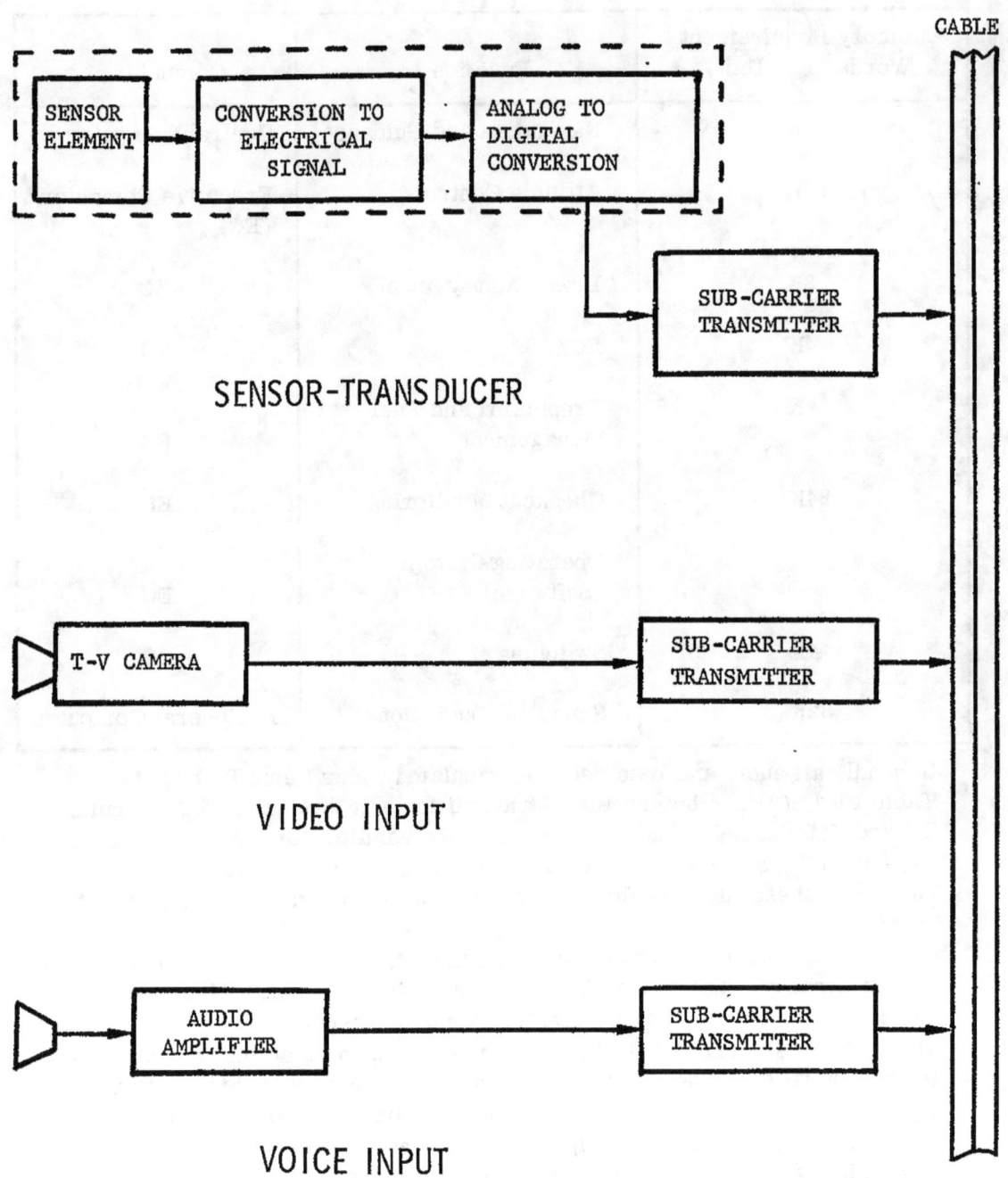

FIGURE III-5(a). SENSOR-VIDEO-VOICE INPUT UNITS

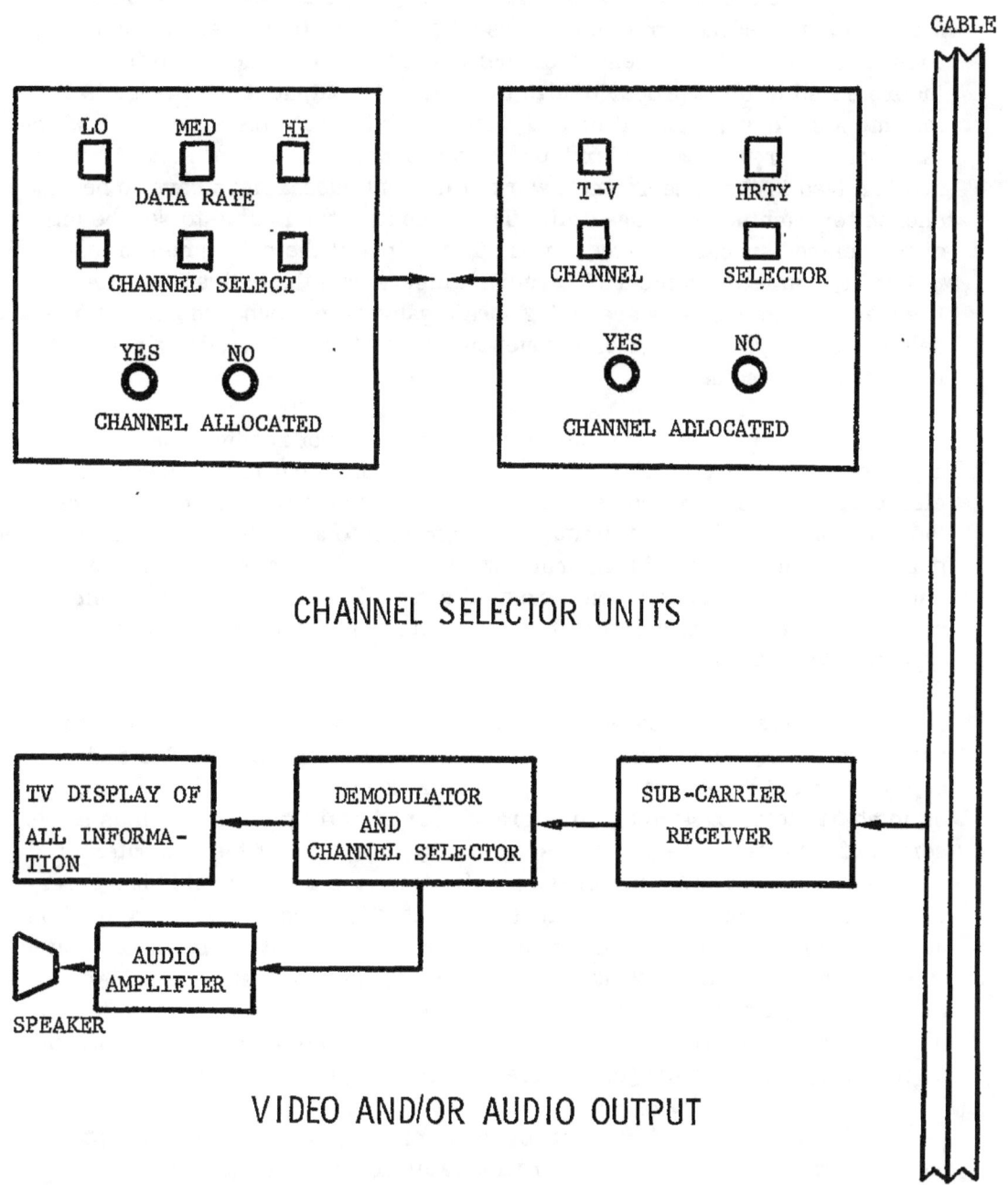

FIGURE III-5(b). CHANNEL SELECTION AND INFORMATION OUTPUT UNITS

3-23

As an example of how the various subgroups are utilized, consider the sensor-transducer unit and suppose that this particular sensor unit is to measure strain. The sensor element is the strain gauge which is incorporated into the proper bridge circuit. The output of the bridge is fed into a solid-state amplifier and then modulated onto the subcarrier which is allocated for this particular type of transducer. If several strain gauges are involved in the experimental work, different subcarriers have to be chosen for each strain gauge unit. The experimenter is able to set the unit at the desired subcarrier frequency. The output of the subcarrier transmitter is then fed into the TSP circuits which serve the entire STARLAB. If several experimenters are using strain gauges, the onboard computer selects the proper subcarrier frequencies so that one signal does not interfere with another.

If the experimenter wants to obtain a series of readings or perhaps a plot of strain versus stress, then some form of output is required. The electrical signals from the sensor are fed into the central computer which then converts the electrical voltage or current into a reading that is proportional to the strain. This format would have to be programmed into the computer. The computer then sends out a signal to the TV monitor which could be in the form of a digital readout or a strip chart plot or some other type of data display.

The type of data to be presented on the TV monitor depends on the nature of the experiment and the desires of the experimenter. The digital computer could be programmed to provide X-Y plots, phase plane plots, strip chart plots, bargraphs, line graphs, or digital displays, such as a series of numbers. If a storage scope is employed, or if the computer is programmed to maintain the image indefinitely, the experimenter has a very versatile type of output device. It is felt that it is almost mandatory for both an all-electronic data processing and electronic read-outs to be incorporated in STARLAB. Space and weight are at such a premium that it is undesirable to attempt to provide much paper for plotting, writing, or storage of information. A writing pad activated by a light pen is used for a scratch pad and a temporary written storage. An electric field is used as an eraser.

It is expected that very little repair of equipment will be attempted on board STARLAB, but that calibration facilities should definitely be provided. It is hoped that enough commonality will be provided so there will be little need to be concerned with repair of the equipment. A defective unit is simply replaced by another similar unit from a stockroom and the item to be repaired is sent back via space shuttle and a new item will be sent up as a replacement. The calibration of the various instrument subassemblies is done by using computer routines and calibration standards. The instruments are plugged into a test jig to determine the operational status of the units.

To simplify the various instrument subassemblies [Figs. III-5(a) and III-5(b)], it is necessary for the STARLAB computer to perform almost all of the data processing necessary to obtain the desired output. Suppose that 10 to 50 voltmeter readings are needed simultaneously. It would be necessary to check out from 10 to 50 voltmeter type sensor-transducer units from the central stockroom. These units would be attached to the desired terminals at which the voltage is to be measured; the output of the subassembly would be fed into a transmitter subassembly, and the transmitter output in turn would be fed into the TSP common cable. The STARLAB computer then converts this signal into an appropriate digital voltage which is fed into the television receiver as either a digital display of digits or perhaps as a plot of voltage versus time. By using time sharing techniques, all 10 to 50 voltages could be monitored simultaneously and very accurate readings could be obtained by using a sophisticated process unit inside the computer. It is anticipated that this technique will reduce considerably the complexity of the individual instrumentation subassemblies and at the same time greatly increase their accuracy and reliability.

Other instrument subassemblies are required [Figs. III-5(a) and III-5(b)] such as the video input unit, the voice input unit, the television output, and the audio output unit. The proper amount of equipment to be in each subassembly has to be given additional consideration, but to maintain the commonality principle and at the same time decrease the complexity and equipment utilization, the number of equipments for instrument subassemblies should be minimized. Selection of proper carriers or subcarrier frequencies are obtained by use of a touch-tone type dialing system which could either be plugged into each unit or mounted permanently with each unit.

In addition to the instrument subpackages shown in Figures III-5(a) and III-5(b), additional subassemblies are necessary to provide the remote control capability for experiments on board the free-flying module and for the remote controlling of equipment and experiments inside the free-flying module, and perhaps the STARLAB itself. Although it is anticipated that many of the remote control devices will perform unique functions, it is expected that much commonality can be provided also; for instance, in an experiment requiring the positioning of devices, a unitized and unified remote control bench could be designed to accommodate the needs of many types of experiments and the desires of many experimenters. For example, it might be desirable to design only a two-dimensional bench in which equipment is placed in a plane. If three-dimensional motion is required, it might be possible to use translation motion only and still have a relatively simple system which would position quickly and accurately. TV monitors could be used with the experimenter providing the feed-back function or the unit could be remote controlled through its own feed-back network and the computers. Additional investigations are needed in this area.

Information Distribution

It is assumed throughout this report that almost all of the data acquisition, processing, and distribution is done by means of electronics. Although man makes almost all of the major decisions, he is assisted by computers and a wealth of information displayed in a convenient form. An overall information flow diagram is instructive (Fig. III-6). The block diagram is divided into two similar areas; in space and on earth. The first step, either in the STARLAB or on earth, is the acquisition of information or data. This information or data can come from many different sources [Figs. III-7(a), III-7(b), III-7(c), and III-7(d)] as the various sensor and display units are found in the laboratory area, crew quarters, FFM's, and shuttle. For instance, in space, the information can come from man using a voice mike or a touch-tone input to the computer. Here, man could be situated either inside the STARLAB, in the EVA configuration, in the space shuttle, or in the manned free-flying module. Information input could also be achieved via the TV cameras or the many data sensors used in the experimental and manufacturing activities. Critical or emergency data input is furnished by the many sensors located throughout the STARLAB, the FFM's, and the space shuttle.

On board STARLAB, all information is fed into the triply redundant systems of the TSP that is conducting frequency multiplexed carrier frequencies from dc to about 275 MHz. All of this information is fed into the onboard computer in a parallel mode; i.e., although the computer performs some sort of function on all of the information, the information is available at any receiving station that is connected to the TSP. The onboard computer performs two major functions; (1) it either switches the signal to another station in the STARLAB, the space shuttle, the FFM's, or the EVA, or it sends the unprocessed data or processed data directly to earth, and (2) the computer performs data processing, either for onboard usage or for transmission to earth. As mentioned in the introduction of this chapter, the onboard computer does not have enough storage or computational capacity

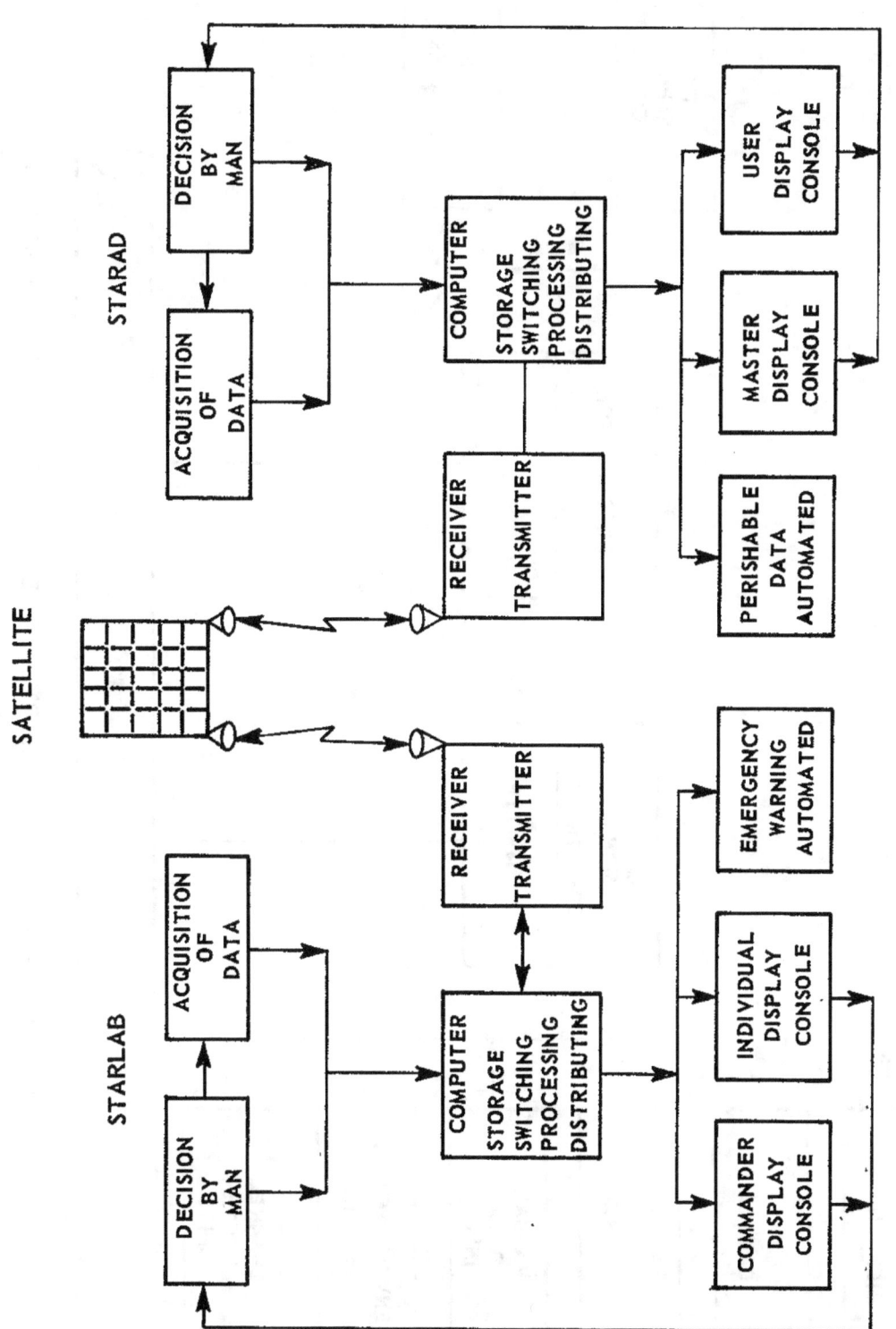

FIGURE III-6. INFORMATION FLOW

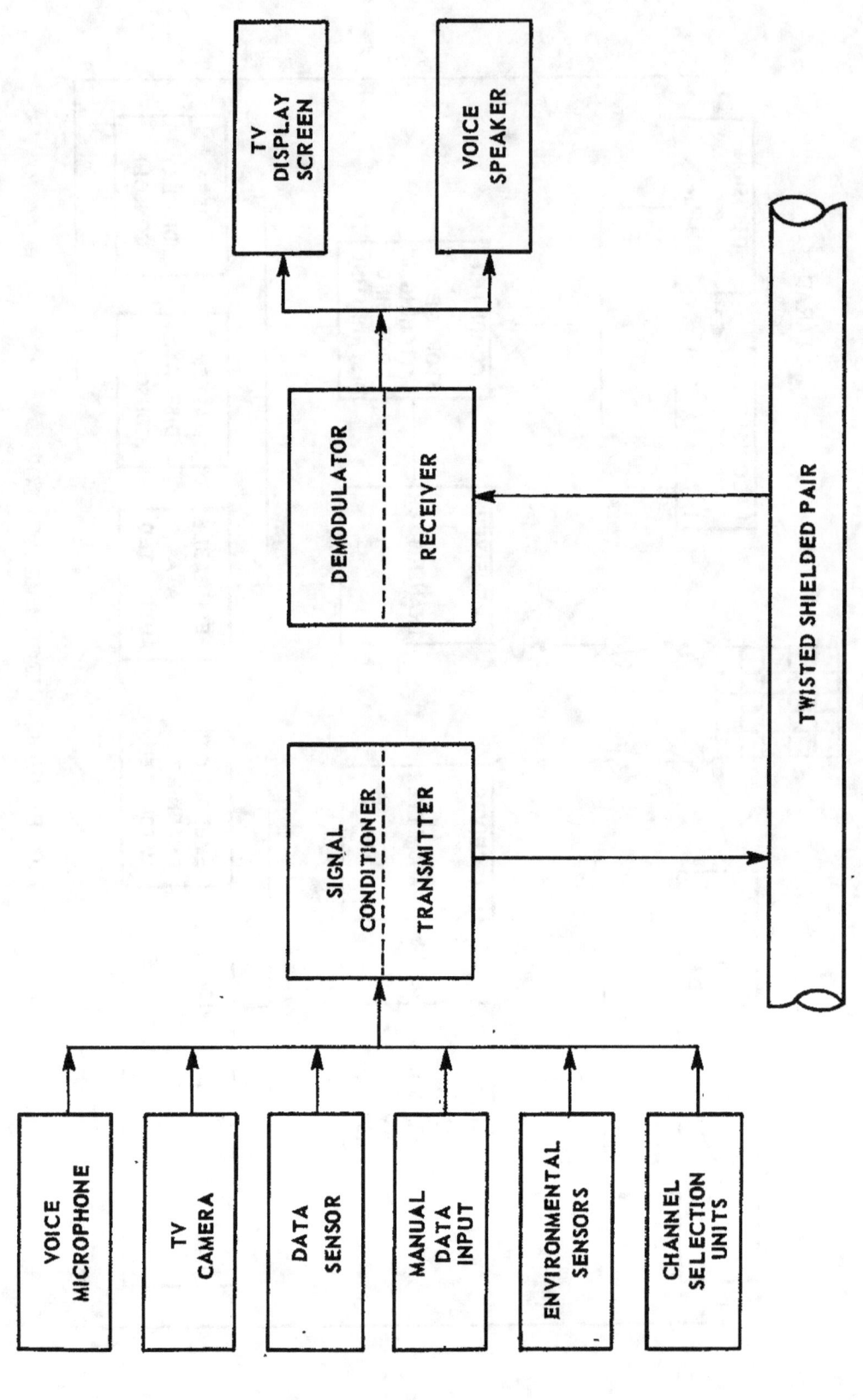

FIGURE III-7(a). LABORATORY AREA DATA ACQUISITION AND DISPLAY

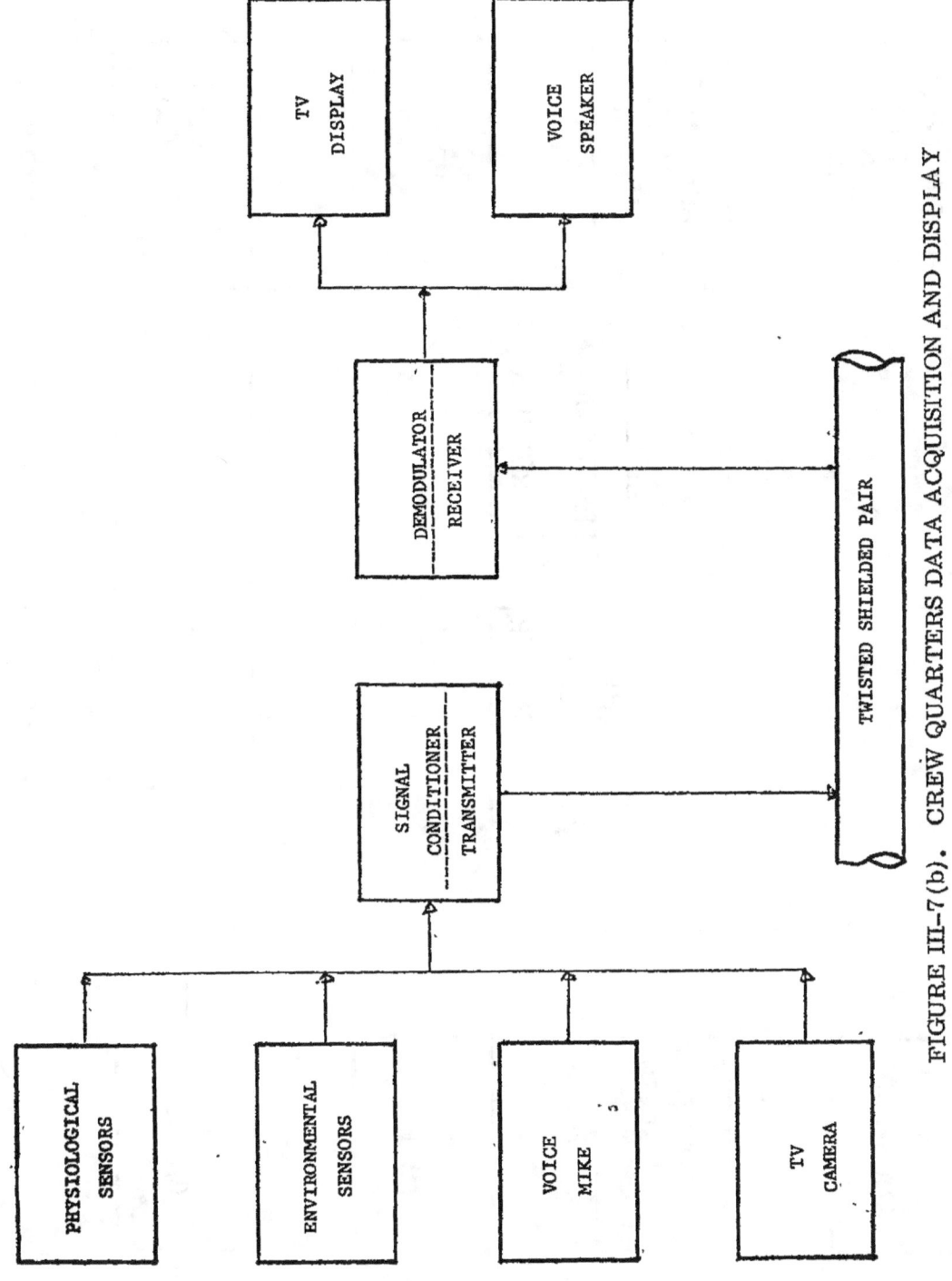

FIGURE III-7(b). CREW QUARTERS DATA ACQUISITION AND DISPLAY

3-29

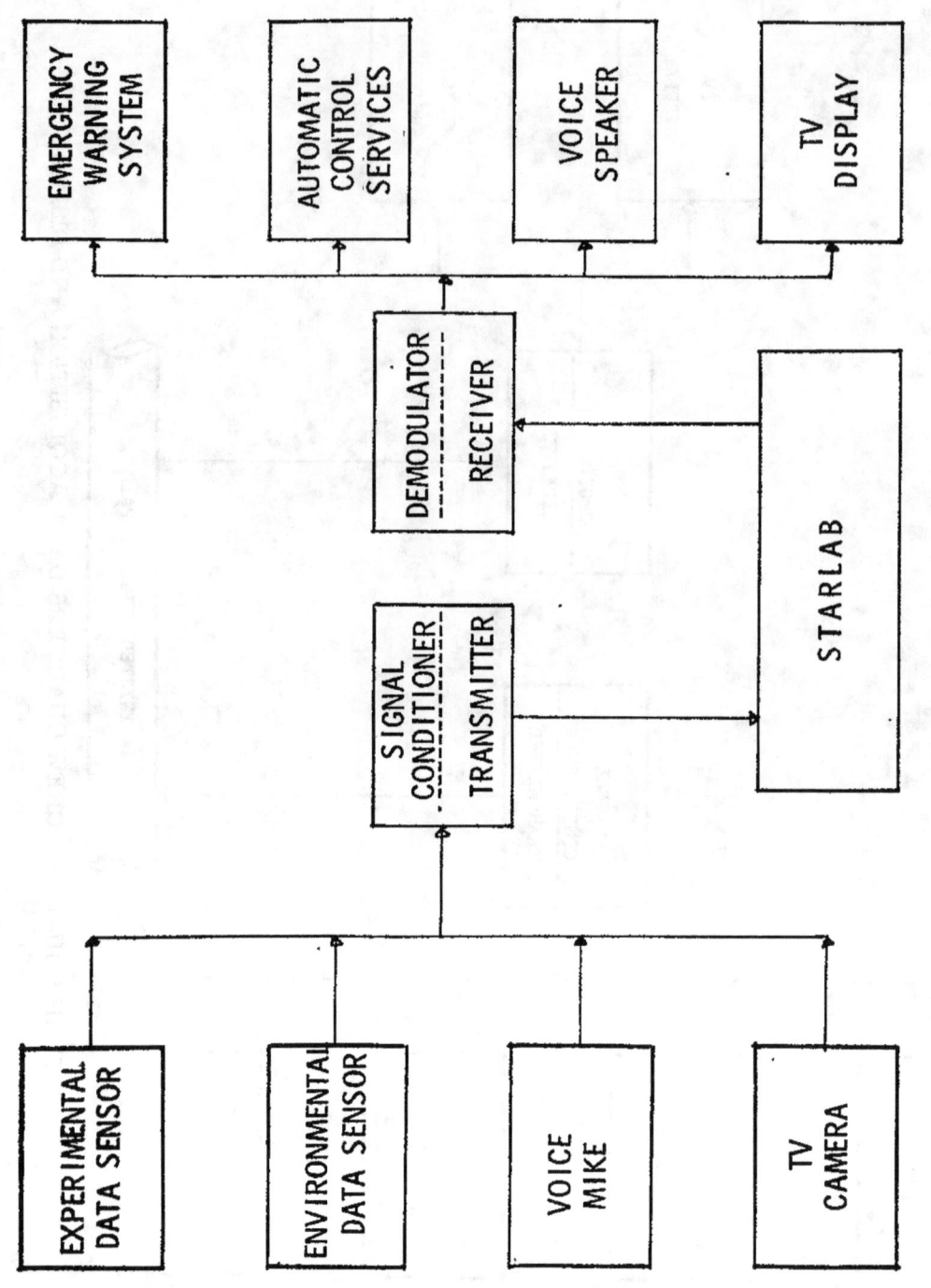

FIGURE III-7(c). OFFBOARD DATA ACQUISITION AND DISPLAY MANNED RFM

3-30

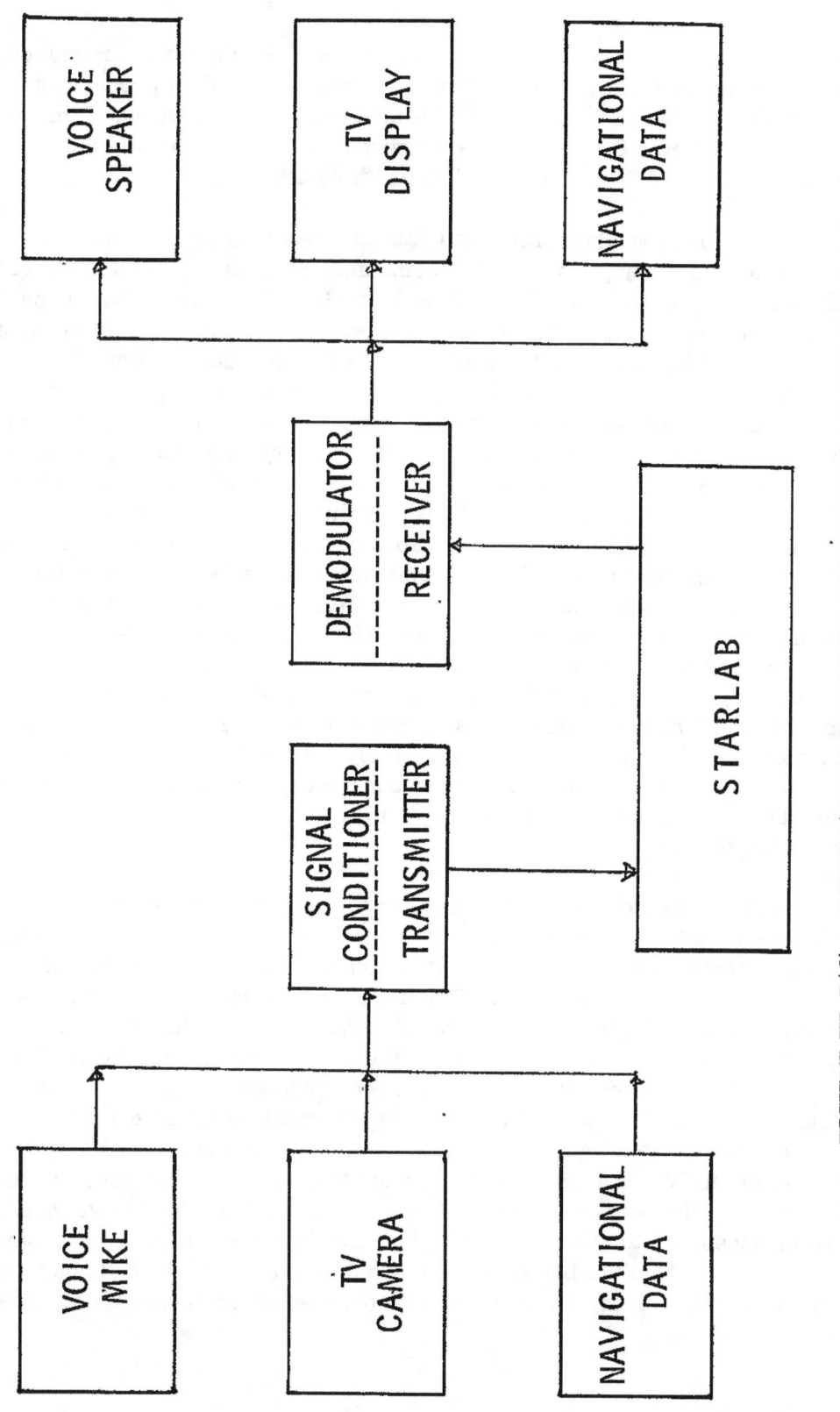

FIGURE III-7(d). OFFBOARD COMMUNICATIONS STARLAB-SHUTTLE LINK

to process all of the data that can be acquired by STARLAB in its research, manufacturing, and observation endeavors. Much of the data processing will be accomplished on earth because of the much larger computation facilities that will be available there, and this processed data will in turn either be returned to STARLAB or distributed on earth as desired.

The unprocessed or processed data are sent to four locations; (1) the storage unit of the computer, (2) the commander's console, (3) the individual console located throughout STARLAB or in the free-flying modules, or perhaps the space shuttle, and (4) the automated emergency warning and action system (Fig. III-6). The commander's console is rather sophisticated and is primarily concerned with the station keeping and housekeeping activities. The individual experimenter's console is one of many located throughout the STARLAB. The equipment in these console will vary depending upon the needs of the experimenter or crew member. If experiments are being performed, the individual consoles will contain TV transmitters and receivers as well as the many data sensing units. If the console is used in its recreational mode, then the TV receivers and a telephone unit would probably be located in the console. Each console would also contain a minimum number of emergency signaling devices as deemed appropriate. The automated emergency warning and action system would not be a console in itself. The emergency functions would already be reflected at the commander's console and, to a lesser degree, in the individual console. It is expected that there will be certain functions that will be performed automatically without decision by man. These might include the sealing off of an entire area because of some catastrophic failure in the area, increasing the oxygen input to a given area, turning on warning lights, etc.

As described earlier, man has a very important part of this closed-loop data management system, because a decision by man must be made for the loop to continue except in a few cases. In many cases, there will be no need to process the data on board STARLAB. For instance, in the case of high resolution TV during astronomical observations, the signal could be switched by the onboard computer directly into the laser modulator and transmission system to be relayed to the earth receivers and processed on the ground. Because of the large bandwidth requirement, it is almost mandatory that lasers be used during high data rate transmissions such as the use of high resolution TV. However, it is expected that an S-band microwave backup will be employed when the use of the laser is not feasible. In all likelihood, it will be necessary to eliminate the high data rate channels when the S-band transmitters are being utilized. Once these data are on the ground, they can either be stored and sent out as either unprocessed or processed data. The unprocessed or processed data can then be distributed to the master console located in the main receiving computational centers or sent out to user consoles that may be located anywhere in the country or perhaps the world. The earth-based computer will have the capability of duplicating all of the onboard computational processes. Of course, it will not be able to perform the switching functions, but will be able to activate the switching functions on board STARLAB. This will provide an added backup in case the STARLAB computer fails.

Consider an experimenter working on board STARLAB who wants to observe the experiment he has just performed in slow motion. The TV picture of his experiment is relayed via the laser network to the storage unit on earth at CIMC. This can be interrogated in such a way as to play back the TV picture in slow motion or stop motion much as we see with today's football game broadcasts. It does not appear that it would be feasible to attempt this on board STARLAB because of the large storage capacity and peripheral equipment that would be required. Note that the data management system on earth is almost identical to that located in STARLAB. Data can be acquired both on earth as well as in STARLAB. For instance, in the earth resource experimental work, it might be necessary to make various measurements on the surface of the earth and transmit this to STARLAB or CIMC for comparison purposes with the data acquired by the STARLAB sensors. Again note that man is in the closed-loop system on earth just as he is in the closed-loop information system on STARLAB.

As mentioned earlier, a little more detail about the equipment that would be contained in the laboratory areas, the crew quarter areas, the command areas, the unmanned FFM's, the manned FFM's, EVA, and the space shuttle [Figs. III-7(a), III-7(b), III-7(c), and III-7(d)]. In all cases, note that information from these various areas is fed into the TSP. As will be detailed in the next section, the TSP represents a triply redundant system in that three separate pairs of twisted wires are located throughout the STARLAB area. Although all three pairs of wires are used during normal activity, if something should happen to one or two of the shielded twisted pairs to cause them to become inoperative, then the final pair could be used for emergency purposes and for emergency operational functions such as maintaining life support and communications between STARLAB, the ancillary vehicle located around STARLAB, as well as communication to earth. This would necessarily restrict the accumulation or taking of data as the bandwidth would be limited by about a factor of three.

The commander's display console is quite elaborate but functional (Fig. III-8). Emphasis is placed on the environmental display panel located in the commander's area and is also duplicated in the emergency shelter area. From this point, the commander is able to obtain all desired information about critical factors in and around the STARLAB as well as the free-flying modules, the EVA's, and the shuttle. This information comes from the various sensors located throughout STARLAB and the ancillary modules. The commander is able to interrogate any of the sensors by pushing the proper buttons with the information displayed on any one of the three TV screens. Notice also the constraining seat in which the commander is sitting. He straddles the restraining bar that is attached to the seat and by kicking with his feet or using his hands he can glide around the console area on the guide rod. This same seating arrangement is recommended for the laboratory and workshop areas.

EQUIPMENT DETAILS

Introduction

The communication subsystem is designed to provide excellent service to the members of the Starsystem. The communications equipment is intimately tied to the computer to provide programmed channeling of all information transmissions.

The main communications link between STARLAB and STARAD via satellite is provided by the integrated laser subsystem. It provides for voice and video communications, telemetry, command, tracking, and ranging. The laser system was chosen because it provides the bandwidth required to support the information flow associated with experiments performed in astronomy, earth resources, manufacturing, and biosciences. This is done within the weight and power limitations imposed on space flight system. Further, it avoids conflicts of use of the frequency spectrum and meets with Federal Communications Commission's approval. It inherently is a secure system since the laser beam illuminates only a small area on earth. An additional reason for choosing the laser system is to encourage the development of the high potential available in this communications system.

The unified S-band system is used primarily as a backup system. It is also used for coarse pointing of the laser apertures. The S-band system is of the Apollo type and is chosen as the backup system since in utilizes

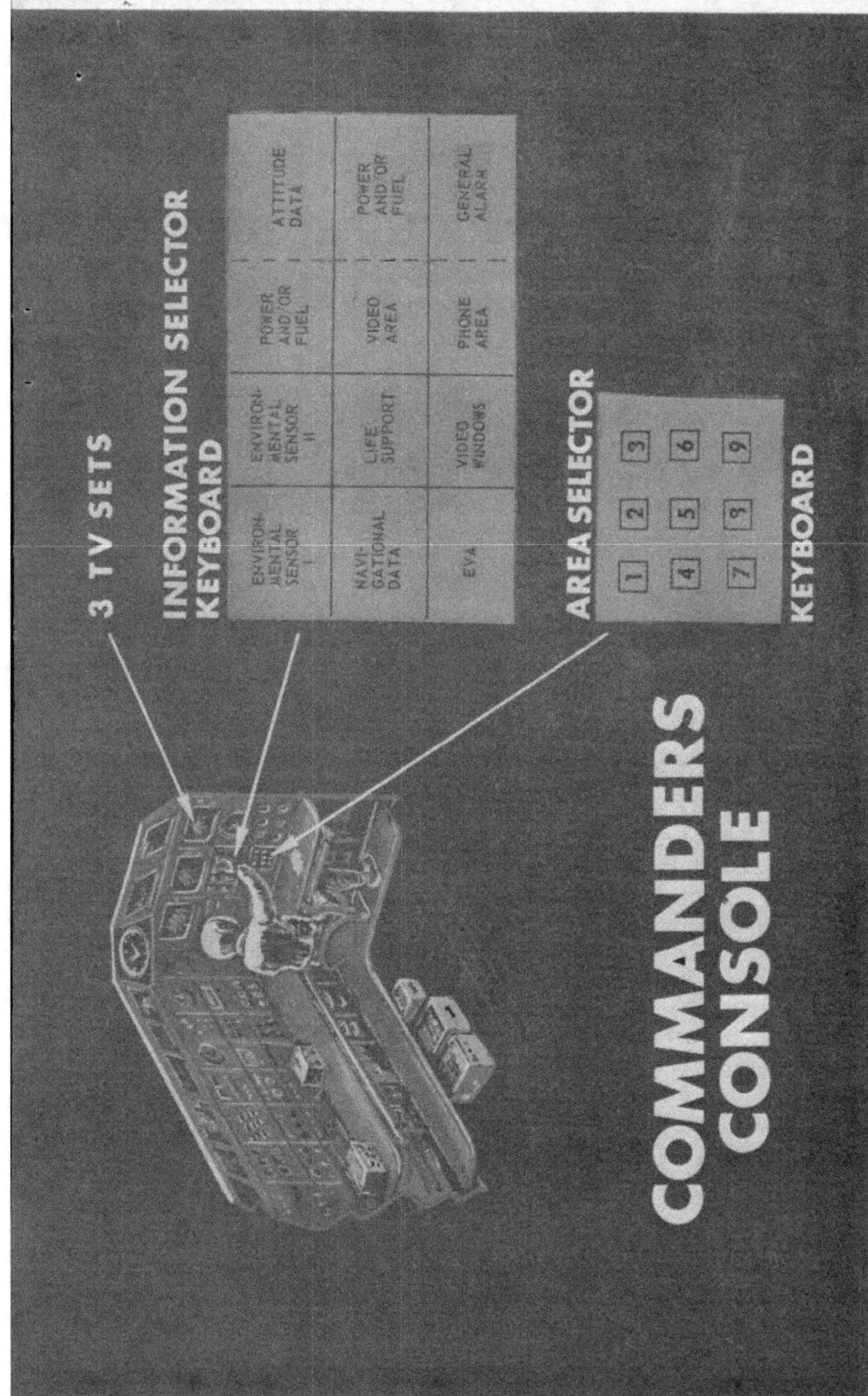

FIGURE III-8. THE COMMANDER'S DISPLAY AND CONTROL CONSOLE

technology which is already developed and equipment which is more or less off the shelf. The X-band radar subsystem is used for docking and tracking of the unmanned ancillary modules.

The low power C-band subsystem is used for voice and video communications, telemetry, command, tracking, and ranging with the ancillary modules. This is a low power system and thus will not affect ground links utilizing the same frequency spectrum.

The distribution subsystem is used to tie the terminals and display consoles into the information exchange system.

STARLAB Information Distribution. To provide for the complex information distribution subsystem aboard STARLAB without using miles of wire, the frequency multiplexing technique is employed to allow a few transmission lines to carry many information channels. This means that information is distributed throughout STARLAB by assigning different bands of frequency to each of various types of information carrying signals. New developments on the TSP type of transmission line make it possible to use the entire bandwidth of 275 MHz for information transmissions. This bandwidth provides for 1000 low data rate channels (defined in Table III-2), 100 medium data rate channels, 30 high data rate channels, 16 TV channels, and 2 high resolution TV channels on each TSP (Fig. III-9). By utilizing three TSP's, only two need to be used full time with the remaining TSP used as a spare. The frequency allocation on the TSP is made for both transmitting and receiving channels. A typical allocation utilizing six colors of a laser subsystem for transmitting and six colors for receiving is shown in Figure III-9.

The first 20 MHz of bandwidth may be used for the S-band critical system (Fig. III-9). The critical functions are defined as those necessary for communications during emergency or checkout periods. A typical frequency allocation for the critical function bandwidth would include a voice and video link to the CIMC, a TV monitor to the CIMC with very rapid sequencing through all computer formated displays, analog and digital telemetry channels for basic life support functions, tracking and ranging information, and emergency control functions (Fig. III-10).

STARLAB Stations. All information is processed before being modulated on a subcarrier for transmission on the TSP at the STARLAB stations. The premodulation processor (PMP) utilized depends on the rate that information is generated by a particular sensor and must be determined by the user. (Example 1: An experiment has several sensors which are to be monitored

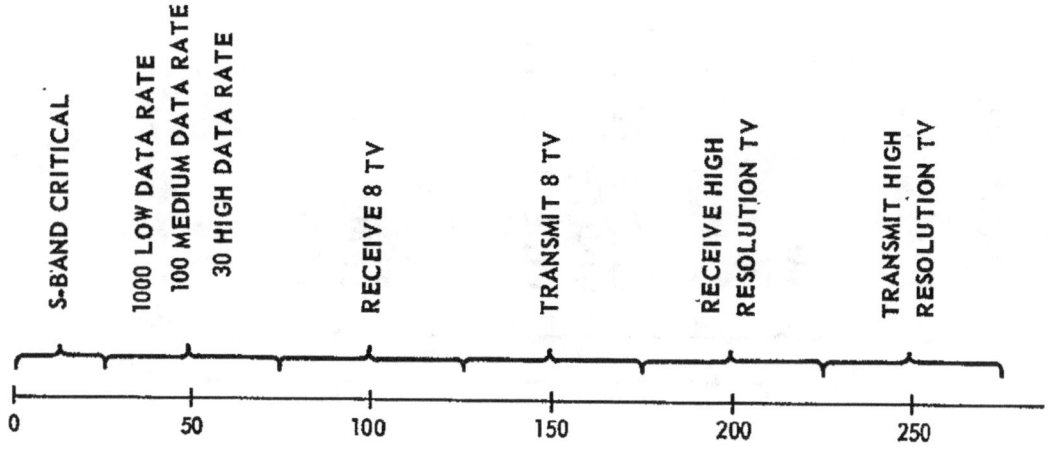

FREQUENCY (MHz)
TWISTED SHIELDED PAIR NOS. 1 AND 2

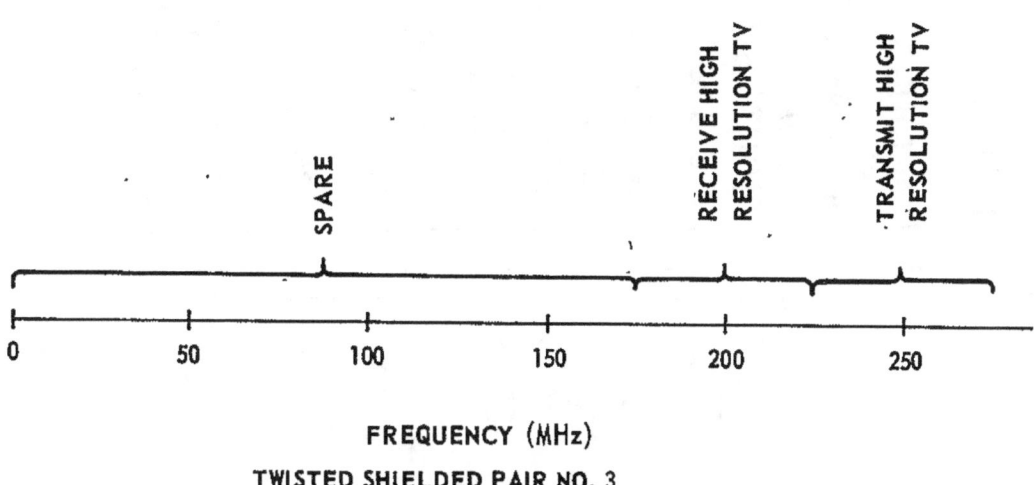

FREQUENCY (MHz)
TWISTED SHIELDED PAIR NO. 3

FIGURE III-9. FREQUENCY ALLOCATION ON TSP

3-37

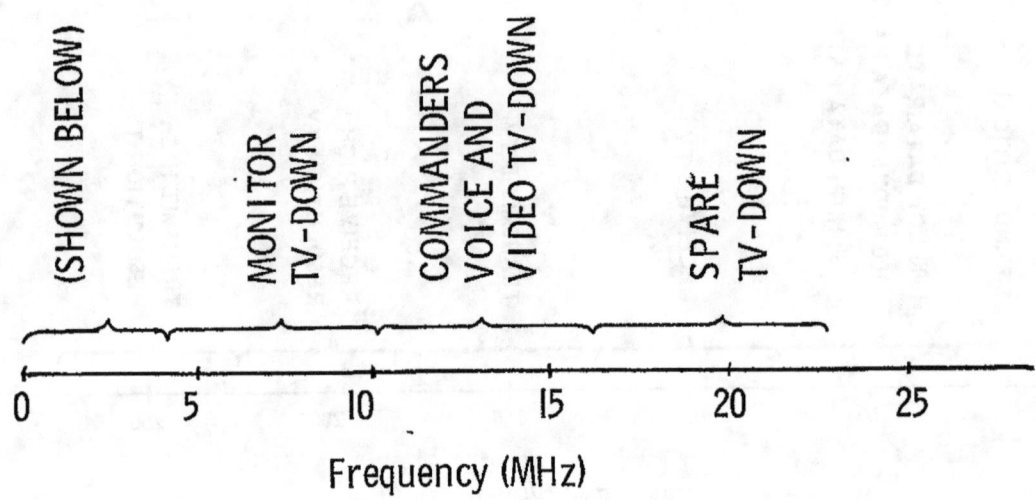

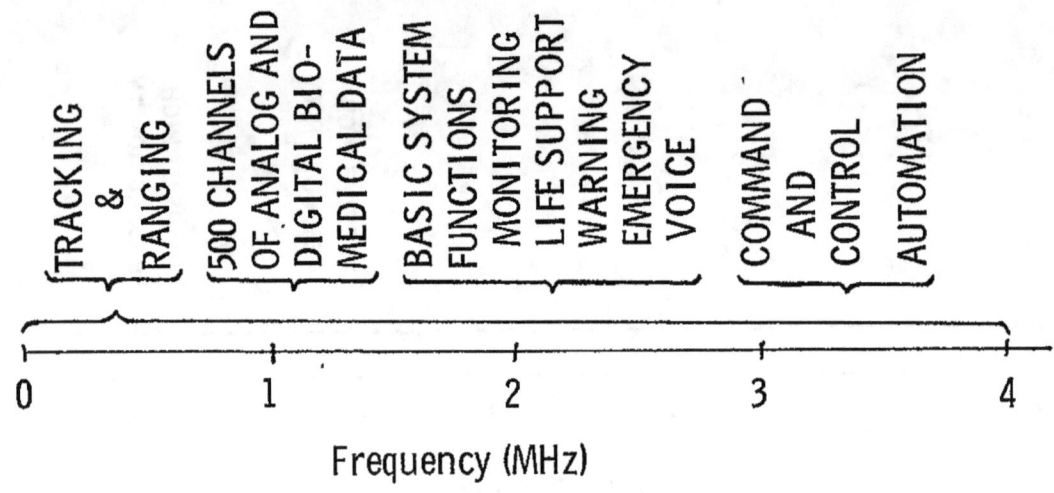

FIGURE III-10. FREQUENCY ALLOCATIONS ON CRITICAL S-BAND CHANNELS

only a few times per second. Therefore, a PMP is selected which time multiplexes their outputs for modulation of a low data rate channel subcarrier. Example 2: Voice has most of its information located in the frequency band from 300 Hz to 3000 Hz and therefore uses only an audio amplifier as a PMP before it modulates a low data rate channel subcarrier.)

Each station is assigned a low data rate channel with a given subcarrier frequency for initiating call signals. A push-button channel selector [Fig. III-5(b)] sends a call signal to the computer information exchange over the call channel. This information exchange is similar in concept to the telephone exchange which is in common usage today. The allocation of frequency bands on the TSP by the information exchange is similar to the connecting function performed by the telephone exchange. Further instructions can be given to the computer using the call channel and vocoders. (Vocoders are devices which change the voice signals into electrical signals that are translated directly into computer instructions.) Since the call channel is critical, it will be assigned on both TSPs to have redundancy.

The Laser Subsystem. Several types of lasers were considered for the main optical communications link. These were a gas laser using CO_2, a helium-neon laser, a double-YAG laser, and an argon-krypton laser.

The carbon-dioxide gas laser has already been considered for space use. The purpose of the coherent laser satellite-to-satellite experiment proposed for the ATS-F and ATS-G program is to establish the feasibility and value of optical space communications using present state-of-the-art laser equipment and technology (Ref. III-3).

The proposed system would utilize 23-pound transceivers consuming 20 watts of primary power. A 30 MHz information base-band would be frequency modulated on a 10.6-micron carbon-dioxide laser beam. Five-inch optical apertures and 400 milliwatts of carrier power would provide for a 28-decibel signal-to-noise ratio for a range of 18 600 miles. The helium-neon laser would be suitable for a narrower bandwidth requirement (Refs. III-4 and III-5).

The argon-krypton laser was selected since it lases up to 11 colors. It stabilizes to provide for modulation rates up to 500 megabits per second per color. The proposed system would utilize 500-pound transceivers consuming 500 watts of primary power. An incoherent pulse/AM type modulation technique operating at rates of up to 500 megabits per second is to be used.

Optical apertures of 50 centimeters (20 inches) and carrier power of 100 milliwatts provide for error rates of less that 10^{-7} errors per bit for a range of 24 000 miles [Fig. III-11(a)] (Refs. III-6 and III-7).

A typical laser communications subsystem is shown in Figure III-11(b). Very coarse aperture alignment is attained using the unified S-band equipment. This places the apertures to a 1-degree field-of-view. The laser telescope is physically oriented using the coarse guidance system to within a 2-arc-minute field-of-view. This allows the fine guidance control package to complete the alignment to within an arc-second.

The double-YAG may quickly overtake the argon-krypton laser as the main communications link. If progress is as rapid as some researchers predict, the double-YAG promises much higher efficiency and mode frequency stability. The double-YAG laser is yttrium-aluminum-garnet crystal that emits ultraviolet light that is frequency doubled using a barium-sodium-niobate crystal. The light frequency is doubled to produce light which can be detected and amplified using photomultipliers operating at 5300 Å. Researchers predict 100-milliwatt output power at 10-percent efficiencies within 1 year (Ref. III-8). Such lasers would greatly reduce the size and primary power requirements of a laser communication system. (Information on the backup and emergency subsystems and the circuit studies is presented in Appendix B of this report.)

LASER TYPE	-	ARGON-KRYPTON
RANGE	-	24 000 MILES
ALL APERTURES	-	50 CM
XMIT POWER	-	100 MW (0.2 % EFF)
MODULATION	-	PULSE/AM (INCOHERENT)
MODULATION RATE	-	500 MEGABITS/SEC/COLOR
SIGNAL	-	280 PHOTONS/BIT
ERROR RATE	-	$< 10^{-7}$ ERRORS/BIT

FIGURE III-11(a). COMMUNICATIONS LINK PARAMETERS

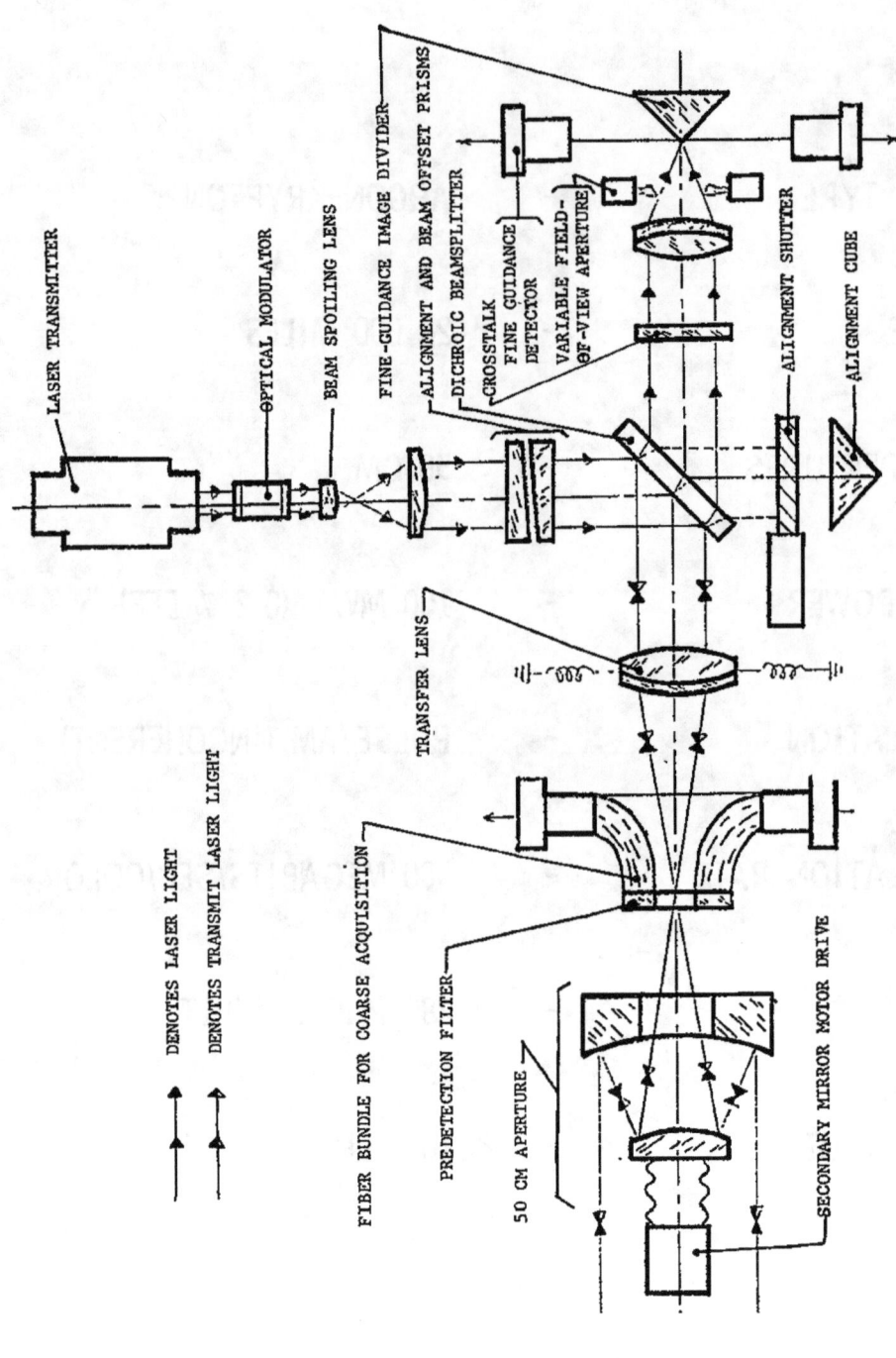

FIGURE III-11(b). LASER COMMUNICATION SYSTEM

3-42

REFERENCES

III-1. Levy, Sanl: A Survey of the Last Decade of Computer Development. 5-9 Computer Technology, RCA Defense Electronic Products, Moorestown, N. J., 1967.

III-2. Narrow, Bernard: Satellite Telemetry Data Processing at Goddard Space Flight Center. X-564-55-336, 1966, p. 37.

III-3. McAvoy, N., Richard, H. L., Plotkin, H. H., Carrion, W. J., McElroy, J. H., and Richards, W. E.: 10.6 Mission Laser Communications System Experiment for ATS-F and ATS-G. Goddard Space Flight Center. Goddard Space Flight Center, Greenbelt, Maryland, April 1968.

III-4. Integrated Laser Communications System. A Set of Viewgraphs From a Presentation Made by ITT Aerospace to NASA, ITT Aerospace, 15151 Bledsoc Street, San Fernando, California.

III-5. Laser Communications Satellite Experiment Report No. 8399. Perkin-Elmer Electro-Optical Division, Norwalk, Connecticut.

III-6. Jasik, Henry: Antenna Engineering Handbook. McGraw-Hill Book Co., Section 33 by Kenneth Bullington, 1961.

III-7. Craiglow, R. L.: Power Requirements for Speech Communications Systems. IRE Transactions on Audio, November-December 1961, pp. 186-190.

III-8. Private Communications with J. Randall, Astrionics Laboratory, George C. Marshall Space Flight Center, Marshall Space Flight Center, Alabama.

BIBLIOGRAPHY

Candidate Experiment Program for Manned Space Stations NASA. Payloads Office, Advanced Planning Missions Program Office of Manned Space Flight, 1 May 1969.

Command and Service Module — Manned Space Flight Network Signal Performance and Interface Specification Block 2. NAS9-150, North American Aviation, Inc., 22 February 1965.

Craiglow, R. L.: A Study of the Effects of Elementary Processing Techniques on the Intelligibility of Speech in Noise. Collins Radio Company, Report No. CRR-201, Cedar Rapids, Iowa, May 1960.

Design Study Specifications for the Earth Resources Technology Satellite ERTS-A and -B. NASA Goddard Space Flight Center, Greenbelt, Maryland, S-781-P-3, April 1969.

Error Rates for Poisson Processes. ADCOM Inc., Cambridge, Mass., Contract NAS8-20001, 30 October 1966.

Henney, Keith: Radio Engineering Handbook, McGraw-Hill Book Co., N.Y., Toronto, London, 1959.

Kendrick, J. B. (editor): TRW Space Data 3rd Edition. TRW Systems Group, TRW Inc., 1967.

Lawton, John G.: Comparison of Binary Data Transmission Systems. IRE 2nd National Conference Proceedings on Military Electronics, 16-18 June 1958, pp. 54-61.

Micron Laser Communications System Experiment for ATS-F and ATS-G. Goddard Space Flight Center, Greenbelt, Maryland, April 1968.

MOL Communication Subsystems. Preliminary Copy of Proposal to DOD by Collins Radio Co., 1965.

Moore, James R.: Apollo Entry Radar Acquisition Study. X-513-65-225, Goddard Space Flight Center, Greenbelt, Maryland, 28 May 1965.

BIBLIOGRAPHY (Concluded)

Narrow, Bernard: Satellite Telemetry Data Processing at Goddard Space Flight Center. Goddard Space Flight Center, Greenbelt, Maryland, July 1966.

Orbital Astronomy Support Facility (OASF) Story Task C — Orbital Astronomy Support Facility Concepts. Vol. IV, Douglas Missile and Space Systems Division, MacDonnell Douglas Corporation, DAC-58144, 28 June 1968.

Quann, J. J., Walton, B. A., and Keipert, F. A.: Aerospace Computer Systems Conference. Paper No. 69-972, Los Angeles, California, 8-10 September 1969.

Recent Advances in Display Media. NASA SP-159 (A symposium Held in Cambridge, Massachusetts, September 19-20, 1967), 1968.

Relay Program Final Report. Goddard Space Flight Center, Greenbelt, Maryland, NASA SP-151, 1968.

Satellite and Rocket Experiments, Data Catalog NSSDC 69-01. National Space Science Data Center, NASA, Goddard Space Flight Center, Greenbelt, Maryland, January 1967.

Snow, Robert M.: Final Report on Optical Communication Studies. ADCOM Inc., Contract NAS8-20001, 15 February 1967.

Sohn, H. M.: Maintainability and Impact on System Design. Lockheed Georgia Company, 16 June 1969.

Television Broadcast Satellite Study. Interim Presentation, Contract NAS8-21036, Convair Division of General Dynamics, San Diego, California, 25 June 1969.

The Operation of the National Space Science Data Center NSSDC 67-41, Goddard Space Flight Center, Greenbelt, Maryland, October 1967.

Waltz, Donald M.: Earth Resources Surveys Conducted by Remote Sensing From a Manned Orbiting Research and Applications Laboratory. Space Vehicles Division, TRW Systems, Redondo Beach, California, 8 July 1969.

CHAPTER IV

MISSION AND ANALYSIS

CHAPTER IV. MISSION AND ANALYSIS

DEFINITION OF SYMBOLS

i	Inclination of orbit plane to equitorial plane
L	Launch site latitude
az	Launch azimuth
τ	Orbital period
m	Number of earth rotations relative to target orbit plane
Ω_e	Earth rotational rate
Δt_{asc}	Time elapsed from launch to injection
n	Number of revolutions relative to launch site
n. mi.	Nautical miles
R_e	Equatorial radius of earth
J_2	Coefficient of potential function
r	Radial distance from center of earth
$\dot{\Omega}$	Change in right ascension of the ascending mode per revolution due to oblateness
μ	Gravitational parameter
h	Altitude

MISSION DESCRIPTION

The operational mission of STARLAB is to provide facilities in earth orbit capable of supporting a long-term scientific and applications research program. The research program is multifaceted, being composed of experimentation in various scientific and technological disciplines. Such a diverse program imposes environmental requirements which cannot be optimally satisfied by one vehicle. The solution is a system of manned and unmanned vehicles which are maintained, controlled, and supported by a manned orbiting laboratory base.

STARLAB is composed of the main laboratory base and five free-flying laboratory modules. The nominal orbit for this system is a circular orbit with an altitude of 308 n. mi. and an inclination of 69 degrees. The selection of this orbit for STARLAB was based on considerations of scientific mission, launch vehicle performance, range safety, earth coverage, radiation environment, orbital lifetime, and rendezvous for resupply. The ground track makes a longitudinal shift of 24.23 degrees per laboratory revolution. The orbits of consecutive days are out of phase with a longitudinal shift of 3.46 degrees. The daily shift coupled with the high inclination provide coverage of all populated land masses, but still allows rendezvous compatibility every seven earth revolutions relative to the laboratory.

The free-flying laboratory modules operate remote from the station, but in its close proximity for the purpose of rendezvous when support is required. The differences in the velocities, altitudes, and ballistic coefficients between the modules and the main station causes relative motion between the vehicles in the nominal orbit. For example, consider a free-flying module which has been serviced by the main laboratory and has been maneuvered into a circular orbit slightly above the main laboratory. The main laboratory maintains its position in the nominal orbit by corrective thrust maneuvers, but the orbit of the module is allowed to decay. The resulting relative motion is shown in Figure IV-1. The module is originally above the laboratory with circular velocity at point 1. The laboratory is also in a circular orbit but at a lower altitude, and thus the module has a lower velocity than the laboratory and slowly falls behind. Drag forces are also acting on the module causing its orbit to decay. The module loses altitude and reaches point 2. As the module continues to lose altitude its velocity will increase, and the module will move to point 3 relative to the laboratory. If it is desired that the module continue its free-flight mode, a velocity impulse must be given to the module by its propulsive system to place it at point 4, and the cycle is repeated. If docking

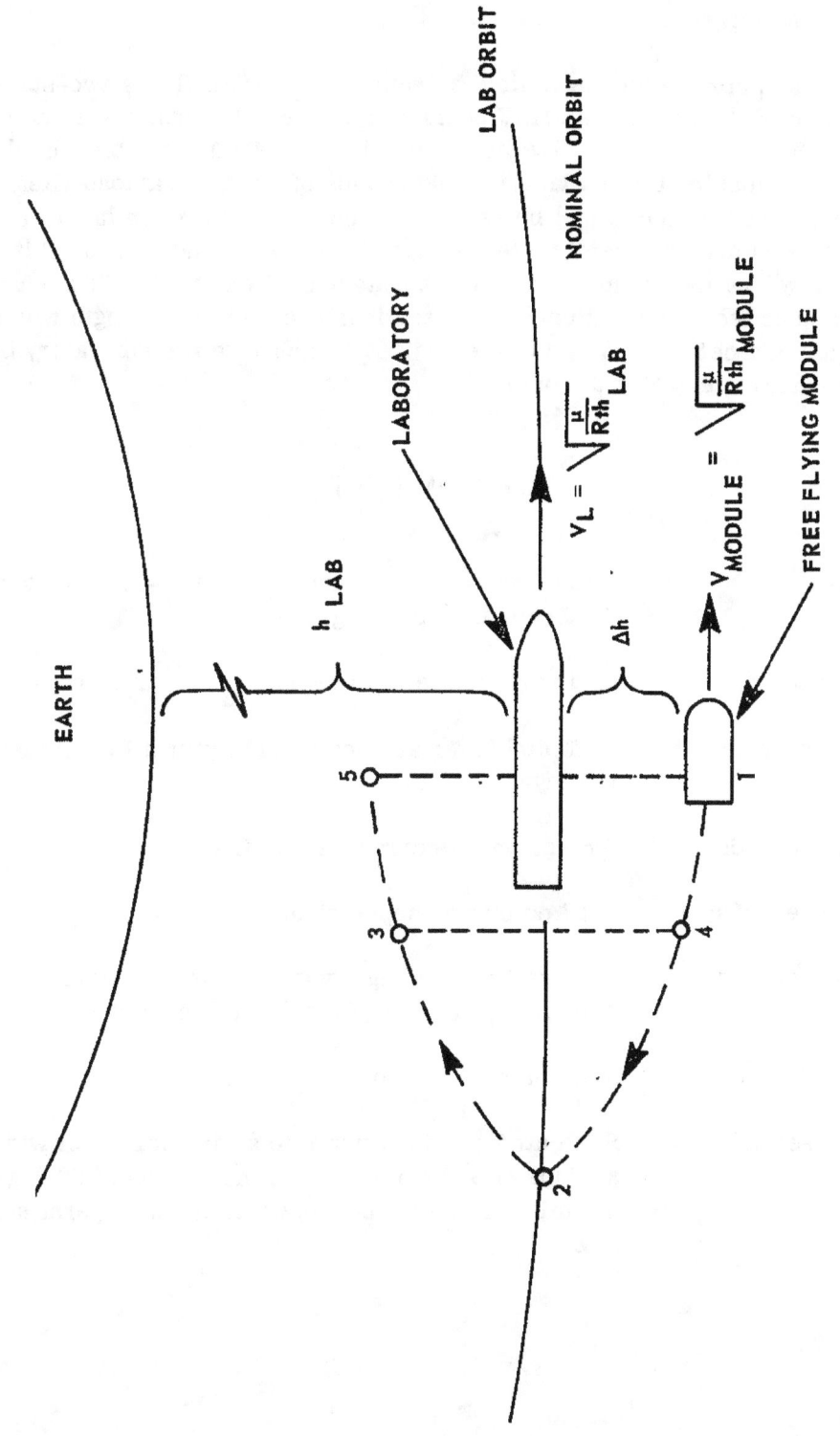

FIGURE IV-1. RELATIVE TRAJECTORY OF FREE-FLYING MODULE

4-3

by the module is required, the module is allowed to move to point 5, and the docking maneuver is accomplished.

The primary launch vehicle selected for STARLAB is a two-stage derivative of the Apollo-Saturn V consisting of the S-IC first stage and the S-II second stage (Fig. IV-2). By direct burn into STARLAB's orbit, this launch vehicle is capable of delivering 145 000 pounds of useable payload (Ref. IV-1). Five unmanned launches will be required to complete the main laboratory. The payloads are completely integrated and are merely docked in orbit. The nominal weight and volume for each laboratory section are 120 000 pounds and 50 000 cubic feet, respectively. The sections are 80 feet in length with a maximum diameter of 33 feet. The desired launch rate for laboratory buildup is one launch every 2 to 3 months.

MISSION PROFILE

T + 0 second:	Launch from Cape Kennedy with two-stage Saturn V. Launch azimuth 45 degrees.
T + 12 seconds:	End vertical rise and begin pitch-over maneuver.
T + 35 seconds:	End pitch-over and begin gravity turn for minimum aerodynamic load.
T + 150 seconds:	Freeze yaw steering for staging.
T + 160 seconds:	S-IC cutoff and separation.
T + 164 seconds:	S-II ignition and begin yaw steering to obtain desired azimuth at injection (dog-leg maneuver).
T + 190 seconds:	Jettison payload shroud.
T + 540 seconds:	S-II cutoff and injection into a circular orbit with an altitude of 308 n. mi. and an inclination of 69 degrees. Terminal rendezvous and/or activation operations begin.

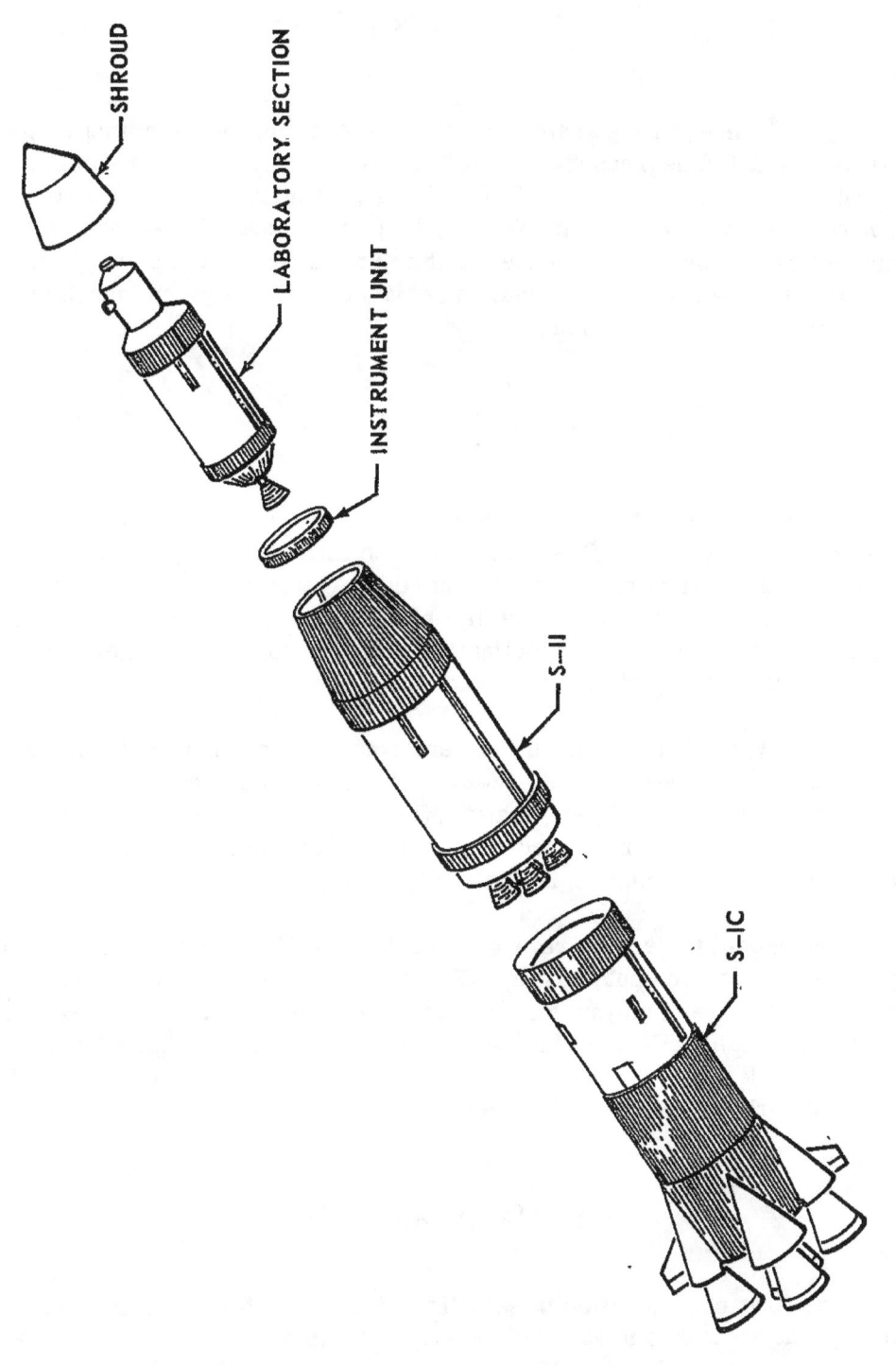

FIGURE IV-2. EXPLODED VIEW OF TWO-STAGE SATURN V AND PAYLOAD

4-5

THE ASCENT TRAJECTORY

Cape Kennedy range safety constraints restrict launch azimuths to between 45 and 112 degrees except for the possibility of a launch azimuth between 140 and 145 degrees (Ref. IV-2). This limits the orbital inclinations that may be obtained from Cape Kennedy by planar ascent trajectories. If it is assumed that the distance and time spent during ascent to the point of burnout are small, the approximate inclination achievable from a given launch site may be found from the relation

$$i = \cos^{-1}\left[\cos(L)\sin(az)\right]$$

where L and az are the launch site latitude and launch azimuth, respectively. For a launch site latitude of 28.5 degrees, the latitude of Cape Kennedy, the launch azimuth restrictions limit the possible orbital inclinations to between 28.5 and 52 degrees with the possibility of an orbital inclination between 55 and 59 degrees. Orbits of greater inclination than this cannot be achieved without the use of yaw steering at the expense of the payload.

A study has been made on the performance capability of the Saturn V incorporating yaw steering (Ref. IV-2). This study indicates that the time at which yaw steering is initiated greatly influences the payload capability. On the basis of these performance curves, a launch azimuth of 45 degrees was selected with yaw steering beginning at S-II ignition.

The ascent trajectory required to achieve STARLAB's orbit with a two-stage Saturn V was computed using ROBOT (Ref. IV-1). A launch azimuth of 45 degrees was given and yaw steering was commanded at S-II ignition. The groung track (Table IV-1) of the ascent profile is shown in Figure IV-3. With this ascent trajectory, 145 500 pounds of useable payload may be injected into STARLAB's orbit (308 n. mi., 69 degrees).

ORBITAL TRAJECTORY

The first consideration in selection of the orbit for STARLAB was support of the scientific mission. The study of earth resources dominates the scientific mission for STARLAB, and therefore its orbital requirements in turn dominate the orbital selection.

TABLE IV-1. GROUND TRACK OF ASCENT PROFILE

t-secs	Latitude (deg)	Longitude (deg)
Launch		
0	28.53	80.56 W
75	28.54	80.55
150	28.85	80.21
189.9	29.29	79.73
243	30.01	79.05
318	31.37	77.98
418	34.07	76.25
468	35.93	75.15
518	38.21	73.82
Injection		
540	39.35	73.15
S-IC Impact	33.5 N	74.13 W

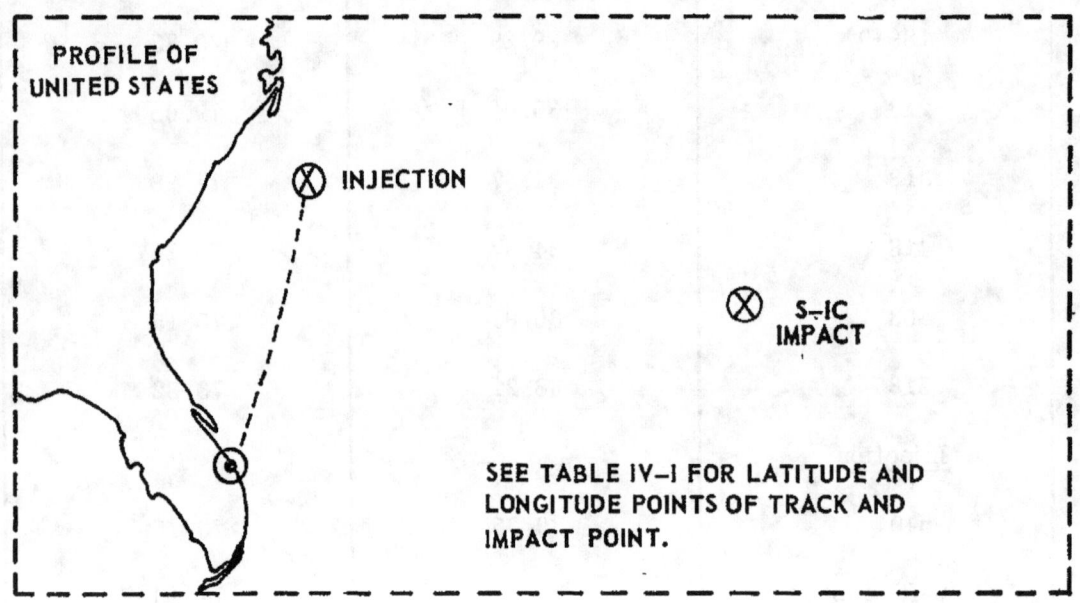

FIGURE IV-3. THE ASCENT TRAJECTORY GROUND TRACK

For the study of earth resources, the altitude of the orbit should be high enough to permit good area coverage and yet low enough to meet sensor resolution requirements. A lower bound on the altitude may be determined from station keeping requirements. At altitudes less than 100 n. mi., orbital lifetimes become so short that the propellant requirement for station keeping becomes prohibitive. An upper bound on the altitude may be determined from the radiation environment. Above 500 n. mi., the trapped particle radiation in the lower Van Allen belt becomes intense enough to make adequate shielding for the crew impractical. A trade-off is required between sensor resolution requirements and area coverage is constrained to the altitude region between 100 and 500 n. mi. with propellant consumption increasing altitude.

For good earth coverage, the orbital inclination should be as high as possible. A near polar orbit can provide uniformity of lighting conditions and almost total earth coverage simultaneously. However, with increasing inclination, the radiation dosage increases because of solar and galactic radiation. At inclinations less than 67 degrees, the radiation due to solar cosmic radiation may be ignored in calculation of radiation dosage (Ref. IV-3).

Another factor which affects the choice of orbital inclination is the phenomenon called the "South Atlantic Anomaly." This is a region where the trapped protons in the inner Van Allen belt dip closer to the earth than at any other longitude. For orbital altitudes of 200 to 300 n. mi., the radiation dosage from Van Allen radiation is a maximum at inclinations of 30 to 35 degrees (Ref. IV-4) and is primarily because radiation accumulated during passage through the anomaly. At inclinations below 25 degrees, the orbit no longer passes through the anomaly, and the dose rate drops sharply. For inclinations above 35 degrees, the dose rate decreases monotonically to half the maximum value at an inclination of 90 degrees.

Introduction of another consideration narrows the range of possible orbits; i.e., frequent rendezvous capability. For STARLAB, rendezvous at specified time intervals is mandatory. For maximum payload capability to rendezvous, the orbit should be selected such that the correct relative position between the launch site and the laboratory for rendezvous occurs at regular intervals consistent with resupply and/or crew rotation requirements. Orbits which satisfy this condition are called rendezvous compatible and are defined by Reference IV-5:

$$\tau = \frac{2m\pi - \Omega_e \Delta t_{asc} + n\dot{\Omega}}{n\Omega_e \left(1 + \frac{\Delta \tau}{\tau}\right)}$$

where $\Delta\tau$ is the orbital period correction due to oblateness

$$\frac{\Delta\tau}{\tau} \approx \frac{3}{2} J_2 \left(\frac{R_e}{r}\right)^2 \frac{1 - 7\cos^2 i}{4}$$

An iterative solution is required because of the interdependence of $\dot{\Omega}$ and τ. Solutions to this equation are given in Reference IV-2, and the orbit for STARLAB was selected from these candidate orbits with trade-offs based on considerations of earth resources observation, launch vehicle performance, range safety, radiation, and orbital lifetime. The orbit selected for STARLAB is circular with an altitude of 308 n.mi. and an inclination of 69 degrees. The period for this orbit is given by

$$\tau = 2\pi \sqrt{\frac{r^3}{\mu}} \left[1 - \frac{3}{2} J_2 \left(\frac{R_e}{r}\right)^2 \frac{7\cos^2 i - 1}{4}\right]$$

and is 96.65 minutes. The ground track of this orbit for the first three revolutions was computed (Ref. IV-6) and is shown in Figure IV-4. The longitudinal shift in track per orbit,

$$\dot{\Omega} = \tau \Omega_e$$

is 24.23 degrees. The ground track for the first day in orbit is shown in Figure IV-5. Note that the orbits of the second day will be out of phase with the orbits of the first day by the amount

$$\dot{\Omega}_{daily} = 15 \dot{\Omega} - 360 \text{ deg}$$

or 3.46 degrees. The phase shift increases the area coverage by allowing new and previously covered areas to be viewed on the same day. The phase shift and the period are such that the ground track repeats every 104 revolutions of STARLAB. This allows rendezvous with STARLAB every 167.53 hours or approximately every 7 days.

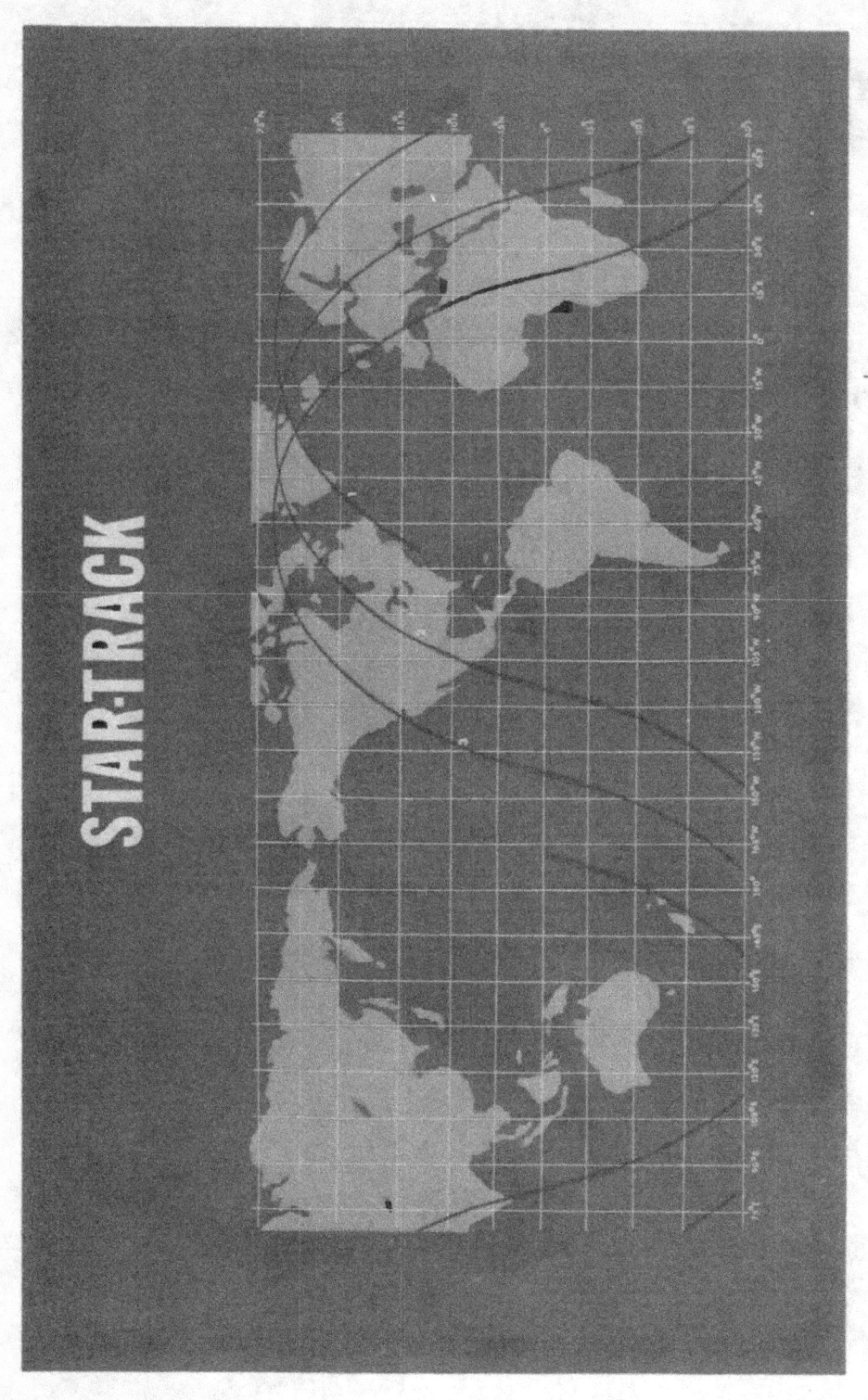

FIGURE IV-4. STAR-TRACK, FIRST THREE ORBITS

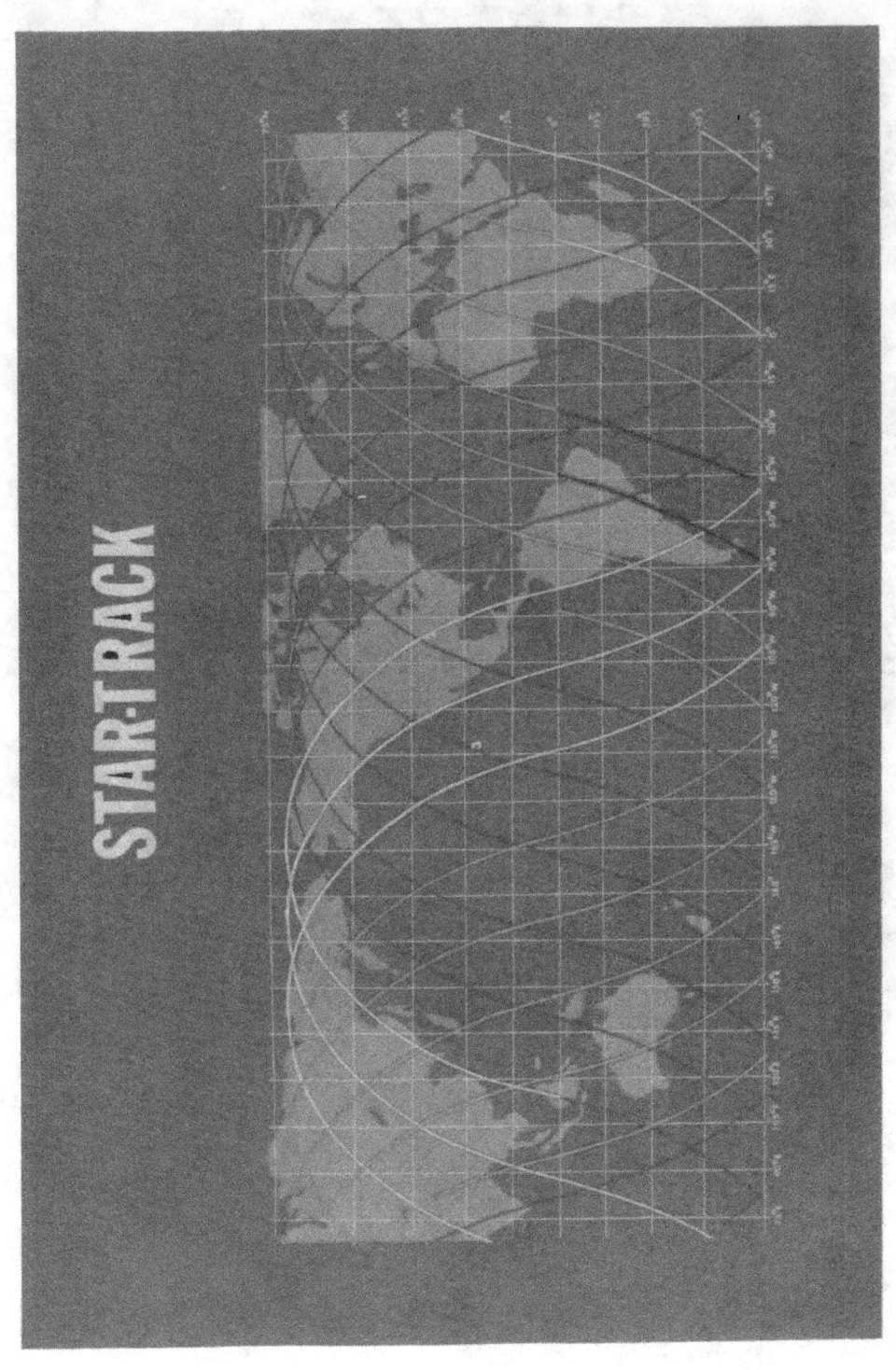

FIGURE IV-5. STAR-TRACK, FIRST DAY'S ORBITS.

REFERENCES

IV-1. ROBOT — Trajectory Profile Optimization Program. Computer Time and Run Provided Through Gordon Solomon, R-AERO-FMS, MSFC.

IV-2. Early Saturn V Workshop. MSFC Report, No Number, 15 May 1968.

IV-3. Manning, H. S.: Annual Radiation Dose for 1975-1980 Space Base Study. MSFC Memorandum (S&E-ASTN-AV-69-57), 13 March 1969.

IV-4. Orbital Astronomy Support Facility Study. Vol. IV, Douglas Missile and Space Systems Division Report DAC-58144.

IV-5. Orbital Flight Handbook Part 2. NASA SP-33, 1963.

IV-6. MSFC Ground Track Computation Program. Computer Time and Program Provided Through Percy Dreher, R-AERO-FT, MSFC.

BIBLIOGRAPHY

Advanced Astronomy Mission Concepts. Martin-Marietta Report ED-2002-795, Vol. II, April 1969.

Battin, R. H.: Astronautical Guidance. McGraw-Hill, 1964.

Compendium of Human Responses. NASA CR-1205 (III).

Ehricker, K.: Space Flight. D. Van Nostrand, 1962.

Jensen, J., Townsend, G., Kork, J., and Kraft, D.: Design Guide to Orbital Flight. McGraw-Hill, 1962.

Purser, P., Faget, M., and Smith, N.: Manned Spacecraft Engineering. Fairchild, 1964.

Space Flight Handbooks. Vol. 1, Orbital Flight Handbook, NASA SP33, 1963.

CHAPTER V

STARLAB VEHICLE AND SUBSYSTEMS

CHAPTER V. STARLAB VEHICLE AND SUBSYSTEMS

DEFINITION OF SYMBOLS

ac	Alternating current
W_e	Watts, electrical
w	Watts
W_t	Watts, thermal
AU	Astronomical unit
cm	Centimeter
SNAP	Systems for nuclear auxiliary power
g	Gravity
kg	Kilogram
lbm	Pound mass
kw	Kilowatts
μ	Micron
m	Meter
M	Million
vdc	Volts, direct current
B	Billion
Btu	British thermal unit

DEFINITION OF SYMBOLS (Continued)

Δp	Pressure differential
ΔV	Velocity differential
psia	Pound per square inch absolute
°F	Degree Fahrenheit
mm	Millimeter
Hg	Mercury
RH	Relative humidity
ID	Inner diameter
OD	Outside diameter
ILM	Integral launch module
S-V	Saturn V
FFM	Free-flying module
EVA	Extravehicular activity
CMG	Control movement gyro
RCS	Reaction control system
RFM	Remote-flying module
LV	Local vertical
CG	Center of gravity
DOF	Degree of freedom

DEFINITION OF SYMBOLS (Concluded)

LB	Angle beta: Angle between Sun, Earth, and spacecraft
GG	Gravity gradient
VV	Velocity vector
POP	Perpendicular to orbit plane
PEP	Perpendicular to ecliptic plan
EP	Ecliptic plane
SPS	Service propulsion system
W-h	Watt-hour
AU	Astronomical unit

PHYSICAL ARRANGEMENT

Introduction

The objective of the laboratory vehicle is to provide project STARLAB with an integral and reliable orbiting base of operations as soon as possible after initiation of the launch sequence.

The scope of project STARLAB requires a laboratory that cannot be launched with a single existing vehicle. Five or more launches will be required. While the advantages of modular construction are recognized and utilized, it is still more efficient to provide certain functions for the entire operation from single specialized equipment complexes; power, communications attitude control, administration, water reclamation, logistics docking, etc. It is important that these services which are sized for the integral laboratory be interfaced with their users as soon as is practicable so that their full potential may be realized and their useful life exploited. These considerations make minimization of in-orbit construction time an important objective.

Launch Package Modules

Large ground integrated units minimize in-orbit construction and interface connections. This advantage is judged to be more important than transportation cost reductions achieved using smaller payload shuttle packages for the orbiting of the base of operations.

The center of STARLAB operations is contained in five ILMs, each having a volume of 50 000 cubic feet. These ILM's are launched weighing 120 000 pounds each. The ILM is the largest unit in both volume and weight that can be constructed and tested on the ground and still be launched into the desired orbit as an integrated payload by a two-stage S-V vehicle. Transportation costs using ILM units, while more expensive than shuttle transport, is still only a fraction of payload development and fabrication cost (Ref. V-1). Comparison of ILM volume to weight ratio with other orbiting laboratory proposals (Ref. V-2) indicates that the major limitation is weight, not volume. To use the volume potential to the fullest, it is necessary to add equipment transported by the shuttle vehicle. Transportation cost of such equipment may be as much as an order of magnitude less than S-V transport, though time-consuming in-orbit transfer and assembly would be necessary.

Each ILM unit is docked to similar units and FFM's at either end. Each can dock smaller FFM units at convenient locations on the sides. Attitude control, power, and life support functions are maintained by each module on a limited emergency basis.

Launch Package Module Integration

The five ILM units will be docked end-to-end to form the basic structure of STARLAB [Figs. V-1(a) and V-1(b)]. This structure is long (430 feet) and cylindrical (33 feet and 22 feet in diameter) and is chosen for several reasons. A large area-to-volume ratio provides more surface on which heat rejection equipment can be mounted. The long shape enables the various functional appendages to be spaced to minimize mutual interference with each other and with the logistics vehicles and FFM's which will be docking frequently. Attitude control requirements for solids of revolution have been studied and attitude control power consumption can be projected into the future. Symmetry of the vehicle should reduce the aerodynamic torque in orbital flight. Further reduction could be achieved by deploying the roll-up type solar arrays in varying amounts to move the drag center to the center of mass. This is probably unnecessary for the altitude of interest (300 n.mi.), as the aerodynamic torques are small compared to the gravity gradient torques. A long cylindrical configuration is easily assembled by four end-to-end dockings of the large ILM's. This reduces the construction time in orbit by minimizing the interconnections that must be made. Finally, if the scope of operations should increase in the future, new ILM units could be added as needed, either to the ends or in the middle by undocking and redocking.

Description of Modules

Each of the ILM units that compose the STARLAB structure will be modifications of a basic module design which are specialized for their particular tasks (Fig. V-2). Module descriptions that follow are in the order of their launch sequence.

The critical services module contains those functions which can be performed most efficiently from a single location and which also have a high

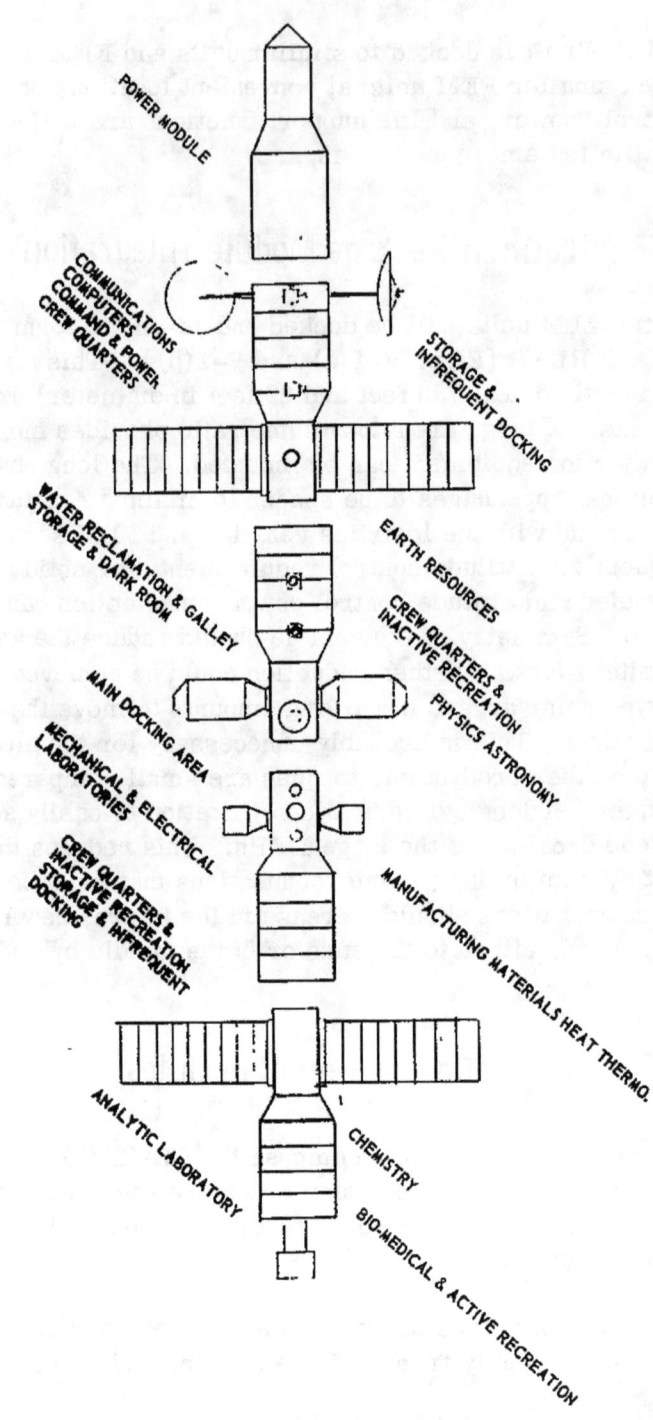

FIGURE V-1(a). PHYSICAL ARRANGEMENT OF STARLAB VEHICLE

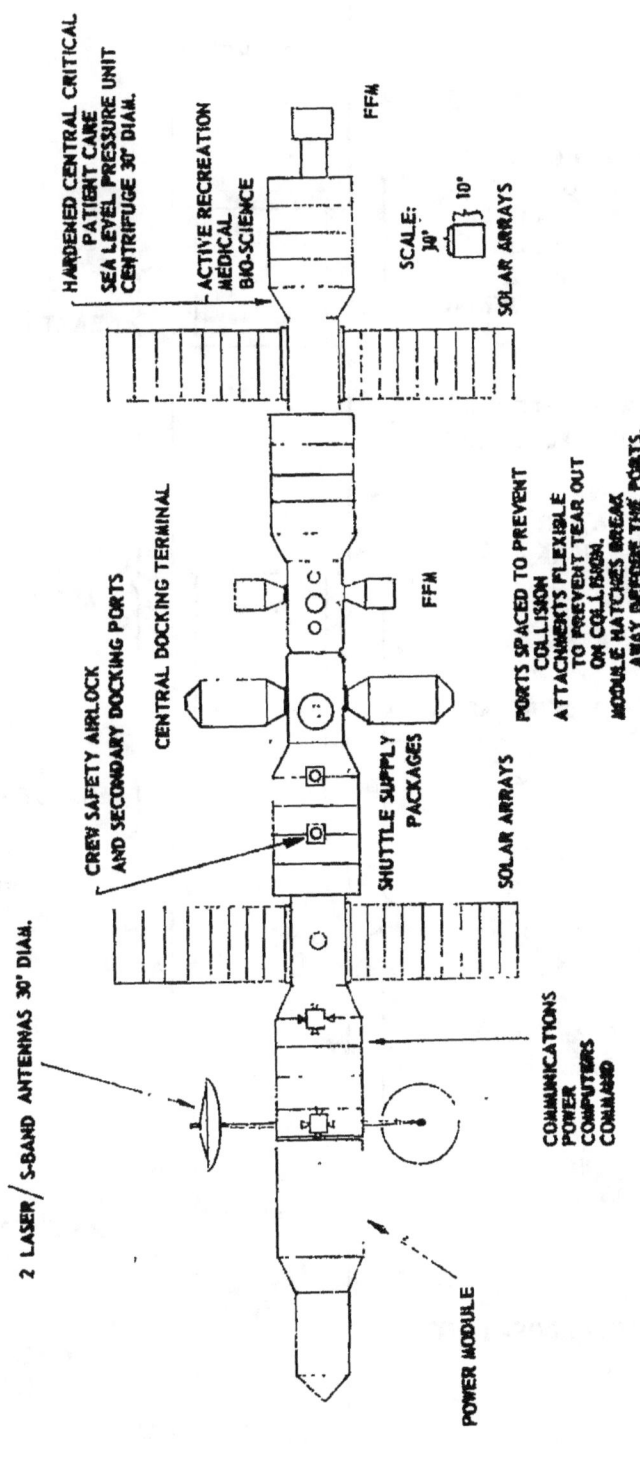

FIGURE V-1(b). PHYSICAL ARRANGEMENT OF STARLAB VEHICLE

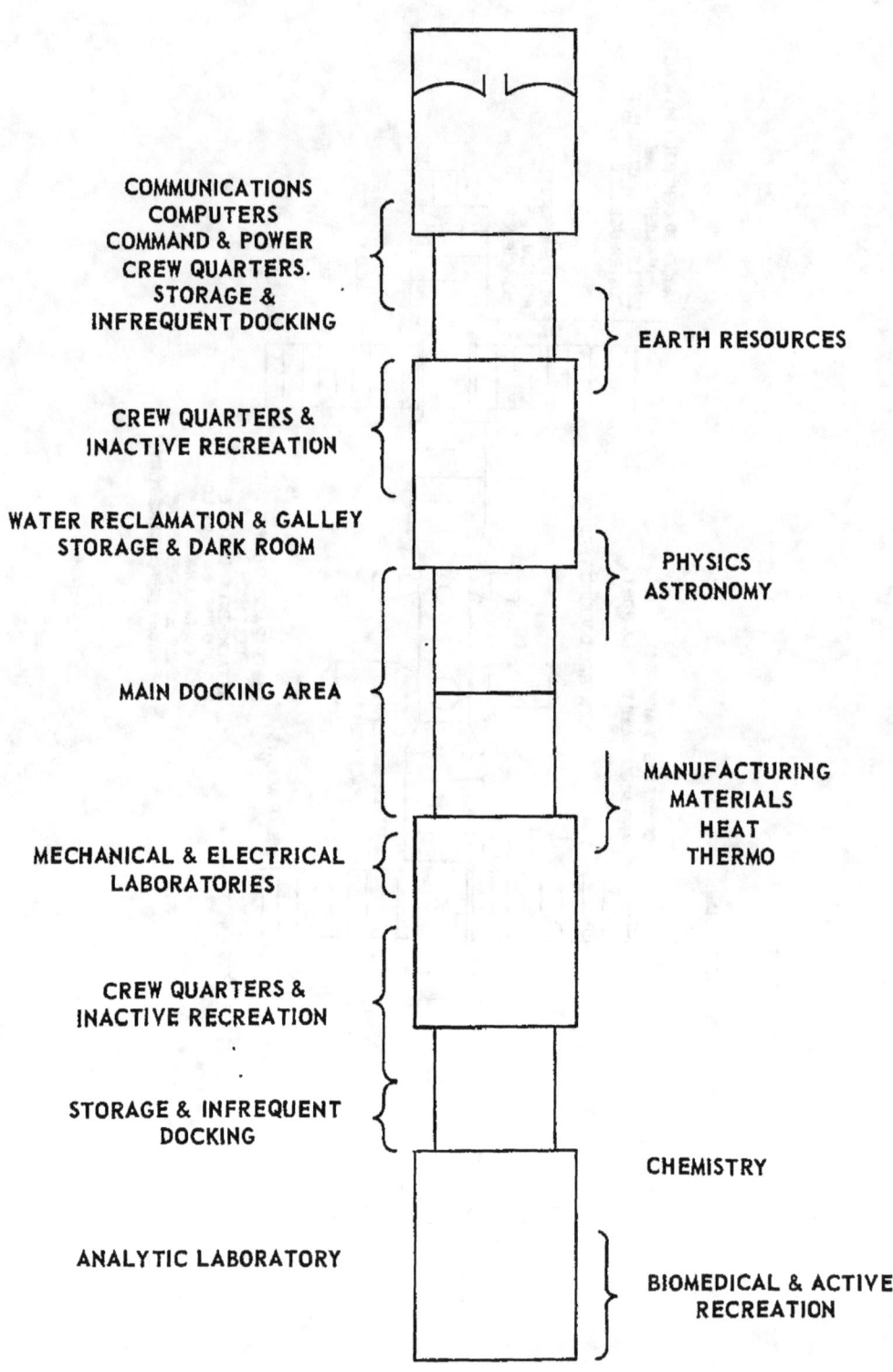

FIGURE V-2. MODULAR ARRANGEMENT OF THE STARLAB VEHICLE

density of mutual interface requirements such as communications, main computer complex, command area, power conversion; and the distribution system.

The power module is launched as soon as possible after the critical services module has been manned and is functioning. The power module contains portions of the power generation system that have to be replaced periodically. Other systems that must be replaced at about the same frequency can be included. CMG attitude control systems, if adopted, are placed in this module. It can also be used to store consumables unaffected by small radiation levels such as foods, fuels, propellants, etc.

The earth resources, physics, and astronomy module are launched at a convenient time after the complete integration of the critical services and power modules. In addition to the function indicated in the title this module contains one of the two main crew-quarters living areas, together with galley and water reclamation facilities. It also contains half of the central main docking area.

The manufacturing and materials module also contains the other half of the main docking facility, the Mechanical and Electrical Support Laboratories, and the second main crew-quarters living area.

The chemistry-medical-biological module also contains the Analytic Laboratory, the active recreation area, the artificial gravity centrifuge for experimentation, and the 14.7 psi experimental area.

FFM units are designed, if possible, to a size that is transportable by the shuttle. FFM units of the payload container size discussed below can share the same systems. Smaller units should serve most purposes (15 feet in diameter by 20 feet long or smaller). These can share systems with the space taxi (discussed below). Maintainability design should place as many repair and service functions as possible internal to a pressurized or pressurizable hull to avoid EVA. External maintenance and repair that would be difficult using EVA or taxi could be handled in two ways; (1) larger units could be returned to earth in the shuttle, and (2) smaller units could be returned or serviced in an in-orbit facility based on the payload container parameters. This hanger could fly free or be docked to STARLAB.

Docking and Resupply

The shuttle vehicle size will be on the same order of magnitude as the STARLAB center of operations structure. Docking would be difficult. The shuttle does not dock but exchanges a payload container for one brought up on the previous visit. This greatly reduces the amount of time the shuttle must spend at the STARLAB operations center. It eliminates a flurry of load and unload activity on shuttle arrival. The smaller and more easily maneuvered payload container flies the short distance from shuttle to STARLAB and is unloaded leisurely, used for storage, and then reloaded with material to be returned to earth. Part of the container may be pressurized so that personnel can be transferred from shuttle to STARLAB without EVA. Some packages brought up in the shuttle are designed for attachment to the exterior of STARLAB; e.g., cryogen containers. These may be attached to the exterior of the payload container during transit. Easy access by the space taxi (discussed below), upon arrival, makes distribution to their functional locations easier.

A centrally located main docking facility has short internal supply routes to the extremities. Equipment needed for resupply of FFM units may be concentrated at a central location. A docking location near the center of mass changes the moments of inertia very little and therefore the attitude control characteristics change little. Docking ports are spaced carefully to prevent collision of one docked module with another. Such a collision could cause large moments at the hatch connections and hatch tear-out and decompression would be a possibility. To prevent this, it is suggested that the hatch connections be flexible, perhaps using a metal bellows connection. The connections would deflect before failure. If failure occurs, the FFM should be the first to fail so that the STARLAB docking port remains intact.

A reusable shuttle vehicle is used to transport logistics packages to the STARLAB. Docking and resupply functions of the payload container logistics package have already been discussed. Shuttle payload to the intended orbit is expected to be 40 000 pounds and 10 000 cubic feet. The shuttle vehicle is still in the concept stage; therefore, the dimensions and shape of the payload have not been determined. Two packages have been proposed; 20 feet in diameter by 30 feet long or 15 feet in diameter by 60 feet long. The first unit is more compact and has been assumed here although the design does not depend upon it.

A number of needs (as listed below) may arise which requires a small multipurpose freeflight vehicle referred to as the space taxi:

1. Distribution and service of small supply (for example, cryogens) or other functional packages attached to the exterior of STARLAB.

2. Visits, scheduled or emergency, to FFM units some distance away from the STARLAB main vehicle.

3. Return of disabled FFM units to the docking facility.

The taxi should have a range approximately the same distance as the maximum flown by other FFM's. It should be capable of sustaining life for at least 24 hours. Addition of external supply tanks would extend both range and duration. Appropriate external manipulators, or serpentuators, are needed.

Imposed Structural Loads

The ability of the STARLAB structure to resist the imposed loads adds an additional constraint to the primary constraints of optimum use of available volume and adaptability to the S-V launch vehicle. The structural problem was approached from the viewpoint that structural considerations should make the smallest possible modification to solutions of the primary problems. This objective appears to be readily obtainable in that the proposed configurations seem to lend themselves basically to simple methods of resisting the loads. A final design would, of course, include an intensive analysis and testing program to assure the structure was capable of withstanding static, dynamic, and vibrational loads. Time did not permit such an analysis in this program.

The expected loads were defined as follows:

1. Total weight above the S-II stage — 120 000 pounds.

2. Axial acceleration during launch — 6 g.

3. Lateral acceleration — 1 g.

4. Rotational acceleration — negligible.

5. Normal pressure differential across hull — 5 psi.

6. Station keeping and attitude control loads — at least an order of magnitude lower than launch loads.

7. Improper docking transient loads — 4-g local, but dissipated due to flexibility.

Load Bearing Structure

The basic member to support these loads is the outer pressure hull. This is assumed to be a cylinder 33 feet in diameter and 50 feet long.

Construction is assumed to be similar to the S-II stage of the S-V; i.e., a 1/8-inch-thick pressure vessel stabilized with internal rings. The S-II stage is pressure stabilized during launch. Since internal structure is required in STARLAB to support the onboard equipment and to help reinforce pressure bulkheads during partial depressurization, it is used to stabilize the pressure vessel during launch without having to resort to internal pressure.

The basic STARLAB module consists of a cylinder 33 feet in diameter and 50 feet long with an additional cylinder 22 feet in diameter and 30 feet long on one end (Fig. V-3). The 33-foot diameter is conically tapered to meet the 22-foot-diameter section over a length of 10 feet. This allows the axial loads in the two sections to be transferred from one to the other.

A 7-foot diameter axial tube is installed the full length of the module. This tube serves to support the internal bulkheads and onboard equipment in addition to providing a number of operational functions. The tube is fitted with pressure-tight hatches at each compartment of the module. This allows access to all other compartments in the event of a depressurization in one or more. The tube is also fitted with a number of sealing rings so that a pair of hatches can be installed turning a short section of the tube into an airlock to enable repairs to be made to any damaged compartment. The tube is the main route for moving men and equipment between compartments in STARLAB. Some additional hatches in bulkheads are provided where it is felt there will be a possibility to move a large amount of equipment rather frequently. The tube also serves as a shelter in case of depressurization of a major portion of the vehicle. It serves either as an escape route to docked emergency vehicles or as a very temporary base of operations to repair some of the

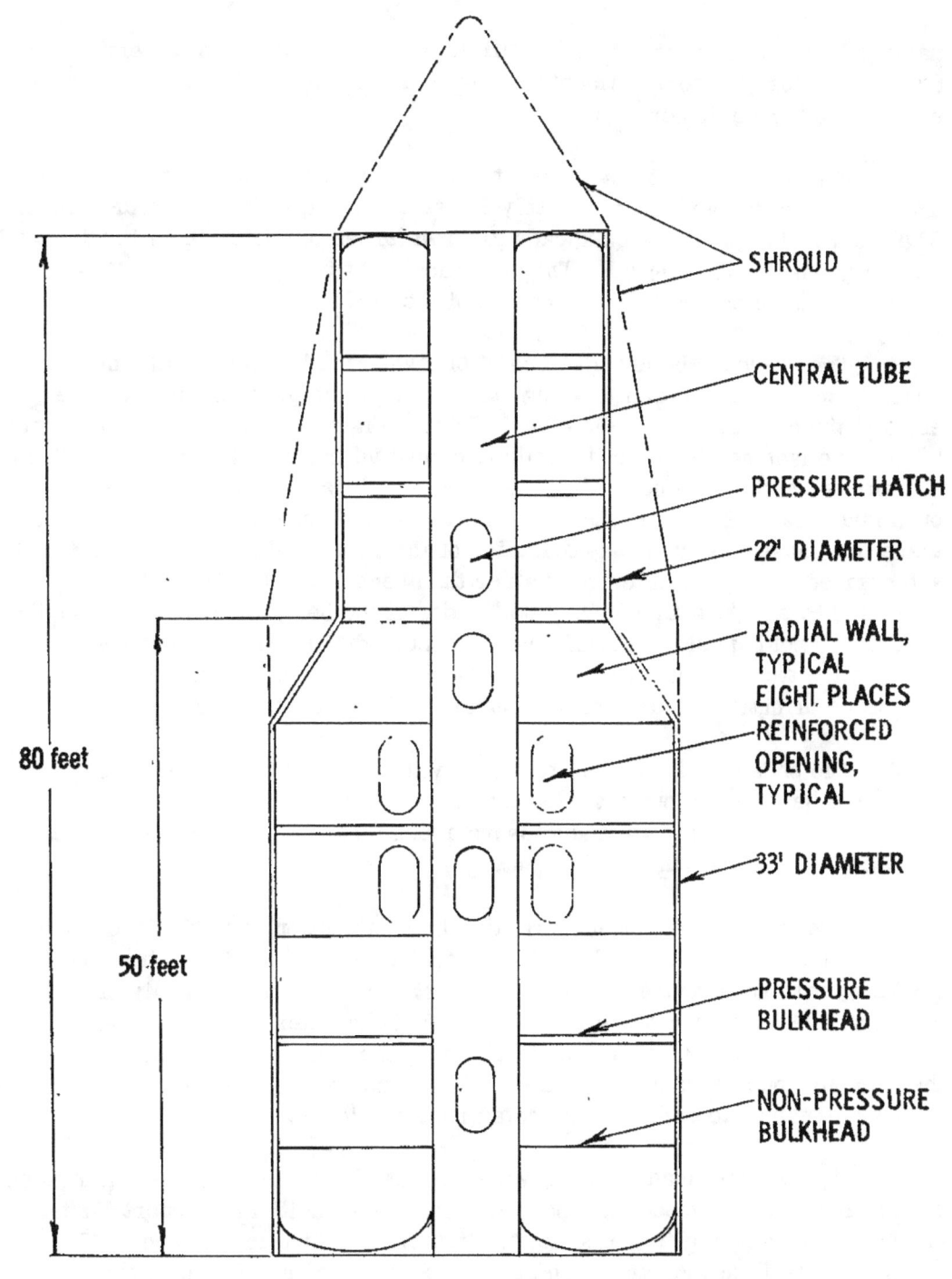

FIGURE V-3. TYPICAL STARLAB MODULE

5-13

damaged areas. Because the tube will not provide much room for emergency equipment storage, compartments are provided in the hatches to stow space suits and other equipment.

Five internal bulkheads are provided in the 33-foot-diameter section. They are disposed and approximately 8 feet apart perpendicular to the axis of STARLAB. Two of these bulkheads are reinforced to enable them to withstand a 5 psi pressure differential. These are located with approximately 16-foot spacing between each other and the end of the module.

The pressure bulkheads consist of eight radial beams, uniformly spaced, supported between the ring structures on the pressure hull on one end and the central tube on the other. Intermediate skin and equipment support beams are fastened between the radial beams and transfer their loads to them. The pressure skin is a normally flat membrane located near the center plane of the bulkhead. An emergency depressurization on one side of the bulkhead therefore causes an elastic deformation of the thin membrane so that the loads are carried through it in tension rather than bending. The membrane is protected by a light skin covering both sides of the beams. Three intermediate bulkheads, not capable of sustaining a pressure differential, are spaced between the two pressure bulkheads. These are a much lighter construction and serve primarily as a reference surface (floor) for the crew.

The ends of the module are fitted with torroidal pressure domes supported between the outer shells and the central tube. The free space provided outside the pressure enclosure around these domes is used to locate pressure bottles and other equipment.

Two pressure bulkheads, similar in construction to those described, are spaced at 10-foot intervals in the 22-foot-diameter section of the module. Docking ports are provided in these compartments so the possibility of a depressurization is greater than in the 33-foot-diameter section. There may even be intentional depressurization in these areas. For this reason, the pressure bulkheads are designed with two membranes and somewhat heavier structure to carry the expected repeated loads.

Eight radial, load-bearing walls are installed equally space and running the full length of the module. These walls provide additional support for the bulkheads, thus permitting a somewhat lighter structure in this area. The walls also stabilize the outer skin against axial loads during launch and improve the vibration performance by reducing the size of unsupported areas. The walls greatly increase the rigidity of the module, thus improving stability of STARLAB. The walls are fastened full length to the outer shell, the central tube and the bulkheads, thus providing the best possible stability. Access through the walls is provided by means of reinforced oval openings as required. These openings provide the minimum strength reduction and weight penalty.

To survive the launch loads, equipment must be mounted to "hard points" on the bulkheads or walls. These areas are reinforced to spread the expected loads over a wide area, thereby reducing the local concentrations. The mounting hardware is as common throughout as possible to provide for future updating of equipment and the possible need to relocate other items.

Materials

The structural materials used in STARLAB are primarily aluminum alloys such as 2219-H1, where weldability is desired, and higher strength alloys such as 7075-T6 as needed. It is particularly important that ductile materials be used where a brittle failure could cause a catastrophic collapse such as in the outer pressure hull or other highly stressed areas. Other considerations are quite important; e.g., fatigue may be the critical factor in preventing failure of the structure. Any material with low fatigue resistance must be avoided or used only in low stress, noncritical areas. Surface erosion and vacuum sublimation are also potential problems because of the long projected stay-time.

Micrometeoroid Protection

The long stay-time planned for STARLAB makes a high order of micrometeoroid protection essential. The long exposure not only increases the number of expected hits by micrometeoroids but increases the probability of hits by larger particles. This makes it important that the shielding system be effective and, if possible, repairable.

Meteoroid shielding is closely related to all the other functions of the outer skin. These other functions are heat absorption and radiation, thermal insulation, and structural strength. The interaction and possible integration of two or more of these functions (Fig. V-4) appears to offer a potential saving in weight, or an increase in protection with no additional weight (Ref. V-3).

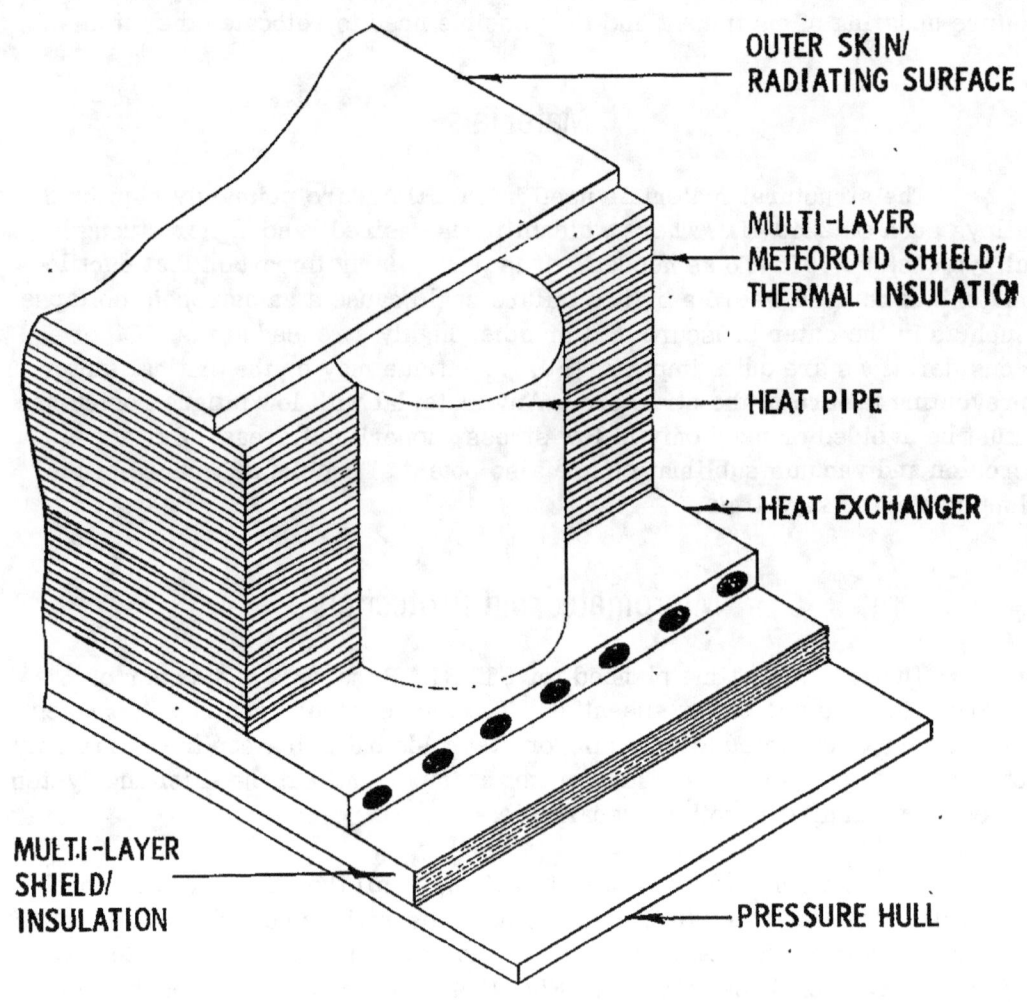

FIGURE V-4. STARLAB SKIN STRUCTURE

5-16

Calculations for heat to be dissipated from STARLAB by radiation indicate that most of the outer surface has to be a heat rejection radiator. The temperature of the radiation surfaces is well above the internal STARLAB temperature, therefore it is necessary that they be thermally insulated from the cabin space. An additional problem arises when a meteoroid puncture of the radiators, which contain tubes for heat exchange fluids, is considered. Any leaks or punctures in these tubes causes loss of fluid and therefore causes loss of at least a section of effective radiator.

The most promising radiator system from the viewpoint of vulnerability to puncture consists of an outer aluminum radiation skin bonded to short heat pipes transferring heat to the skin from a heat exchanger between the skin and the pressure hull. The space between the skin and exchanger is then filled with an insulation material and the exchanger is in turn insulated from the pressure vessel. This insulation is a multilayer type bonded into the radiator-exchanger heat pipe structure thus providing structural rigidity to the unit and integrating the structure as an efficient meteoroid shield. A high degree of meteoroid protection for the exchanger is provided as the tubes are located so that both the outer skin and multilayer structure shield them. Penetration of any heat pipe will only eliminate that one unit and will have an almost negligible effect on overall performance. An additional benefit to the life of the tubes in the heat exchanger is that the multilayer structure prevents spread of debris from a meteorite; thus, only a small penetration is made compared to the free-standing bumper concept where a number of tubes could be punctured by the spreading debris stream. This benefit extends to the pressure hull which has the advantage of the exchanger and to an additional multilayer structure to further reduce the mass of any penetrating particle. A small penetration of the hull is easier to patch and less likely to cause catastrophic failure of the hull than the typical split area caused by a diffuse shower of debris.

The integrated shield-radiator-insulator is rigid enough to withstand flight and launch loads, including aerodynamic loads, and should be rigid enough to prevent flutter. With proper design, this unit could be capable of supporting and stabilizing the pressure hull during launch, thus reducing the weight required to withstand these loads.

POWER SYSTEMS

Overview and Cost Summary

Project STARLAB is to be an orbiting laboratory facility with a design lifetime of 10 years. Because of the large crew size of 35 men and the mission of the space station, 100 kW_e of power must be supplied to the vehicle in a package which has flexibility for future alteration. Primary power is supplied by a 100 kW_e SNAP-8 nuclear heat source, coupled to lead telluride thermoelectric converts and organic-Rankine turbugenerators. A 30 kW_e solar array system with nickel-cadmium batteries and roll-up solar panels supplies the required auxiliary and emergency power to back up the primary power source. In the event that all system power is lost, the batteries which are interconnected into the main distribution system will supply emergency power for life support, communications, and repair maintenance.

The weight of the nuclear power source, including the thermal radiator, is 33 000 kg with an estimated recurring cost of $10 M. The solar array and batteries increase spacecraft weight by 2560 kg and this system has an estimated cost of $50 M. For a 10-year lifetime, the power system including logistics support is expected to cost a total of $1.61 B, based on 1968 costs.

Introduction

Various sources of electrical power find use in space vehicle power systems (Fig. V-5); batteries, fuel cells, solar arrays, mechanical cycles, and thermoelectric converters. Consideration must also be given to available heat sources including solar radiation, and nuclear and chemical reactions. A space power-generating system has numerous constraints not encountered on earth. The following are noted as pertaining to an earth-orbiting laboratory:

1. It must be small enough to be successfully placed in earth orbit along with other equipment.

2. It must be light in weight because of limited rocket payload capabilities.

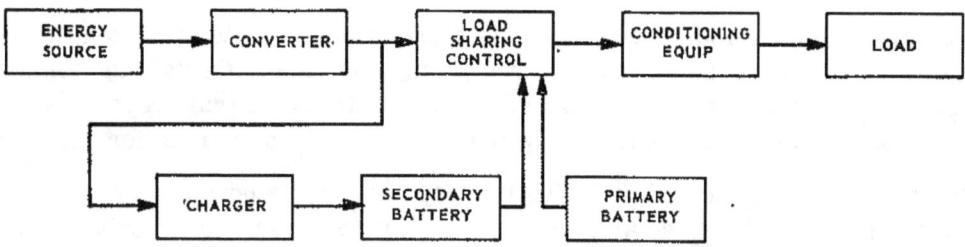

FIGURE V-5. BLOCK DIAGRAM OF A TYPICAL SPACECRAFT
POWER SUPPLY (After Scull)

3. It must consume little fuel for long service life.

4. It must operate in an airless environment, in zero-g, and be reliable for programmed periods of the mission.

Additionally, constraints are defined by the sense of the mission. The most important of these constraints are (1) that the STARLAB system be launched in the year of 1977 and have a serviceable lifetime of 10 years, and (2) that the laboratory is capable of supporting a minimum crew of 35 members.

Batteries are on board the space vehicle. Nickel-cadmium units, which have excellent life-cycle characteristics, have been selected as part of the solar array system. These cells have specific energies of 2.2 W-h/kg and a specific volume of 0.062 W-h/cm^3.

Fuel cells are not recommended as part of the STARLAB manned orbiting laboratory power system since their contribution to overall power is small and their by-product, potable water, can be more efficiently realized in a closed life support system.

An auxiliary/emergency power source is composed of a 455 m^2 solar array with nickel-cadmium batteries. The system is capable of supplying 20 kW$_e$ power to the space station. The solar array is of the roll-up type that is extended or retracted according to the power situation in the laboratory. If technology permits, light arrays made from 100 μ thick silicon cells with a specific weight of 20 kg/kW and a specific area of 17 m^2/kW will be utilized.

Mission Power Requirements

The space station requires a total usable power of 100 kW_e in the 1977-1987 time period. Because of its experiment-oriented mission, this total power need not be fully demanded at any one time. Certain systems such as environmental control, life support, attitude control, repair, and operations require a minimum power of 64 kW_e. Hence, mission oriented power is used as required up to the full capacity of the power source. In the event of power failure, solar cells and battery power take over critical operations. The power reserved as a margin of safety is at the discretion of the space station commander.

Power requirements for the mission are subdivided into four distinct areas. These are prelaunch (ground), launch, propulsive maneuvers, and the earth orbit. Each of these areas is further categorized into such areas as communications, experiments, guidance and attitude control, environmental control, and life support. The individual allotments listed in Table V-1 are detailed in later sections.

The prelaunch phase is principally a checkout phase. Since prime and auxiliary power systems cannot be used during ground operations, it is necessary to rely on external ground power sources to operate and test the various space station subsystems. Power is required to operate lighting (20 kW_e), thermal control (3 kW_e), and communications (6 kW_e). Other systems are operated at checkout as necessary.

The launch phase operation requires that power be supplied for communications, flight control, and thermal control. From the standpoint of value, these are estimated to be 100 W_e for a computer and sequencer, 10 W_e for tape recorders, and 500 W_e for amplifiers and power supplies. A primary battery supply of silver-zinc cells is located on board the launch vehicle to supply this quantity of power.

Available Power Systems

Physical potentials of various candidate power systems for spacecraft electrical power were considered for the project STARLAB mission. Those reviewed are batteries, fuel cells, solar arrays, radio-isotopic thermoelectric/ dynamic generators, and nuclear thermoelectric/dynamic generators; all

TABLE V-1. POWER ALLOTMENT

Mission Phase		Power Requirement (kW$_e$)
I. **Prelaunch**		System Checkout (External Power Source)
Communications		–
Guidance and Control		–
Environmental Control and Life Support		–
Thermal Control		–
	Total	50
II. **Launch**		
Communications		0.8
Guidance and Control		0.5
Electronics		0.7
Thermal Control		1.0
	Total	3.0
III. **Propulsion Maneuver**		
Communications		0.8
Guidance and Control		1.0
Thermal Control		5.0
Experiments		13.0
	Total	19.8
IV. **Earth Orbit**		
Communications		6.0
Experiments		36.0
Guidance and Control		1.0
Thermal Control		5.0
EC/LS		18.0
Docked Modules		4.5
Operations		2.0
	Total	72.5

except the latter system have been successfully utilized on spacecraft missions. The selection of a particular power system was made on the basis of mission constraints, expected system life, reliability, safety, payload weight, and cost for a planned mission duration of 10 years.

There are two classes of batteries; primary and secondary (Ref. V-4). Primary batteries are those which deliver energy in a relatively short period of time. They are short-lived and have high-energy density, high current capabilities, and good reliability. Secondary batteries, otherwise, have long-life characteristics and can withstand numerous charge-discharge cycles. Because project STARLAB has a long-term mission, batteries cannot be considered as a prime source of power; however, they are required to support a system of solar arrays.

A battery is essentially an inseparable combination of energy source and converter, where the reactants are contained within the cell. Batteries are well-developed items with specific energies for primary cells in the range of 22 to 220 W-h/kg, and for secondary cells in the range of 4.4 to 88 W-h/kg. Operational problems of batteries include a restricted operating temperature range (0°F to 150°F), a relatively short shelf life after energizing (for certain types of primary cells), and the necessity of keeping certain configurations nearly upright during ground and launch operations. Battery types that are available are made of silver oxide-zinc, nickel-cadmium, and silver-cadmium.

A nickel-cadmium battery has the highest survivability as well as the longest life cycle, but it is heavier than a silver-cadmium battery. Nickel-cadmium batteries have been used in the past and their characteristics are well known. It is anticipated that in the future, advanced development of these cells will achieve greater uniformity of operating characteristics among cells that presently exist. Currently, a specific energy of 100 W-h/kg is available for spacecraft application (Ref. V-5).

Silver-cadmium cells for a secondary battery have twice the advantage of an energy density of nickel-cadmium cells and a life cycle of one to two orders of magnitude greater than silver-zinc cells. These advantages are offset by a low-cell voltage and the resultant-poor system voltage regulation. Although silver-cadmium cells can withstand rapid recharge rates, their cyclic performance has been inconclusive. There have been reports ranging from 10 000 shallow cycles to poor cyclic performance. Silver-zinc cells have been developed into highly reliable, low capacity primary systems with energy densities up to 176 W-h/kg, and they are often used in launch vehicles.

Fuel cells were considered as a power source during early phases of this study, particularly since a cell using oxygen-hydrogen reaction agents is capable of supplying water as a direct by-product of operation. Unfortunately, fuel cells are more practical for short-time missions where life support is a necessity. In these devices, streams of reactants from separate tanks are passed into a conversion device. A hydrogen-oxygen cell has a specific energy of 330 to 660 W-h/kg depending on storage methods. There is a fixed weight for the conversion cell and auxiliary equipment. Fuel cell operation is limited by membrane reliability, reactant storage, and heat removal problems. For example, power systems of 4 kW_e capacity require an active coolant loop to remove waste heat; fuel cell systems with proven life capacities in excess of 2000 hours are not yet available (Refs. V-6 and V-7).

Fuel cells have also been considered as a replacement for batteries in a solar array battery system. This becomes a possibility by utilizing an electrolytically regenerative oxygen-hydrogen cell. In this configuration, water recovered from the fuel cell in the power generating process is recycled and electrolyzed into H_2 and O_2 reactants. Since recharging of the fuel system is rapid, this type of system finds use in orbital missions with large ratios of dark-to-light time periods. Such systems have potential advantage over battery systems which experience reduced life caused by accelerated charging rates. Moreover, battery weight becomes excessive in large power solar arrays.

Because of the availability of solar energy in space, solar arrays were used on many satellite space missions. They were constructed on cylindrical bodies, set out in flat panel arrays, and placed on spherical surfaces. As a consequence of having to obtain energy from the sun, these arrays must be placed in an unobstructed line-of-sight position with the sun and the orientation maintained by controlling either the entire space vehicle attitude (when the arrays are rigidly mounted) or by physically moving the arrays with respect to the orbiting vehicle.

Solar energy is distributed over the electromagnetic spectrum with most of the energy located in the infrared and visible regions (Ref. V-7). Electrical power is converted from this form of radiation by photovoltaic or thermal conversion, the former being the most commonly used. Many semiconductor materials are sensitive to solar radiation, and they can be used to convert solar radiation into electrical power direct. Semiconductor converter materials include silicon, cadmium sulfied, and gallium arsenide. Cadmium sufide arrays are lighter per unit area than those of thin silicon; but specific weight, excluding deployment and support structure, is about the same.

Moreover, such an array has twice the area of a comparable silicon array. Hence, deployment, support, and orientation mechanisms are heavier. Cadmium sulfide cells require a net conversion efficiency of 6.25 percent to match the specific power achievable with 200 μ silicon cells; an even better performance is required with respect to cells of 100 μ thickness. Consistent and stable performance at such levels in the aforementioned material is unlikely in the forseeable future. Therefore, silicon has a specific weight advantage. Although polycrystalline cells are more flexible than those of silicon, silicon arrays can be safely wrapped around a 3-inch-diameter roller. Multikilowatt solar arrays weighing 20 kg/kw using 100 μ thick single crystal silicon cells are possible in the near future (Ref. V-8). This type array would have an approximate area of 12 m^2/kw. From cost considerations, silicon solar cell prices have decreased by 40 percent since 1965. With increased usage of these cells, prices will tend to lower.

A large area solar array development program was carried out by Jet Propulsion Laboratories during 1966-68 at Pasadena, California, to advance the technology of lightweight electrical systems capable of producing 50 kW$_e$ of power (Ref. V-9). It was shown in that program that significant improvements in solar panel capacity are possible. Frames for the solar array were made of built-up beryllium-bonded rectangular sections connected by titanium fittings where structural loads were concentrated. Results of that study for the 50 kW$_e$ array are: a specific area of 0.0093 m^2/w and a specific weight of 21.4 kg/kw; also, a specific output of 46 w/kg and a specific weight of 2.3 kg/m^2. These values are based on a 465 m^2 array at 1 AU, air mass 0° and 55°C.

One problem encountered with lightweight systems is that they are not satisfactory for surviving a high-acceleration launch environment. For large arrays, there are additional problems to consider. These arrays contain expansive surface areas which cause manufacturing and deployment problems. Since they must be properly oriented and kept out of equipment and vehicle shadow zones, additional attitude control systems are required aboard the space vehicle. Solar arrays cost approximately $2 M/kw. Moreover, they are subject to performance degradation caused by space particles and solar radiation. This necessitates an increase in deployed solar array area which is considerable for large power arrays. Further, these arrays have to be supported by a system of rechargeable batteries to offset the changes in solar intensity encountered in traversing an orbit. Batteries are considered the weakest link in such a power source. Figure V-6 shows solar array area requirements as a function of electrical load and solar degradation.

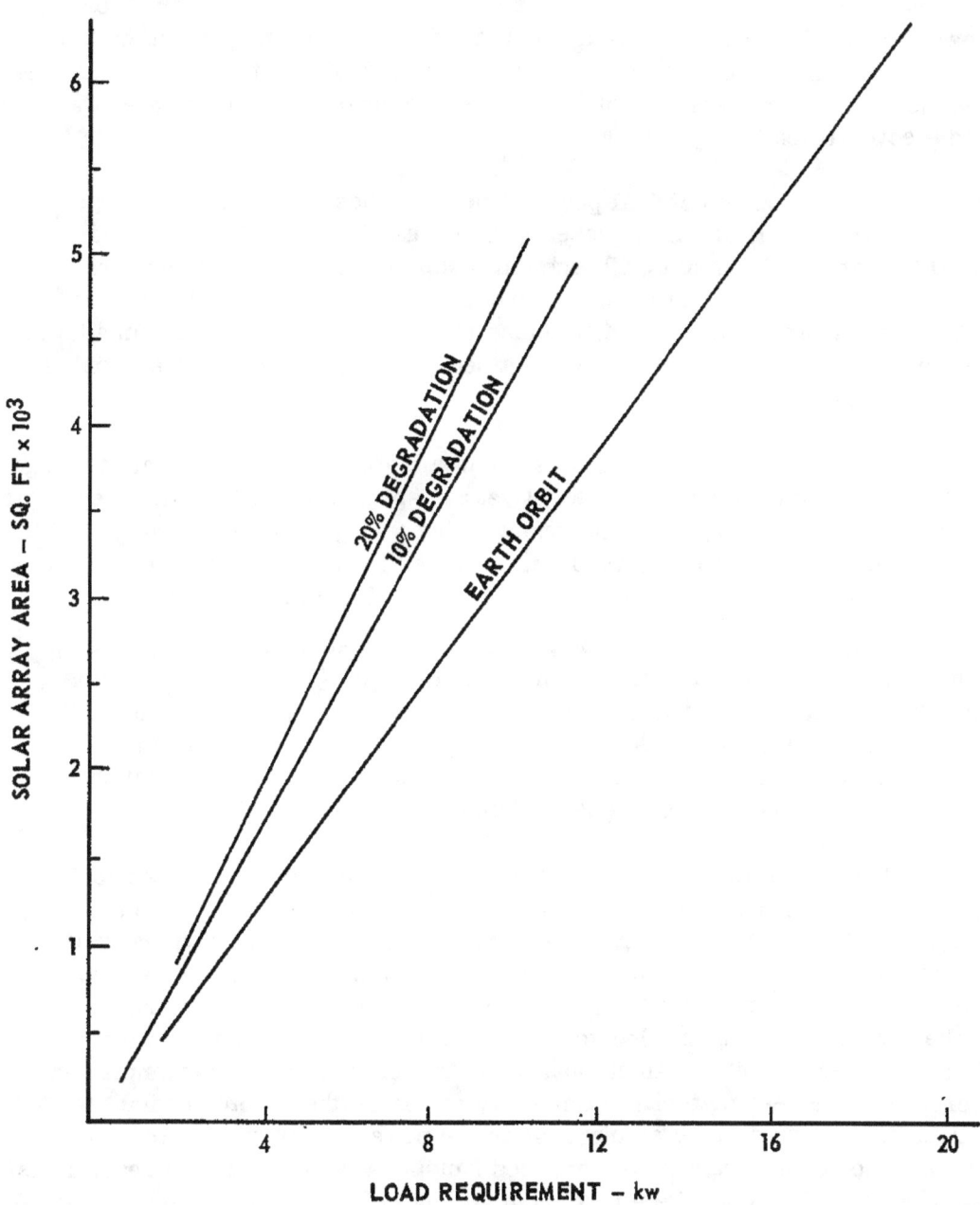

FIGURE V-6. AREA REQUIREMENTS FOR SOLAR CELL ARRAYS

A solar array must be oversized to provide auxiliary power while recharging the secondary batteries. Under some earth-orbital conditions, this means a solar array almost twice as large as would be required for direct power load. The maximum voltage output of the array is dependent on the temperature of the solar cells. Arrays must be designed to limit this voltage. Furthermore, the operating voltage range is dependent on both temperature and electrical load.

Radioisotope electrical power generation has been under development for quite some time and it is presently being used in the TRANSITE and NIMBUS satellite programs. These units supply heat for power conversion by radioactive decay of an isotope. Presently, isotope materials are not readily available in large quantity; these materials are expensive. In addition, radioisotope-powered systems have not been developed to a large electrical capability such as 15 kW_e.

Nuclear reactors are considered as a favorable souce of heat for large power generation systems. In recent years, specific weights of these systems decreased from 365 kg/kw to 238 kg/kw in the 1 kw power range using Pu 239 fuel. Cost of this fuel in the 1970's is expected to be 20 to 70 percent of the equivalent in isotope power in the 25 to 125 kW_e thermal power range (Ref. V-10). One problem to consider is that a reactor has a critical size and weight to maintain a controlled nuclear reaction (Ref. V-11). Further, radiation shielding upwards of 20 000 pounds is required for shielding of the larger power systems. With the present state-of-the-art, SNAP-8 nuclear reactor systems offer many advantages for a manned orbiting laboratory. Information data concerning these systems are given in Table V-2.

Recent reactor systems for thermoelectric conversion use zirconium hydride as a moderator of uranium 235 fuel. Beryllium oxide is used as a neutron reflector around the reactor core. Reactor coolant is typically a eutectic mixture of sodium-potassium (NaK) which can be circulated by an electromagnetic pump. Such a pump could operate from an independent bank of thermoelectric elements located on the heat rejecting coolant structure. This arrangement acts as an automatic power controller. For example, an increase in the reactor fuel rod temperature raises the reactor coolant temperature, thereby causing an increase in the power output to the circulation pump. The coolant mass flow increases causing a decrease in the reactor fuel temperature. The opposite effect occurs when fuel rod temperature decreases, since this is accompanied by a decrease in coolant flow rate and, hence, an increase in the reactor fuel rod temperature.

TABLE V-2. CANDIDATE NUCLEAR-ELECTRIC POWER SYSTEMS FOR
THE 1975 – 1985 TIME PERIOD (After Brantley)

Heat Source	Year	Power (kw)	System Eff. (%)	Unshielded Weight (lb/kw)	Shield Weight (lb × 10³)	Radiator Area (ft²/kw)	Expected Life (yrs)	Status
SNAP-8/TE 1200° F Hot Junction	1975	25-50	4 (PbTe)	400[a]	10 – 20	60	2	Flight Ready
	1980	50-100	5 (PbTe)	350[a]	10 – 20	50	3-5	Flight Ready
SNAP-8HgR 1270° F Turb. Inlet	1980	35-50	9	300[b]	15 – 20	35	2	Flight Ready
	1985	70	12	175[b]	15 – 20	25	2-3	Flight Ready
SNAP-8/B 1250° F Turb. Inlet	1980	100	20	200	10 – 20	65	2-3	Conceptual Design

a. 20 percent power conversion system redundancy
b. 100 percent power conversion system redundancy

Note: TE – Thermoelectric
HgR — Mercury-Rankine
B — Brayton

Thermoelectric systems using lead-telluride converters are now capable of operating with 1200° F at the hot junction. Derating of this temperature level extends converter life. Converters using germanium-silicon alloys were under investigation in an effort to increase system efficiencies. Systems with radioisotope heat sources were placed in operation with specific weights of 2.2 kg/w. The isotope heat generator appears to be the most advantageous heat source for thermoelectric conversion at power levels up to 2 kw. It was suggested that reactor/thermoelectric power systems may find applications in space, depending on advanced technology at power levels of 10 to 25 kW_e (Ref. V-12).

In recent reports by Atomic International, favorable support was given to the use of reactor/thermoelectric power systems. Their study of a 25 kW_e power unit for 1975 deployment concluded that a design lifetime of 2 to 5 years can be expected. The power unit investigated was composed of individual converter modules of 262 W_e output each at 5.25 vdc. Converter material was lead telluride. A test program initiated in 1959 has tested a total of 200 modules with an accumulated test time of 20 000 hours at temperatures in excess of 1150° F. Derated temperatures of 1000° F to 1050° F produced extended service lifetimes. System circuit efficiency was 4 to 6 percent. Advanced research on manufacturing and operational problems, such as bridging of the PbTe during fabrication and heat degradation of the material, have greatly increased the potential of these semiconductor systems. The greatest advantage of thermoelectric converters as compared to dynamic systems is the absence of moving parts. This will be weighed against the inherently low conversion efficiencies which are not expected to exceed 10 percent in the future. Moreover, heat rejection temperatures require relatively large radiator surface areas. It was concluded that reactor/thermoelectric power systems are feasible for a 1977 launch with power capabilities up to 50 kW_e.

Thermionic conversion offers a possible method for generating power levels of the order of megawatts with conversion efficiencies of 15 percent. Either radioisotope or nuclear reactors can be used for heating. However, the reactor has the greatest potential for generating high power levels. This type system requires an advanced technology to solve the many formidable problems that prevent its application to practical use. Brayton cycle conversion utilizes an inert, heated gas such as neon or argon to drive turbomachinery for electrical power generation. Gas temperatures range from a high of 1600° F to 100° F at heat rejection. This also requires a large radiator surface. Additionally, it permits high thermal efficiencies; however, it increases system weight. The inert gas creates no corrosion or erosion problems. But, as a gas, it requires larger structural components and additional weight.

High power level systems favor use of the Rankine cycle because specific weight is reduced by operating the turbine and radiator at increased temperature levels. The nuclear reactor is presently the preferred heat source for Rankine cycle conversion with power capacities in excess of 25 kW_e, mainly because at these power levels the reactor is able to generate more power per pound than radioisotopes. To date, 35 kW_e and 600 kW_e power generating reactor/Rankine systems have been built and tested by NASA and AEC, respectively (Refs. V-13 and V-14). With higher conversion efficiencies, practical systems will be capable of power output levels in the 50 to 75 kW_e range with a heat rejection radiator area of one-half that required for thermoelectric converters. However, dynamic-type units have moving parts which present operational difficulties during a 2- to 5-year mission. Heat is rejected in the Rankine cycle at an average temperature of 575° F. This is much higher than the 100° F associated with the Brayton cycle system. A consequence of this temperature level is a smaller radiator size (Fig. V-7). A Mercury-Rankine system having a power output of 25 kW_e requires a radiator area of 75 m^2, approximately. Although this area may seem large, it is relatively small when compared to a comparable nuclear/thermoelectric power system, which requires 130 m^2 of radiator area. For very small space vehicles, such as those of the Gemini or Apollo programs, these radiator requirements are a definite constraint. It has minimal influence on the large structure considered for the STARLAB mission.

Summary of Power Systems

A review of system power requirements (Table V-3) and the available sources of heat and electrical converters rules out batteries and fuel cells as sources of primary power, mainly because of the mission time involved. This same reason applies to a consideration of fuel cells instead of batteries in a solar array power system. When the operational life of fuel cells is extended to 10 000 hours, they will become a definite competitor to the battery systems now in use. Like batteries, they can be regenerated in a closed system process; but total weight for such a closed system would be far less.

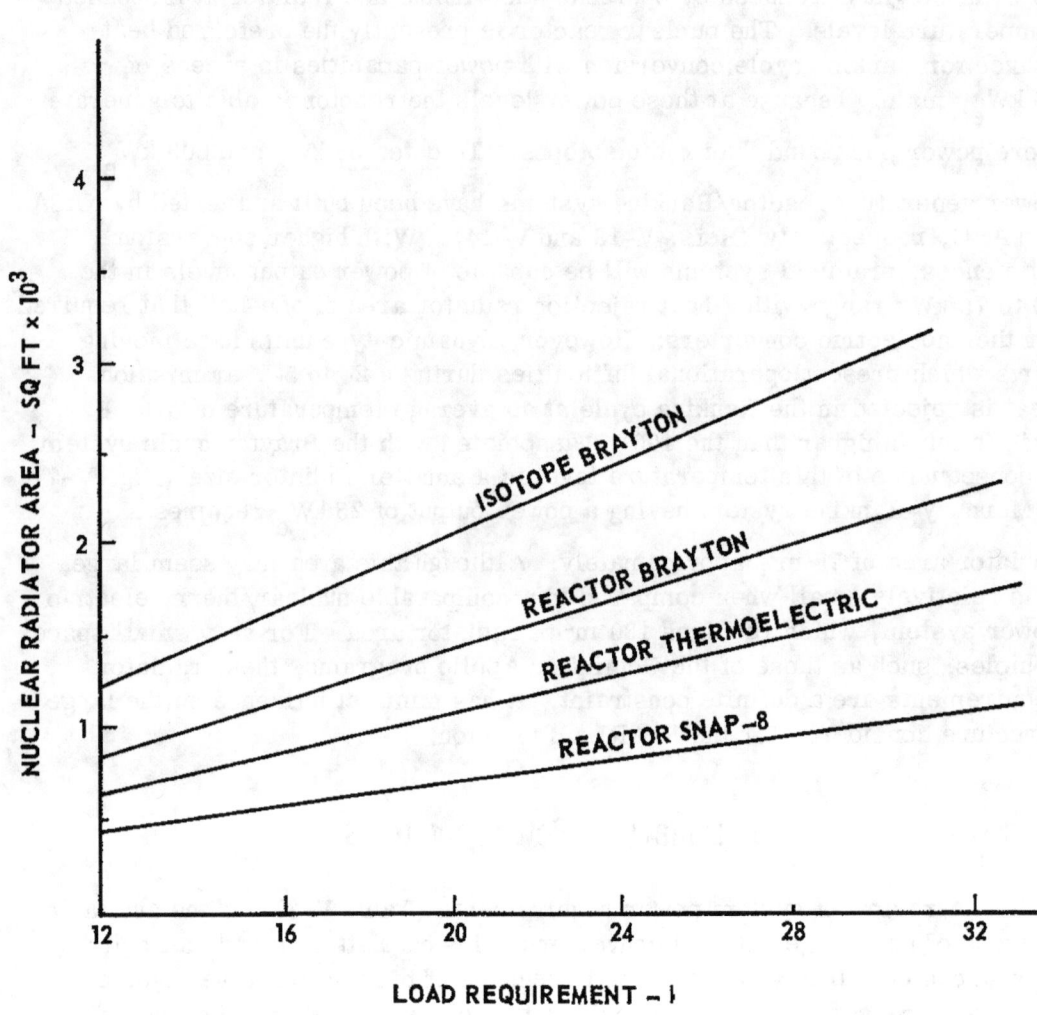

FIGURE V-7. AREA REQUIREMENTS FOR NUCLEAR
HEAT SOURCE RADIATORS

TABLE V-3. SYSTEM POWER REQUIREMENTS

Subject	1977 Orbital Checkout	1977 Operation
Crew	6 men	35 men
EC/LS	6 kW_e	18.0 kw
Lighting	1.5	20.0
Thermal Control	1.0	5.0
Attitude Control	2.0	1.0
Instrumentation	1.0	2.0
Communications	2.0	6.0
Maintenance	2.0	1.5
Airlocks	1.0	4.0
Dock Module	0	4.5
Operations	0.5	2.0
Experiments	3.0	34.0
Shuttle	0	2.0
Total	20 kW_e	100 kW_e
Power Source	Solar Panel	SNAP-8/TE/Rankine
	20 kW_e	100 kW_e
	4900 ft^2	5000 ft^2

Thermionic converters require more advanced technology before use. High-temperature materials have to be developed to make these systems compatible with a nuclear-type heat source for long life application. Consideration was given to a prime power system composed of multiple radio-isotope/thermoelectric units. However, disadvantages outweighed advantages when compared with the system finally selected.

In view of the total capabilities of the various candidate systems, it is proposed that a combination system consisting of the following be used: for primary power, a 150 kW_t nuclear reactor containing uranium 235 fuel moderated with zirconium hydride and coupled to thermoelectric and organic-Rankine converters; and for auxiliary/emergency power, a solar array system using 400 μ thick silicon cells supported by a system of nickel-cadmium batteries.

Description of the Solar Array/Battery System

Mission life, required power levels, and reliability of operation are serious considerations for a system to power an orbiting laboratory. A combination of batteries and solar array satisfies, in part, all three of these requirements; it provides auxiliary power to complement the prime power source during periods of peak demand. But more important is the fact that this power source of 20 kW_e is available immediately after launch for life support, flight control, and other functions which must be in operation for deployment and checkout of the orbiting laboratory. A further advantage is the availability of the battery system for emergency operations. The energy storage capacity of the battery system powers life support systems, communication apparatus, and repair equipment during emergency situations and at such times that primary power is unavailable. This occurs during an emergency reactor shutdown or while changing or upgrading the power source.

The state-of-the-art of solar cell technology is well developed. Systems have been successfully operated in space with power levels of milliwatts to watts. Current technology is being expanded into the design of solar array systems having 50 kW_e power output. These systems are light in weight, and simple and reliable in operation. The expected degradation in an earth orbit of 308 n. mi. is small; hence, no real penalty is involved in increasing the array area. With a 69-degree inclination, maximum dark time amounts to 35 minutes per orbit, and it is during this time that the battery system is used for backup power. The design orbit results in 15 orbits per day which is equivalent to 5500 cycles per year.

If the operating life time of the nickel-cadmium batteries is to approach 5 years, they should not be discharged below 80 percent of capacity. At 33 W-h/kg, this requires a total battery weight of 1510 kg. Associated with this is a volume of 1.58 m^3. A specific cost figure of $1.5/W-h results in a recurring battery cost of $15 000. The silicon cell has been constantly upgraded in performance and environmental resistance. The present 10-percent efficiency will persist until there is some new breakthrough in the art of manufacturing silicon materials. Thin-film cells are presently being investigated, and they hold promise for future low-cost, lightweight solar cell arrays.

The solar array for the STARLAB vehicle has the following subsystems: (1) solar cell panels; (2) deployment drive and control; and (3) orientation drive and control. Each solar cell panel consists of individual cell modules housing appropriately connected 400 μ thick silicon cells attached to the supporting panel structure. The solar cells can be treated with lithium doping if necessary to reduce degradation caused by radiation effects.

A beryllium or titanium beam structure can be fabricated for mounting the individual panels with a fiber-glass substrate into a completed array. Frames are built-up beryllium-bonded rectangular sections connected by titanium fittings where structural loads are concentrated. It is expected that such panels will have a specific output of approximately 44 w/kg for the total array, based on a thickness of 0.004 m.

Because of the large area of the solar panels it is recommended that they be mounted in a roll-up configuration and attached to the sides of the laboratory vehicle. On command, the panel drive motor and clutch mechanism deploy the proper amount of irradiated panel area whenever auxiliary power is needed. During emergencies, maximum surface area can be deployed. Roll-up arrays have the advantage of being easily stowed both during launch and when they are not in use during orbital flight.

As thinner and lighter weight solar cells are developed, lighter substrates can be used for mounting the cells. This results in lower specific weights for future panels. The following data are indicative of design criteria for the solar array:

Solar cell cover glass absorptivity	0.8
Solar cell failure rate per hour	10 × 10-10
Solar cell efficiency	10-12 percent
Solar array assembly loss	3.4 percent
Panel specific weight	0.8 kg/m^2
Panel thickness	0.004 m
Reliability	0.95-0.99

For a 20 kW$_e$ solar array using a specific area of 0.0093 m^2/w, an area of 186 m^2 would be required. However, this area must be increased by a factor of 2.3 to account for 10-percent degradation of the energy output and power for recharging batteries. The resulting net area is 455 m^2. At 2.3 kg/m^2, the solar array weighs approximately 1050 kg in addition to the weight of the deployment and orientation mechanism. Brantley has suggested a solar array cost figure of $2 M/kw. Therefore, the total solar array costs $50 to $60 M.

Reactor Power System

With a requirement for large quantities of power, it quite naturally falls to the nuclear systems to supply the STARLAB power demand (Ref. V-15) Radioisotopes fall short of the necessary level of power. More important is the problem of isotope materials. These materials have always been in short supply, and it is not anticipated that supplies will increase significantly during the foreseeable future. On the other hand, nuclear materials, such as U235, are available in sufficient quantity to power the large 25 to 50 kW$_e$ systems with lifetimes in excess of 10 years (Refs. V-16 and V-17).

The reactor-type that is recommended for the STARLAB power system is a 150 kW$_t$ SNAP-8. SNAP systems are the outgrowth of over 13 years of design and development of zirconium-hydride reactors. The reactor fuel has a hydrogen content of 6.3 by 10^{22} atoms/cm^3 for long life. As a cladding material for the fuel element, either Hastalloy-N or Incoloy-800 suffices, the latter exhibiting greater ductility after irradiation. The control drum contains tantalum — 10-percent tungsten alloy for nuclear poison. In addition, beryllium oxide is required for reflecting radiation. These latter materials can operate for extended periods at 1800° F.

The shield for the reactor uses a combination of lead and lithium hydride placed about the reactor package. A 4π shield for a 25 kW$_e$ power unit (Fig. V-8) results in a weight of 6820 kg, approximately. Lead is used along the critical axis as the first gamma shield to reduce the secondary gamma-ray production that is associated with most other materials. This shield is followed by a lithium-hydride neutron shield to prevent activation of the NaK heat transfer liquid metal. Next, there is a shield of depleted uranium and then another shield of lithium hydride. Tungsten and LiH shields are used in the noncritical directions.

Because of the use of nuclear power, consideration was given to the installation, replacement, personnel safety, and radiation fields as well as the launch weight and configurations. The installation of a large reactor on board the space vehicle creates adverse conditions with respect to all the above-noted considerations. In addition, docking and replacement of a massive unmanned module onto the structural vehicle must be considered as a hazardous operation. If a locking collar or guide fails because of mechanical, structural, or radiation-induced failure, it would cause a critical condition aboard the space station in the event of an inoperative or runaway reactor. The nuclear source for the STARLAB vehicle is in an independent module that is placed into orbit on an S-V launch vehicle and docked automatically to the orbiting laboratory. The power module contains attitude and guidance control including a supply of boost propellant. When the power system is no longer serviceable it can be boosted into a 450-mile storage orbit or deboosted to a deep ocean area. Considering current emphasis on earth environmental conservation, the former approach is probably the more acceptable.

The thermoelectric converter is composed of individual tubular modules which contain lead telluride elements. Each element is capable of producing 262 w at 5.25 v. The modules are 3.8 cm OD by 38 cm long and weigh approximately 2.61 kg. The lead telluride is set into the tubular module in the form of washers isolated from each other by mica. The IDs are in contact with a tube carrying hot NaK; the ODs are cooled by NaK circulating through the dynamic system heat exchanger. Output power of the module is a function of the temperature difference across the elements. The individual modules are combined in various arrays to satisfy the overall system power requirement. Selection of lead telluride follows from extensive background test

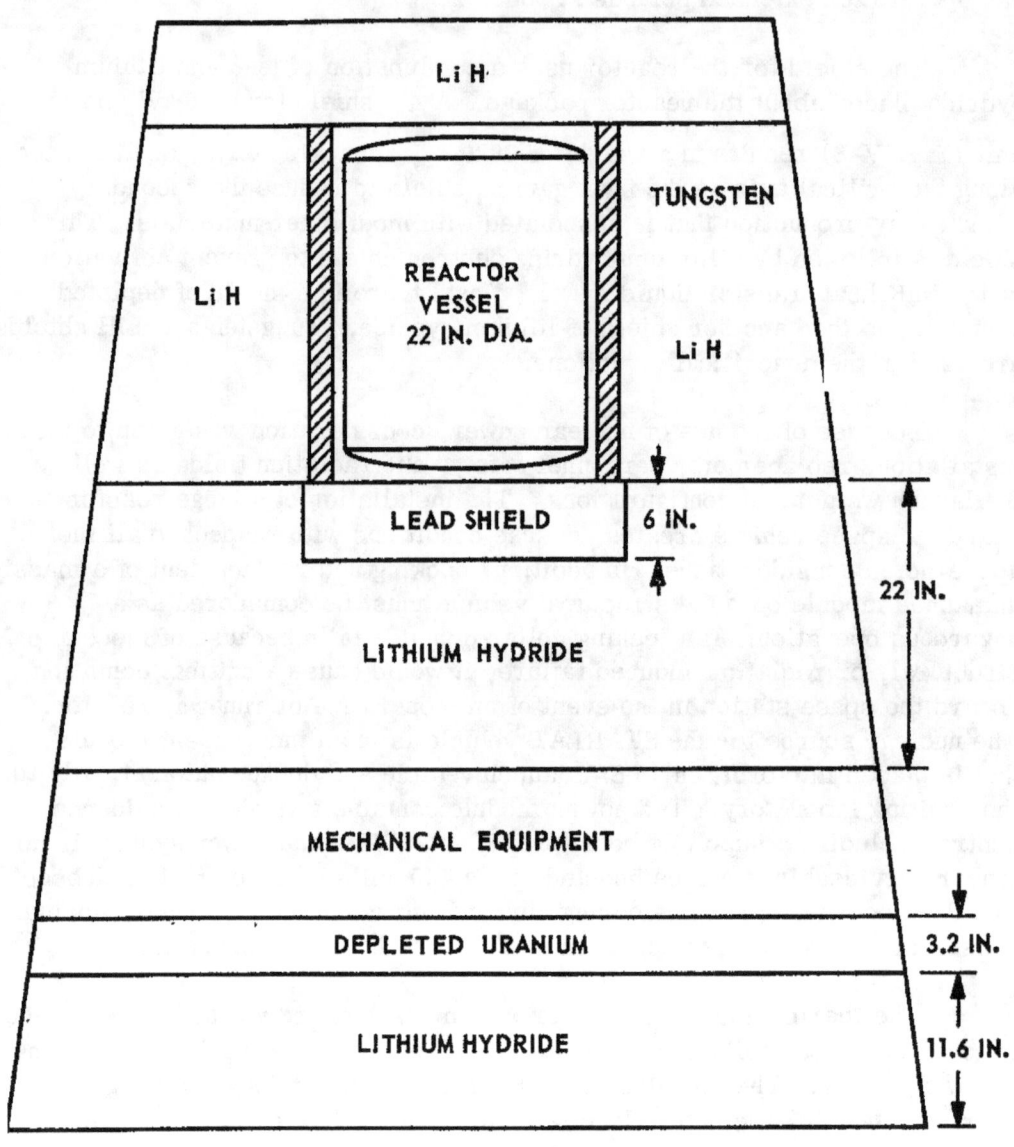

FIGURE V-8. NUCLEAR REACTOR 4π SHIELD FOR A 25 kW POWER SYSTEM

experience. An assembly of four individual converter sections, each delivering 6.3 vdc will supply a total of 25 kW_e to the power system.

The remainder of the power is generated by three organic-Rankine powered generator systems which derive their energy from heat rejected by the thermoelectric converters. At 23.6 kW_e per generator, the total output power of the nuclear reactor/thermoelectric/organic-Rankine system is 100 kW_e (Fig. V-9). The first loop in the overall system contains liquid NaK, pumped by a thermoelectric powered pump. Heat energy is utilized by the thermoelectric converter in addition to being transported in a second loop to a NaK-organic heat exchanger. In the last loop, an organic liquid is vaporized for use in driving the turbogenerator equipment. Waste heat is rejected to the space environment at 270° F. This system is presently under investigation by NASA. It has potential for producing a large quantity of power in the near future with some sacrifice in reliability, but overall efficiency is high. An expected service life of 1 to 2 years is anticipated for launch readiness in 1978.

In order for this power system to be ready for the STARLAB mission, a development program will need to be started in 1970. Based on recent cost figures, the development program for flight readiness will cost $144 M. Additionally, there will be a recurring powerplant cost of $10 to $12 M. In support of a short lifetime system budget, allowances must be made for approximately five replacement systems during the life of the space station. This includes for each replacement a S-V vehicle at a revised cost of $200 M. Thus, the reactor energy system has a total 10-year cost of $1.61 B, assuming that a development program proceeds for a luanch plus a 5-year upgraded reactor system with a rated life of 5 years. The 100 kW_e reactor system and its man-rated shield weighs 27 200 kg. With a radiator requirement of 420 m^2 as part of a redesigned third-stage vehicle, total system weight is 61 200 kg.

During ground and prelaunch operations, standard handling procedures such as those used in the nuclear industry must be observed. For example, when handling, transporting, or storing the reactor, control drums are replaced by neutron poison material to prevent any possibility of accidental criticality. Other than this, it is necessary only to keep the power system in a warm, dry, and clean environment to prevent environmental degradation. No significant fission-product inventory is present in the reactor during prelaunch and launch periods; hence, no severe safety problem exists. Should the reactor fall into the ocean on an abortive launch, the worst possible

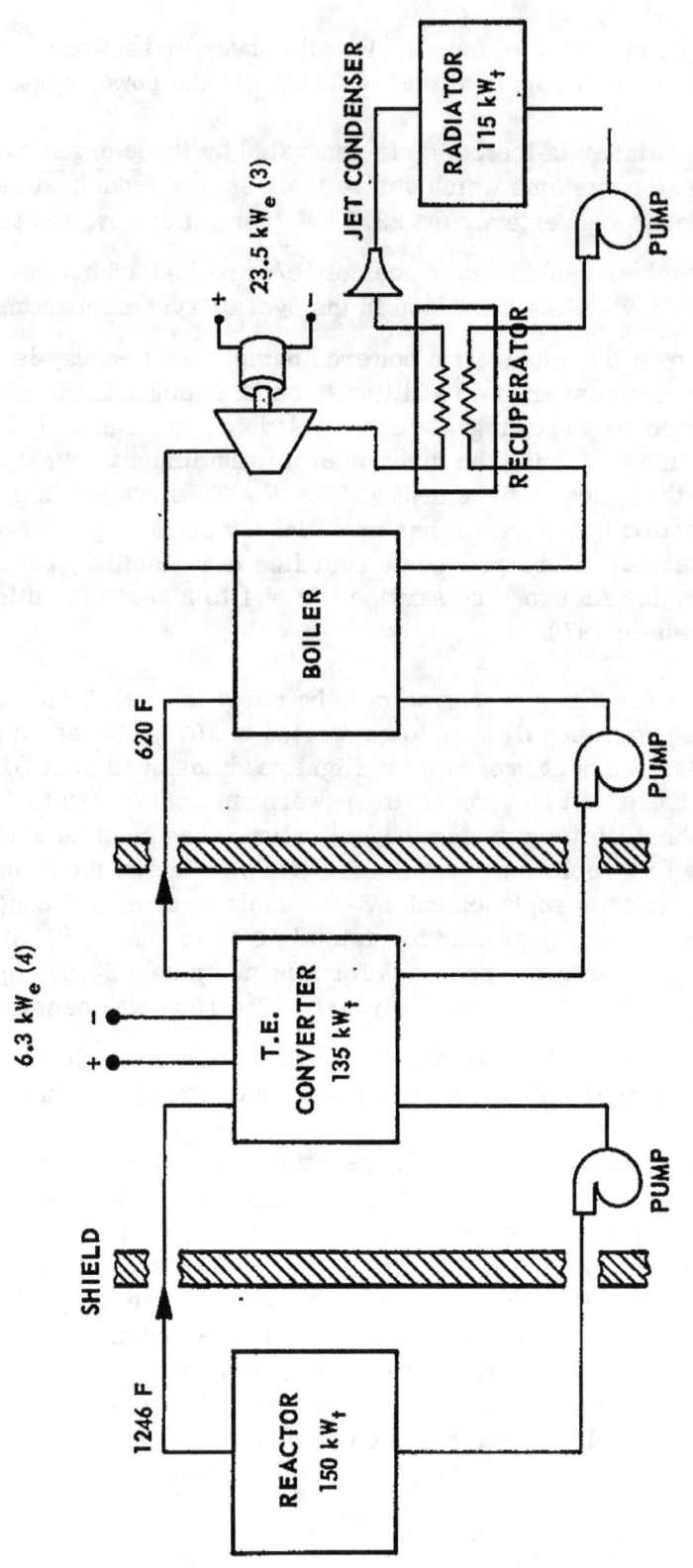

FIGURE V-9. BLOCK DIAGRAM OF SNAP-8/T.E./ORGANIC-RANKINE POWER SYSTEM RATED AT 23 PERCENT EFFICIENCY OVERALL

5-38

result would be an explosion equal to 1 pound of TNT. Accompanying this would be a small localized release of fission products. After the power module is placed into orbit and integrated with the laboratory facility, remote command can initiate reactor start-up. The reactor goes critical at about 90 minutes into the start-up and full power is reached in 4 hours. From prelaunch to orbital connection, power must be supplied to heaters and pumps on the reactor system to maintain the NaK in a fluid state at a 5-percent-rated flow rate.

Power Distribution System

Electrical loads present in the spacecraft present widely differing demands in terms of power output; such as voltage level, regulation, total demand, switching profile, and frequency stability. Hence, a power conditioning system must be provided to transform converter output into forms suitable for consumption by various devices (Fig. V-10).

With the anticipated complexity of the STARLAB vehicle in terms of the number of individual loads, it is considered advantageous to transform the output of the primary converter into a small number of standard distribution voltages. The voltage and frequency of these standard values are regulated by the central power subsystem of the spacecraft in such a way that the majority of loads can be supplied by further simple conversion in the load circuit. The electrical power of the space station is distributed in a way which (1) has maximal efficiency and (2) has flexibility such as providing the several voltages needed to perform the functions on the space station. Figure V-10 shows an arrangement of the power distribution system with the two power sources, solar panels, and nuclear reactor generation interconnected to supply energy to the space station. The output voltages of the solar panels and the dc converter of the nuclear reactor generator system are not constant. The voltage regulator following the solar panels and the dc converter maintains the input voltage to the booster regulator at a level voltage, which is anticipated to be about 42 v. A commonly used voltage regulator system is shown in Figure V-11(a). This is a servo system whose details of operation are readily available (Ref. V-7). The purpose of the "booster regulator [Fig. V-11(b)] is to maintain the output voltage at a predescribed regulated level voltage, where $vmin < v_{R_1} \leq v$, where vmin is a lower limit voltage which is a function of the regulator parameters.

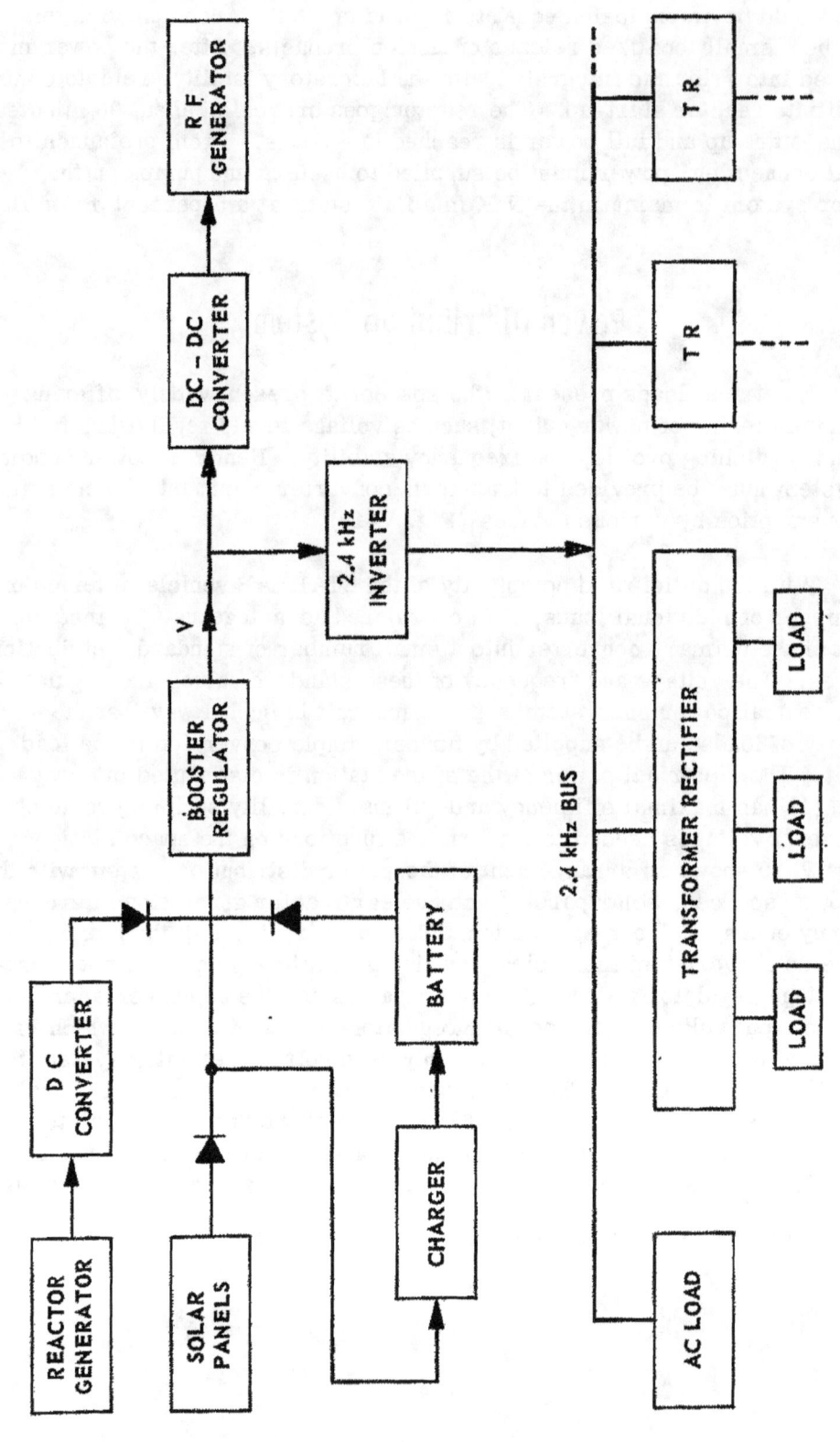

FIGURE V-10. POWER DISTRIBUTION SYSTEM

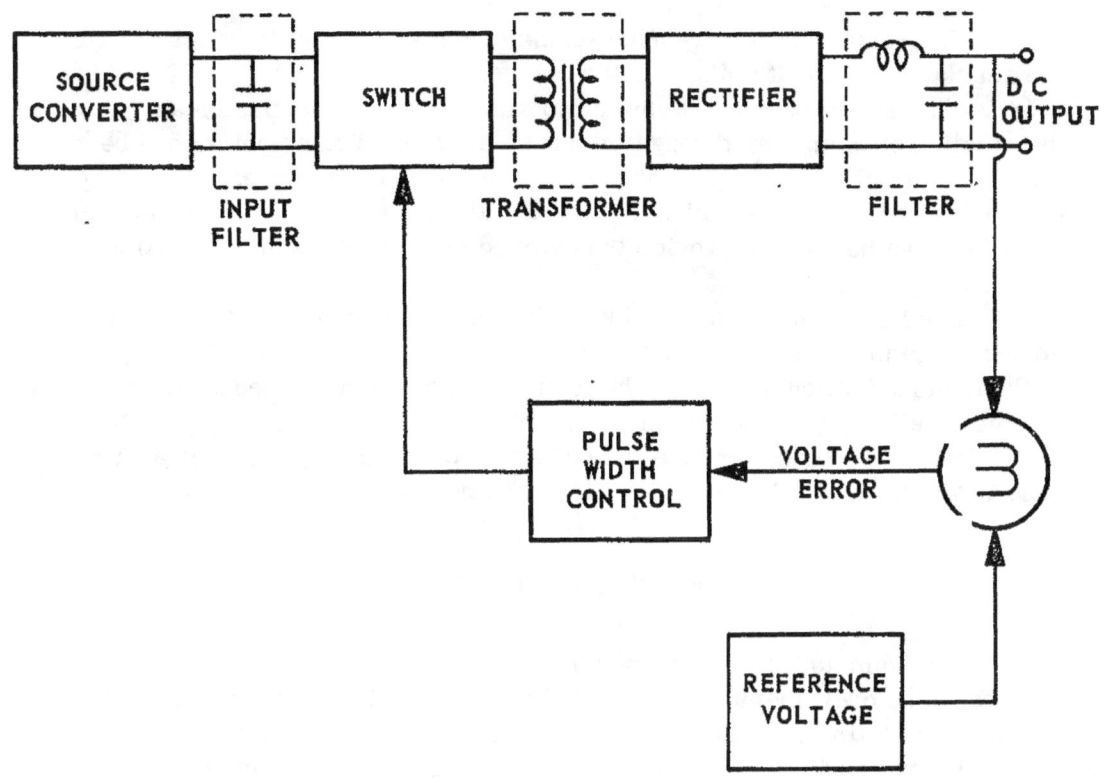

FIGURE V-11(a). VOLTAGE REGULATOR FOR THE
STARLAB POWER SYSTEM

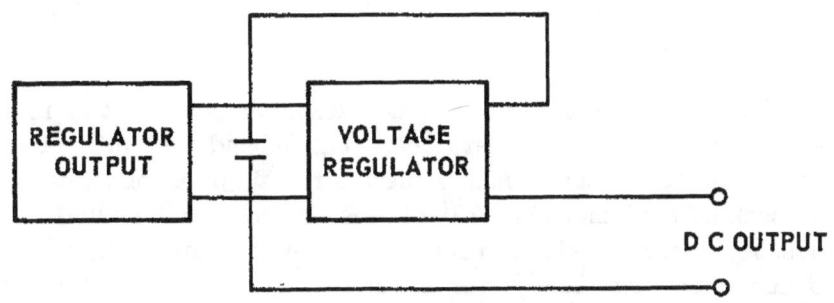

FIGURE V-11(b). BOOSTER REGULATOR FOR THE
STARLAB POWER SYSTEM

5-41

The remaining power-conditioning devices (Fig. V-10) operate from the regulated dc output voltage of the booster generator. The 2.4 kHz inverter is used to convert dc into ac to simplify the transmission and distribution of the power through the space station. The ac power can be distributed in either a single-phase or three-phase arrangement depending on the desired voltage flexibility. Suitable arrangement of transformers and rectifiers can be used to provide the desired voltages through the station.

The RF generator driven by the high dc voltage from the dc-to-ac converter can be a large load such as an induction heater of the RF amplifier of the communication system. The load-sharing control is necessary to avoid draining the battery system when the sources are capable of supplying the entire load. The diodes isolate the battery, solar panels, and nuclear reactor generator system so that no energy is consumed by any of the sources.

Onboard Communications

The communications subsystem is designed to provide excellent service to the crew members of the STARLAB vehicle. Communications equipment is intimately tied to the computer to provide programmed channeling of all information transmission. The main communications link between STARLAB and STARAD is via satellite and is provided by the integrated laser subsystem. It provides for voice and video communication, telemetry, command, tracking, and ranging. A laser system was chosen because it provides the wide bandwidth required to support the information flow associated with experiments performed in astronomy, earth resources, manufacturing, and biosciences. This is done within the weight and power limitations imposed on a space flight system. Further, it avoids conflicts of use of the frequency spectrum and meets with Federal Communications Commission's approval. It inherently is a secure system since the laser beam illuminates only a small area on earth. An additional reason for choosing the laser system is to encourage the development of the high potential available in this type communications system.

A unified S-band system is used as a backup system. It is also used for course pointing of the laser apertures. The S-band system is of the Apollo type and it was chosen as the backup system since the technology has already been developed and equipment is available that is more or less "off-the-shelf." The radar subsystem is used for docking and tracking remote flying modules which are unmanned. The low-power C-band subsystem is used for voice and video communications, telemetry, command, tracking, and ranging with the remote flying modules. This is a low-power system which will not affect ground links utilizing the same frequency spectrum. The distribution subsystem is used to tie the terminals and display consoles into the information exchange system. The information exchange is similar to a telephone exchange which provides a channel for communications.

NAVIGATION, GUIDANCE, AND CONTROL

Navigation

The tracking and orbit determination of the spacecraft can be accomplished with either an earth-based system and an earth-based computer, or with a complete onboard system. The use of an onboard computer for trajectory determination and corrections does not appear feasible since weight and power considerations for onboard computers make it impossible to provide a capability equal to that for an earth-based computer. There will be several RFM's operating in conjunction with STARLAB. Navigation capability for rendezvous, docking, and other maneuvers associated with the RFM's must be available on STARLAB along with computer monitoring of the orbit and operational status of the RFM's. The computer monitoring should include a periodic-automatic checkout of the RFM systems and appropriate display of any present or potential problem areas.

STARLAB will be a semipermanent installation and thus a stationkeeping requirement exists to prevent orbit decay. The orbit will be periodically updated through the navigation system and the onboard Service Propulsion System. Based on the requirements to update the orbit as determined by the navigation system, commands are relayed to the STARLAB and stored in the central computer and sequencer. At appropriate predetermined intervals, these commands are sent to the attitude control system which positions STARLAB to the maneuver mode and aligns it so that the thrust vector points in the direction of the required ΔV. With STARLAB oriented in this manner, the appropriate engine is fired and continues to burn until an accelerometer-time response indicates that the desired magnitude of ΔV has been achieved. At this time, the engine is shut off and STARLAB is returned to its normal cruise mode by the attitude control system.

Guidance and Control

Attitude control is defined as the attempt to constrain the attitude of the vehicle to a desired orientation. This implies that the attitude control system must be able to generate the necessary restoring torques whenever the vehicle's attitude deviates from the desired orientation.

The four major problem areas for attitude control of an earth-orbiting vehicle are (1) choice of attitude reference axes and the dynamics of motion relative to these axes, (2) analysis of the disturbing torques acting and the orbiting vehicle, (3) sensing methods, and (4) control actuation methods. Attitude control systems analysis must then trade the various factors against one another. Attitude control concepts can generally be classified as passive, semipassive, or active. Passive control systems make use of either spin stabilization techniques or environment fields such as gravity, solar radiation, pressure, or magnetic fields. Active control systems use either external control moments or momentum storage devices, or both. Semipassive control systems make use of both passive and active control techniques. For use with STARLAB, pointing requirement, response time, and docking considerations will dictate the use of an active attitude control system. Several active systems have been considered and some of these are discussed in following sections. These systems can be broadly classified as momentum exchange devices and external moment systems.

Vehicle Orientation

Several vehicle orientations were considered for STARLAB (Figs. V-12 and V-13). Basically, these are either earth fixed reference modes or solar reference modes. Some of the characteristics of these reference modes are tabulated in Table V-4.

The configuration, power generation system, and experiment program for STARLAB greatly influence the operational modes. The dynamics of vehicle motion for a given orientation and orbit are determined from the configuration characteristics. These parameters dictate the size and propellant requirements for the attitude control system and service propulsion system. Solar panels must be aligned perpendicular to the sunline for efficient operation. This requires either a complex gimballing system for the solar panels or a severe restriction on the orientation of the entire vehicle. The experiment program is the prime justification for STARLAB

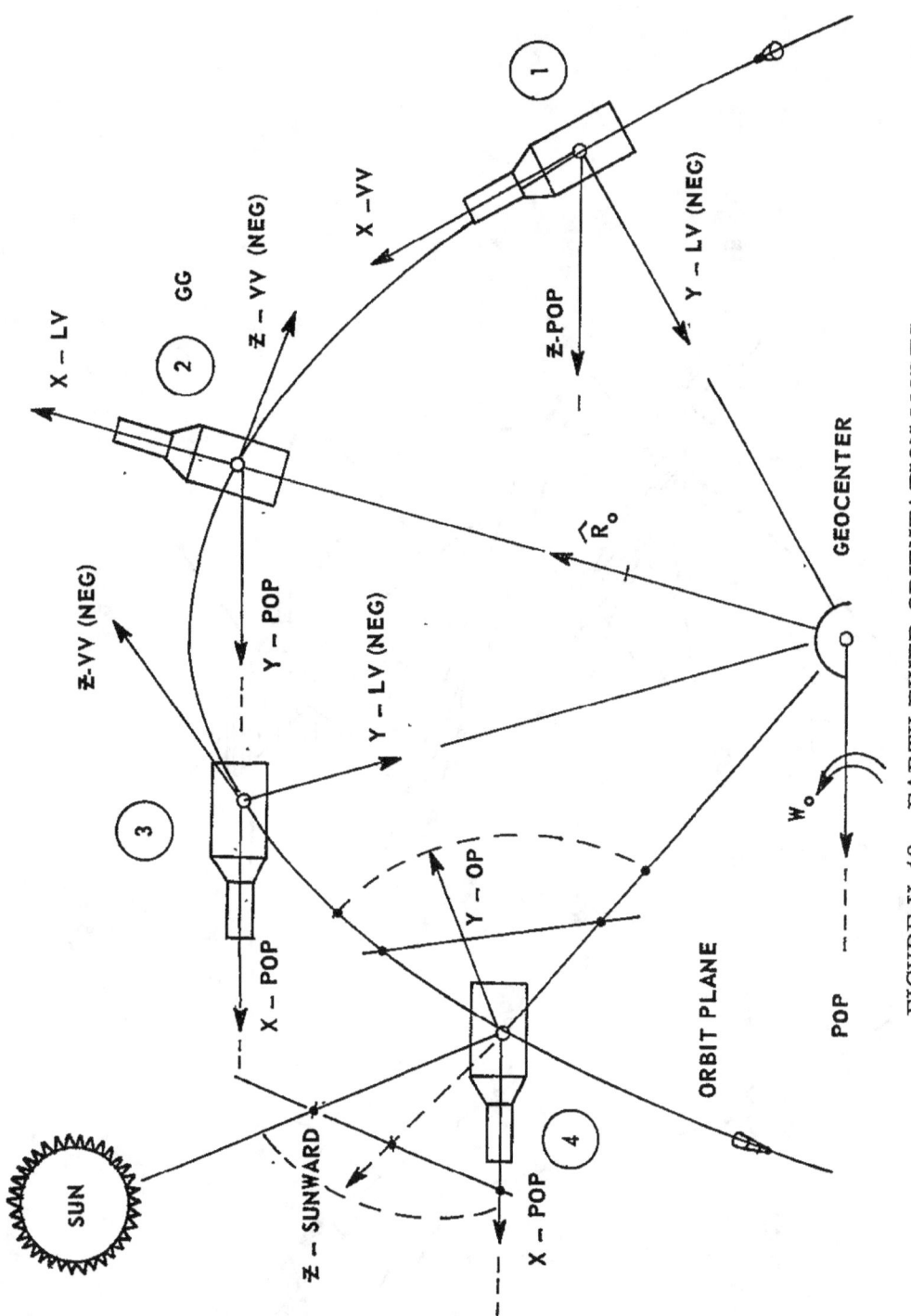

FIGURE V-12. EARTH FIXED ORIENTATION MODES

5-45

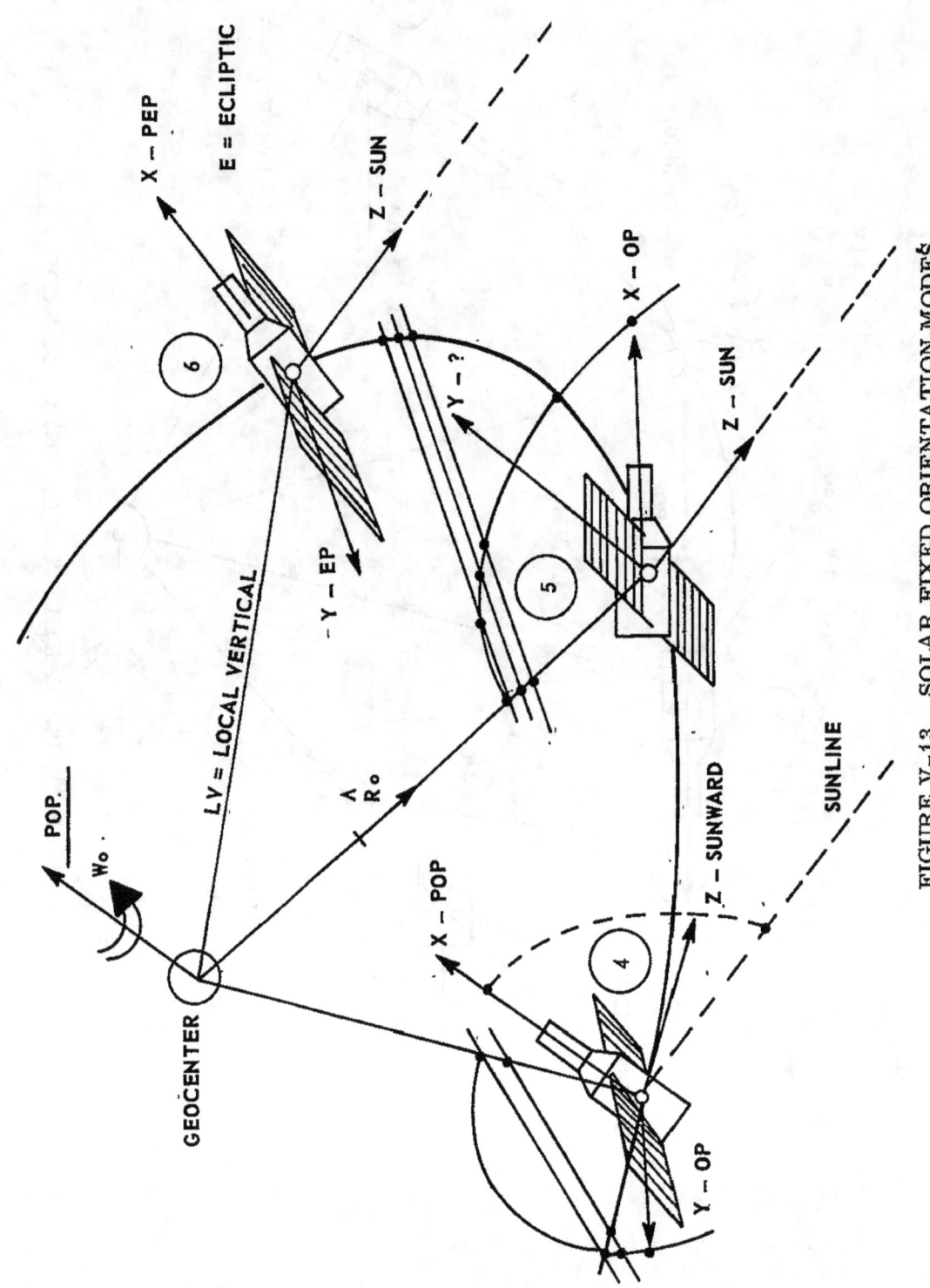

FIGURE V-13. SOLAR FIXED ORIENTATION MODES

5-46

TABLE V-4. CHARACTERISTICS OF ORIENTATION MODES

Characteristics of Earth Fixed Modes
• Two axes are in the orbital plane, one axis is POP.
• The vehicle is in an equilibrium position when the principal axes are aligned with the LV coordinates. The GG torques are zero.
• Axes misalignments such as control-principal and principal-reference misalignment produce attitude errors about equilibrium positions — GG torques build up.
• Attitude errors about an axis aligned with the LV do not produce GG torques. Hence, docking should be along an axis perpendicular to the LV such that any change in principal axes is equal to a rotation about the LV.
• GG torques due to attitude errors are minimized by aligning the axis of intermediate inertia with the LV.
Characteristics of Solar Modes
Mode 4: Two axes are referenced to the orbital plane. The vehicle is in an unstable equilibrium position which minimized GG torques. Only 1 DOF is required for solar panel or experiment module pointing.
Mode 5: One axis is restricted to the orbital plane which results in much less GG torques than an inertial mode. Gimbals are not required for pointing of solar modules. The vehicle angular rate commands are about 6 deg/day for orbital regression and 3.5 deg/day for solar look-angle variation.
Mode 6: All three axes are referenced to the ecliptic plane for the solar inertial mode. GG torques attain their maximum values with at least one biased component.

and requirements dictated for support of this program must be satisfied whenever possible. The orientation requirements desired by individual components are conflicting so tradeoffs are necessary to establish the most acceptable overall system.

Disturbance Torques

The principal disturbance torques that operate on an earth-orbiting vehicle are (1) magnetic, (2) solar pressure, (3) aerodynamic drag, and (4) gravity gradient. Magnetic torques arise from the interaction of the body of the station and its magnetic field with the earth's magnetic field. Solar pressure results from the particle flux from the sun and is parallel to the sunline. Aerodynamic torques result from the vehicle passing through the remnants of the atmosphere found at the altitude under consideration. Gravity gradient torques arise because of the variation in the earth's gravitational field with altitude.

For the altitudes of interest for STARLAB, earlier studies have shown that gravity gradient torques will be the dominant environmental forces acting on the vehicle and for preliminary design purposes will determine the control system energy requirements for attitude control. The combined effect of the other environment torques is only about 10 percent of the gravity gradient contribution.

Momentum Exchange Devices

High-pointing accuracy and certain long-time duration missions make control systems attractive by utilizing momentum exchange devices. The principle-of-momentum exchange is based on Newton's second law of angular motion which states that the total external momenta, $\overline{M}_{ext}$, acting on a system, is proportioned to the time rate of change of angular momentum with respect to inertia space and can be written as

$$\overline{M}_{ext} = \left\{ \frac{d \overline{H}_{system}}{dt} \right\} \qquad (V-1)$$

If the system consists of two momenta, $\overline{H}_{vehicle}$ and $\overline{H}_{control}$, then integrating equation (V-1) yields

$$\int \overline{M}_{ext} \, dt = \overline{H}_{ext} = \overline{H}_{control} + \overline{H}_{vehicle} - \overline{H}(o)_{system} \qquad (V-2)$$

which provides the equation defining the principle of momentum exchange. If the controller does not expel mass, then changes in controller momentum can be used to balance external torque as well as change the spacecraft attitude by varying the vehicle angular momentum.

Momentum exchange is desirable for the following reasons:

1. Continuous control.

2. Efficient control of cyclic disturbances.

3. Ease in management of propellant expulsion for continuous disturbances.

Continuous control results from the continuous exchange of momenta between the vehicle and the control system rather than a periodic exchange such as obtained from a pulse-modulated reaction control system. It is obvious that the momentum exchange device can handle cyclic torques on a continuous basis over long time periods without propellant expulsion. Biased torques applied to the vehicle will cause the controller to eventually reach its maximum capacity and thus saturate the momentum exchange device. This saturated condition requires the expulsion of propellants to remove some momentum from the momentum device; however, this propellant expulsion can be arranged at times in the mission profile when other activities are at a minimum or when pointing requirements for the vehicle can be relaxed. It is possible to economize on propellant expulsion in some cases if pointing requirements for the vehicle can be relaxed for a portion of the orbit and utilize environmental torques to partially desaturate the momentum device.

The CMG is basically a wheel that rotates at a constant speed which provides a constant angular momentum magnitude capable of variable orientation relative to the spacecraft. A moment is imparted to the vehicle by causing a change in direction of the constant momentum magnitude. A single constant momentum magnitude device is not usable since equation (V-2) shows that this is capable of producing only a change in direction of the vehicle's angular momentum. It is necessary to have two or more CMG's to effect both a magnitude and orientation change on the vehicle. With two CMG's, each constant in magnitude and both with two degrees of freedom, it is possible to control the magnitude and direction of the momentum vector in three degrees of freedom.

The maximum momentum possible in the system is obtained when all CMG momentum vectors are parallel and thus momentum exchange is possible for all CMG cluster momenta within a given momentum volume. The shape of this volume depends on the particular CMG cluster configuration and the condition of saturation is the surface of this volume, thus a momentum magnitude outside of this volume is unattainable. Typical state-of-the-art physical characteristics and performance characteristics of a 2000-foot-pound-second CMG are presented (Tables V-5 and V-6). For large vehicles where several such CMG's are required, the weight and electrical power requirements become rather large.

One method of CMG desaturation is to use RCS propellant in which case fuel weight may pose a problem for long lifetime missions. A more practical method of CMG desaturation is to interchange momentum with the spacecraft's natural environment forces. Such an interchange requires that the attitude pointing requirements be relaxed during the desaturation period; however, the total energy requirements for long lifetime missions can be minimized.

An attitude control system with CMG actuators in a standard or typical configuration uses CMG's without undue complications and is basically a position servo with attitude rate stabilization. Some of the major functions of the control system are:

1. Position sensor.

2. Vehicle control law.

3. CMG control law.

4. CMG configurations.

5. Vehicle dynamics.

6. Momentum management.

A block diagram of the standard control system is shown in Figure V-14.

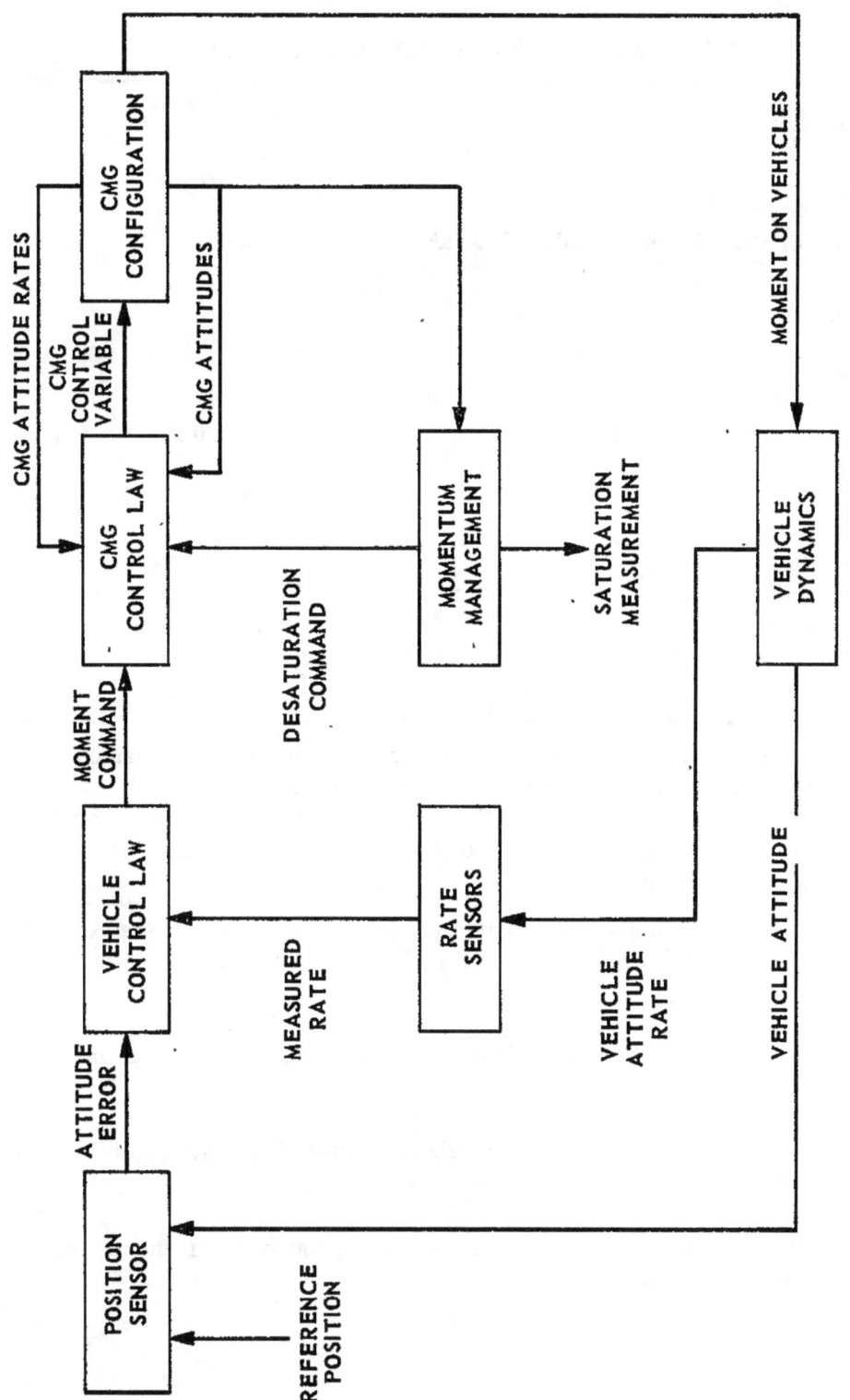

FIGURE V-14. STANDARD CONTROL SYSTEM BLOCK DIAGRAM FOR CMG SYSTEM

TABLE V-5. CMG PHYSICAL CHARACTERISTICS

Size (Approximate)	
CMG	39 by 39 by 40 in.
CMG Electronics Assembly (CMGEA)	8.6 by 9.8 by 3 in.
Weight	
CMG	
Inner Gimbal and Wheel	242.0 lb
Outer Gimbal	75.3 lb
Frame	94.7 lb
CMGEA	8.2 lb
Total	420.2 lb

The position sensor, which is an optical device, will measure the attitude error between the actual vehicle attitude and the desired attitude and this error is combined with the vehicle angular rate in the vehicle control law. The vehicle control law is basically a proportional plus differential controller which determined the desired command moment, $\overline{M}_{com}$, to be applied to the vehicle. Depending on the mode of operation of the vehicle, the position sensor for STARLAB will include one or more of the following sensors:

1. Sunfinder — Locates direction toward sun from any initial orientation.

2. Sunsensor — Indicates angular deviation from the direction toward center of the sun.

3. Planet sensor — Indicates angular deviation from the direction toward center of radiation of a planet.

TABLE V-6. CMG PERFORMANCE CHARACTERISTICS

Momentum Storage	2000 ft-lb-sec
Degrees of Freedom	2
Life	10 000 hr at run speed
Output Torque	
Range	0 to 120 ft-lb
Threshold	0.25 ft-lb
Gimbal Rates	
Range	0 to 3.5 deg/sec
Threshold	0.0056 deg/sec
Gimbal Rate Servo	
Bandwidth	70 rad/sec
Phase Margin	70 deg
Gimbal Pivot	
Actuator (dc Motor) Torque (max)	7.0 ft-lb
Tachometer Gain	1 vdc/rad/sec
Tachometer Amplifier Gain	1.013
Gear Ratio	56.55:1
Resolver Speed	Single (4.8 kHz Excitation)
Wheel	
Operating Speed	7850 rpm
Acceleration Time	9 hr Maximum
	7 hr to 90% Full Speed
Deceleration Time	4 hr
Power Requirements	
Bearing Heaters	
-60° F to 50° F	240 w
50° F to 70° F	48 w
During Wheel Runup	170 w
Wheel at Run Speed	56 w
Gimbal Freedom	
Inner	
Electrical	± 75 deg
Mechanical	± 80 deg
Outer	
Electrical	+215 deg, -125 deg
Mechanical	+220 deg, -130 deg
Gimbal Caging Positions	
Outer	0, 45, and 90 deg
Inner	0 deg

4. Horizon seeker — Indicates angular deviation from the direction toward the true center of a planet.

5. Star sensor — Indicates angular deviation from direction toward a star.

The command moment is processed by the CMG control law and the output of this law provides the input command to the CMG gimbal servos.

The command variable to the CMG servos drives the individual momentum vectors of each CMG and this change in angular momentum causes a reaction moment on the vehicle.

The reaction moments on the vehicle are forcing functions which act on the vehicle dynamics and change the vehicle attitude and attitude rate, thus closing the control loop.

The individual properties of the CMG control law and CMG configuration can vary radically, depending on the applications.

Momentum management is the method of maintaining the total momentum of the momentum controller within its momentum volume and is subdivided into the following three functions:

1. Saturation measurement.

2. Desaturation computation.

3. Desaturation implementation.

Saturation measurement involves the measurement of the controller momentum components along a given reference coordinate system (usually the vehicle principal control axes). Combining these components provides for the determination and display of the state of the momentum controller relative to the controller momentum volume.

Desaturation computation contains the logic and computations necessary to affect the desired momentum state to which the controller is driven and the determination of the intial and final desaturation commands. Desaturation implementation is the actual mechanization of forcing the momentum controller to attain the momentum state determined by the desaturation computations.

Flywheel attitude control is another momentum exchange method and employs the reaction torques of accelerated flywheels as a restoring agent for the space vehicle. The system consists of three cylindrical flywheels, one in each of three mutually perpendicular axes. Each flywheel is powered by a separate motor.

There will be large quantities of water and other liquids on board STARLAB. It may be practical to utilize these liquids in a momentum exchange device. The liquids can be pumped in paths around the three-vehicle axes, with the pumping velocity as a variable. Since angular momentum of the complete system must be conserved, an angular momentum can be imparted to the vehicle.

Reaction Control System (RCS)

The same basic equation, equation (V-2), governs the RCS. In this case, the momentum required is supplied by a jet system ejecting mass from the vehicle. Two basic types of pulse-modulated jet control systems are possible. One system uses jets that are pulse-rate modulated, and the other uses jets that are pulse-width modulated. Each of these methods can be optimized with respect to fuel consumption.

The typical arrangement of attitude control nozzles consists of four nozzles on each of three axes. Torques generated by the gas jet system are nominally applied to the vehicle as pure torques. With 12 nozzles, there are a total of six possible couples, one plus and one minus about each of the three control axes. One of these thrusters will belong to one system, and the other will be part of an identical system. In case of failure of either jet to operate, the remaining jet will provide torque (plus a force) although at half the design value.

Actually, two separate attitude control systems are required if a pure RCS system is used for attitude control. One having relatively high thrust level engines for maneuvering, docking, and change of orientation where fast response and turning rates may be desired and a low level thrust system for maintaining orientation where low burning rates would be desirable.

STARLAB is to be a semipermanent installation so provisions must be made for maintenance, repair or replacement of the attitude control thrusters. This should be done without EVA if possible. Procedures for accomplishing this task have been considered previously (Ref. V-18). A standard contro system for an RCS system is outlined in Figure V-15. This system is simpler than that for a CMG system since no momentum management and saturation measurement must be considered.

Control Considerations

In all these methods, the system begins to compensate when a certain deadband is exceeded. The "on-time" required to compensate a disturbing impulse can be determined either by the angular velocity of the vehicle which exists at the time of motion activation or by employing instantaneous attitude and velocity information.

All of the methods eventually require the expulsion of mass from the system to maintain the attitude control system in an operational mode. The attractive feature of the momentum exchange systems is that this mass expulsion can be scheduled at convenient times and slightly better pointing accuracy of the vehicle is possible. With the momentum exchange devices, it is also theoretically possible to use environmental gradients to desaturate the CMG's. However, this concept still requires development and the extent of the application to STARLAB is not well defined. The major disadvantages of a CMG system is weight, electrical power requirements questionable reliability and maintainability over a long time period, and the necessity for relaxing the pointing requirements for the vehicle during a portion of the orbit if any significant propellant economy is to be obtained.

Using a CMG attitude control system, it will be possible to maintain STARLAB to ±0.5 degree about all three axes. A reaction control system can reasonably expect to maintain STARLAB to ±1.5 degrees. In any case very stringent requirements for experiments will have to be supplied by having the experiment mounted on its own stable platform.

Service Propulsion System (SPS)

An SPS must be available for maneuvering, docking, and station keeping. This system will operate in conjunction with the central onboard computer and the ground-based navigational system. Propulsion thrust-to-weight ratios of less than 1×10^{-6} are generally adequate for these operations for orbits above 200 km.

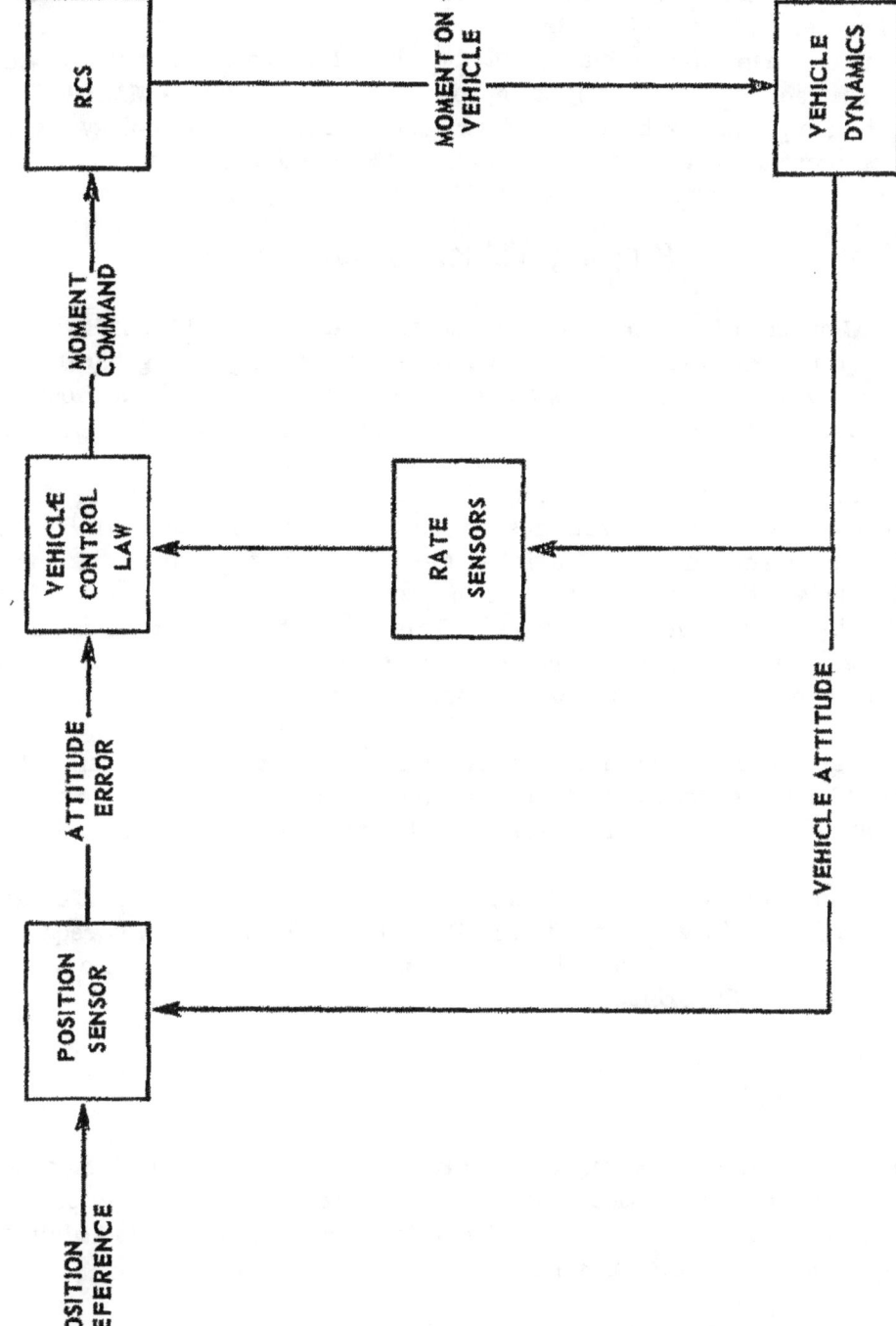

FIGURE V-15. STANDARD CONTROL SYSTEM BLOCK DIAGRAM FOR THE RCS SYSTEM

The propellant systems for STARLAB must be safe, reliable, possess good long term storage characteristics, and be easily refueled. Cold gas systems can meet these requirements but their low specific impulse make them unattractive. At present, a bipropellant system such as MMH + N_2O_4 might be the prime candidate, however, advancements in technology may produce more attractive systems before STARLAB is operational.

Proposed Vehicle Orientation

Consideration of possible attitude orientations for STARLAB was based upon requirements to support the experiment program, power generation, logistics, propellant requirement to maintain orientation, and other operations. Therefore, operational requirements make it necessary that STARLAB operate at various times, in four attitude control modes. Three of these orientations are earth oriented and one is solar oriented. These orientations are shown in Figure V-12. The first and most frequent mode of operation will be X-POP, Y-LV. Here the long axis (X-axis) of the vehicle is perpendicular to the orbit plane and the vehicle is rotated one revolution per orbit to maintain the Y-axis aligned with the local vertical. This orientation allows the earth resources experiments a continuous ground view and is also attractive from propellant requirement and docking considerations.

The second orientation is the X-POP, Z-sunward mode. This orientation may be necessary for certain onboard experiments and will also be important when power from the solar panels are important considerations.

The third mode of operation is a maneuver required for orbit updating. In this mode, the long axis of STARLAB is aligned in the direction required to update the orbit. The service propulsion is then fired to yield the desired ΔV for updating the orbit.

The attitude control for these three orientations will be an active control system.

The fourth orientation is a gravity gradient mode with the long axis aligned with the local vertical. This will be utilized if STARLAB is placed in a dormant or storage mode. Here, the attitude control is passive utilizing gravity gradient torques for stabilization.

Proposed Attitude Control System

Many control system studies have shown that from standpoints of economy and control accuracy, the CMG system offers many desirable features for stabilization of manned space stations. This is particularly true when a relatively high-accuracy attitude stabilization system is required. However, for a very large semipermanent vehicle, such as STARLAB, there are problems associated with a CMG system. These problems become apparent when consideration is given to the physical characteristics of the system required for STARLAB.

Using outlined equations and procedures (Ref. V-19 and V-20), estimates were made of the CMG system and RCS propellant required for STARLAB in the X-POP, Y-LV mode. These calculations are only order of magnitude calculations since the physical and dynamic characteristics of STARLAB are not well defined. Only conceptional designs of the system have been completed. The results of the calculations have been tabulated (Table V-7) for a 300-mm, 70-degree-inclination orbit.

The CMG system and propellant requirements are relatively small if body axes of the vehicle coincide with the principal axes and sensor axes are exactly aligned. However, there are certain errors in both knowledge of instantaneous principal axes (docking, movement of personnel and equipment, use of expendables, etc.) and the orientation of sensors. Reasonably expected errors of this type introduce biased perturbations into the system and must be considered in sizing the attitude control system. These misalignments greatly increase the size of a CMG system and the propellant requirements.

The CMG systems are ideal for handling torques of cyclic nature and lead to considerably smaller RCS propellant requirements. However, for biased type torques, CMG systems yield no propellant savings unless the pointing requirements of STARLAB for a portion of each orbit can be relaxed and utilize gravity gradient torques for desaturating the CMG system. This operating procedure would probably be undesirable for STARLAB. In this case, the major advantage of the CMG system is scheduling of RCS burn at convenient times when other activities are at a minimum.

For a reasonable expected misalignment of axes and sensors, the propellant requirements and requirements for a CMG system for STARLAB increase drastically. Calculations were carried out for a 5-degree misalignment (Table V-7). Based upon these calculations and present technology, a CMG system for STARLAB would weigh approximately 18 000 pounds and

TABLE V-7. SUMMARY OF PHYSICAL CHARACTERISTICS OF STARLAB ESTIMATES OF PROPELLANT REQUIREMENTS AND REQUIRED CMG SYSTEM

PHYSICAL CHARACTERISTICS	
Weight (lb)	500 000
Volume (ft^3)	225 000
I_{XX} (slug-ft^2)	1.86×10^6
I_{YY} (slug-ft^2)	5.86×10^8
I_{ZZ} (slug-ft^2)	5.86×10^8
I_{SP} (sec)	220
L_X (Moment Arm, ft)	17
L_Y	160
L_Z	160

SYSTEM REQUIREMENTS		
	No Misalignment of Axes	5-deg Misalignment
Propellant Weight (lb/orbit)	16.4	2.4
CMG Weight (lb)	3000	18 000
CMG Electric Power (kw)	0.5	2.5

require 2.4 kw of electric power with no essential decrease in RCS propellant requirements. There is also some question about long-term reliability and maintainability of CMG systems. The only advantage of the CMG system in this case is slightly better pointing accuracy for STARLAB.

A CMG system does not appear attractive for initial operations of STARLAB. A simple RCS system is initially proposed for STARLAB. However, after STARLAB is operational, it is also recommended that a study program be instituted to determine more about the principal axes of the vehicle under various operational conditions and proper alignment of sensors. For such a study, a scheme for maintaining a favorable relationship between principal axes, vehicle axes, and sensor axes may be developed. Once this has been done, it may be possible to go to a CMG system of much smaller weight and power requirement which will yield better pointing accuracy and smoother operation from the attitude control system while minimizing the weight and electrical power penalties.

CMG systems may be attractive for some of the RFM's depending primarily on the printing accuracy required for the module and the availability of electric power for this operation.

ENVIRONMENTAL CONTROL AND HABITABILITY

Introduction

The objective of the environmental control and life support systems (EC/LS) is to produce and maintain an environment on STARLAB which has an optimum, not minimum, level of habitability. The crew will be unable to function at a high level of efficiency unless the environment is comfortable and convenient (Ref. V-21). Also, the EC/LS systems must be reliable. A person who is concerned about the quantity and quality of his next breath or his next drink of water is not likely to be able to function at a high level of effectiveness. Also, the operating crew for the station must be kept at a minimum to keep the number of scientific personnel as high as possible, the EC/LS will be automated as far as possible. Also, since the scientific crew members will be untrained astronauts, the backup systems must be highly automatic, to maintain confidence in the system. This is primarily true of the atmosphere mixture, oxygen supplies, and atmospheric pressures.

The EC/LS system must be able to maintain good living conditions for the normal crew of 35 scientific personnel and station operating staff plus any visitors who may be present for a short visit (i.e., shuttle pilots). The system will be resupplied at regular intervals, but must be capable of operation for at least 4 weeks if a resupply flight should be aborted. Also, STARLAB may have to supply service and power to RFM's.

Adequate emergency supplies of food and water will be stored throughout the station. Also, the atmospheric control system will have adequate redundancy to maintain a breathable atmosphere if the primary atmospheric system should fail, or if collision or meteorite penetration should cause loss of pressure and some of the atmosphere supply (Ref. V-22).

The system must be able to operate on the available power. Power is the primary constraint. The EC/LS systems have to be efficient users of power and also not place an excessive load on the logistics system. In the event of a main power source failure, the EC/LS systems must be able to function using only the solar-battery backup system.

The capabilities of the logistics system are described later in this report. The next sections describe the EC/LS systems that were selected and show how they are a compromise between available power, system requirements, and logistics system capabilities.

Laboratory Atmosphere

The air-conditioning system will be designed to handle each of the four modules individually that will be connected to form the habitable areas of STARLAB. This will then allow for redundancy during emergency operation in the event one or more modules are put out of operation caused by meteoroid penetration, fire, or are not in use.

Small air blowers (10 w, 65 cubic feet per minute at 0.2 inch H_2O Δp) will be used to draw air from each living area to the main ducts and across the heat exchangers. For the AAP proposal the heat exchangers will use "Coolanol-15" (Monsanto Chemical Co.) as the heat transfer fluid which pumps the fluid to the radiation heat exchanges that dump 15 000 Btu per hour, at less than $B = 0$ degree or 17 000 Btu per hour at less than $B = 78$ degrees with a pump power use of 240 w. These surface heat exchangers are calculated to remove 33 Btu per square foot-hour.

After part of the return air has been filtered, cooled, humidified or dehumidified, the air will be redistributed to the compartments. Return air from areas where odors will be most likely to be obtained will be passed through charcoal filters before being mixed with the other return air volume. Only 15 percent of the air will be filtered each pass because of a high power penalty. Fans to each room will be controlled by the room light switch and will shut off when the lights are turned off. Temperature override switch could be used to turn the fans back on for a high or low temperature condition. Results from the AAP will govern the design of STARLAB in this area.

The actual air environment composition will show 5 psia total pressure, 2.7 psia O_2 partial pressure, and 1.3 N_2 partial pressure. Room air will be designed for 70° F, 45° F dew-point minimum and below 95-percent RH. The water vapor pressure in the air will be controlled to a minimum of 8 mm Hg and a maximum CO_2 partial pressure of 7.6 mm Hg. Toxic gases such as CO will be kept below 0.76 mm Hg partial pressure or as normally specified.

Two sizes of cryogenic storage containers have been designed, tested, and/or proposed by McDonnell Douglas (Ref. V-23). The smallest vessel has an ID of 39 inches and the largest has an ID of 60 inches and operates at 1000 psia. The size and capacity of liquid cyrogenic storage vessels are given below.

Tank Size	O_2	N_2	Para H_2
39-in. ID	1200 lbm	850 lbm	75 lbm
60-in. ID	4500 lbm	3150 lbm	275 lbm

This system of storage vessels has an in-space heat leak of less than 9 Btu's per hour and a fluid boil-off rate of approximately 0.09 lbm per hour at equilibrium for the 39-inch-ID LOX storage vessels. The values were obtained by test measurements. An insulation for the ground hold condition is a closed-cell polyurethane foam, and a second insulation of multiple layers of randomly perforated, crinkled, aluminized polyester film of approximately 60 layers per inch is applied over the first layer of insulation for the low-vacuum space operation. The perforations in the insulation are to allow the hard vacuum to be obtained. Thus, the compartments where the cryogenics are stored will be open to space vacuum and closed off from the space laboratory environment. A recommendation is made that 60-inch-ID containers go up instead of the four large space ship launches, and that either the 60-inch-ID or 39-inch-ID resupply containers go up as needed and then be attached on the outside of the hull of the STARLAB for refill using a bellows type diaphragm operated by helium gas pressure.

Regulation of the oxygen pressure and the total pressure in the atmosphere of STARLAB will be through the use of pressure sensors where the total pressure is checked and if low, then the O_2 partial pressure is read and either O_2 gas or N_2 gas is added as indicated. This method can also be used in closed-off areas where other pressures may be required.

A red light system in the command module could operate showing when normal operating design limits were exceeded. If danger levels or near danger levels are obtained, a flashing red light and alarm could then be operated for the module or the STARLAB as indicated.

Food and Water

The initial food supply and the emergency food supply will be the Apollo-type food packages. These weigh approximately 1.4 lbm per person per day and thus can be stored compactly with little weight penalty. Water is necessary to be added to this type food. Apollo-type food will provide 2800 calories on a balanced-type diet.

Once the STARLAB becomes operational, food will be prepackaged on the ground, and frozen and brought to STARLAB at a rate of 5 lbm per man per day. This food will be thawed and cooked by infrared ovens. Because of the high percentage O_2 atmosphere, vitamin E may have to be specifically added to the frozen food and Apollo diets.

Body water must be supplied each day at a minimum rate of 4.4 lbm per person. Other water necessary for personal hygiene will amount to about 6 lbm per person per day. Therefore, it is a requirement that a closed water cycle be put in operation as soon as possible. Figure V-16 shows a studied method of waste water that is processed twice before being reused as drinking water. Control and check-point locations are necessarily located throughout the entire system.

To prevent any serious problems with the water systems, each of the four integral launch modules will have their own closed waste water systems with a direct on-off connection between each of the modules. This will allow a maximum safety for the system and a more uniform module construction

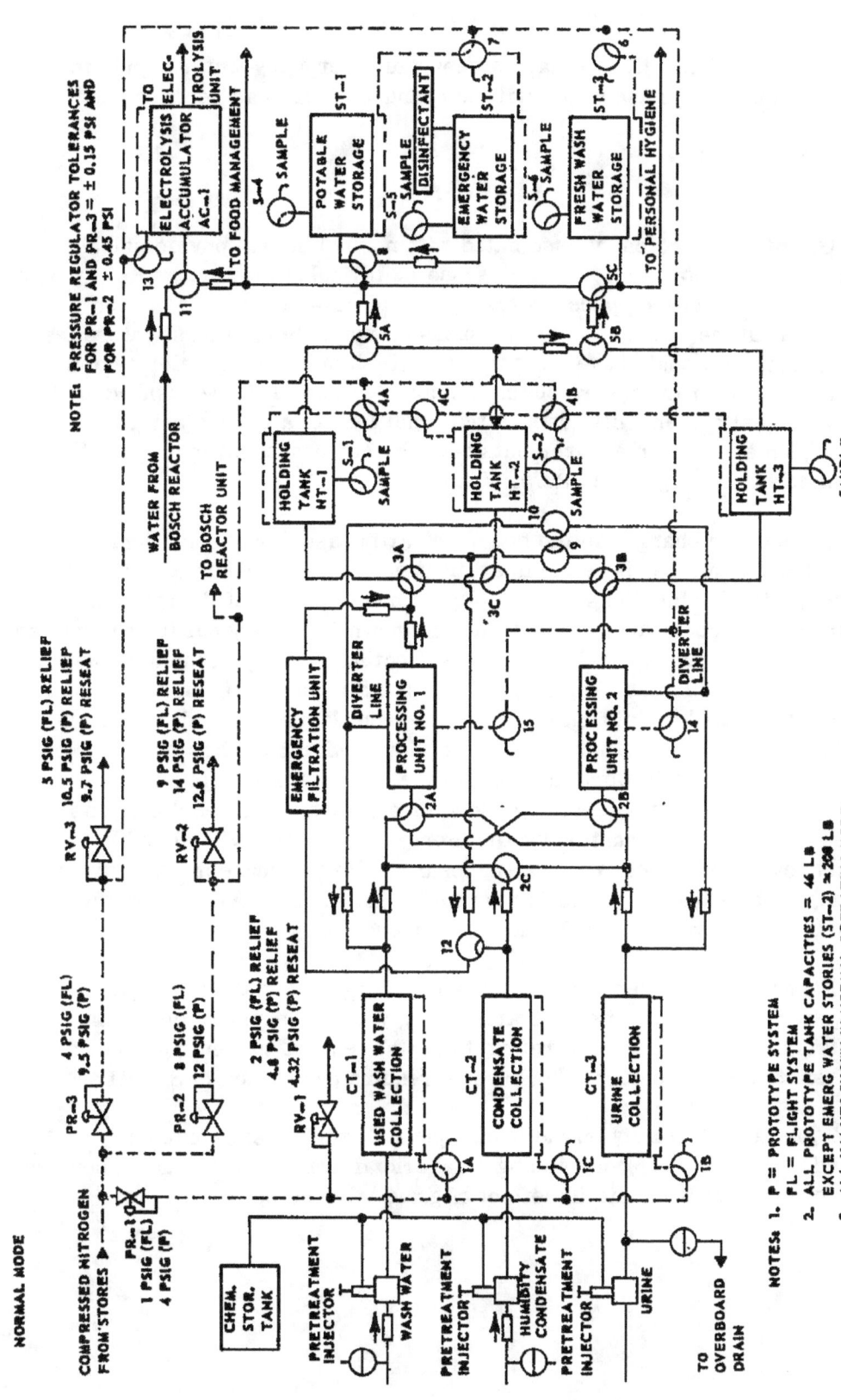

FIGURE V-16. INTEGRATED WATER MANAGEMENT SYSTEM REVISED SCHEMATIC (Ref. V-24)

design. Each module will also have a separate emergency water supply for both drinking and wash use which will be changed or used as necessary.

Gravity

Gravity is so strongly associated with normal human physiological activities that the total effect of its absence is difficult to evaluate (Ref. V-25). Since the inception of the space program, considerable concern has been expressed about the effect of weightlessness on man. Some investigators feel that weightlessness could have a significant effect on human physiology and limit the ability of man to work during a stay in space. It is well known that the physiological system will adapt to the conditions of a zero-g environment. This has been shown by measurements taken on astronauts after a space flight and by simulations on ground (Refs. V-26 and V-27).

Some of the changes in the body are cardiovascular deconditioning, decreased blood volume, bone demineralization, muscle denitrogenation, and disturbed vestibular functions. Probably the most serious of these is the cardiovascular deconditioning. It is not yet known if the deconditioning process levels off to a new plateau or if the process continues. The ground simulations indicate that a new plateau level is reached, but this has not been borne out by flight data. The longest flight to date has been Gemini 7 which lasted 14 days. It was not clear if a new level of cardiovascular condition had been reached. However, it is worth noting that the crew were able to return to a normal 1-g environment without any undue discomfort. The results of this flight led to the recommendation by Dr. Berry et al. (Refs. V-25 and V-28), that a man can safely stand a 30-day exposure in larger spacecraft. Further research, such as that proposed for the AAP (Ref. V-29), will determine if man can safely live in space for longer periods.

Based on the present recommendation, longer stay-times in space will require the provision of either artificial gravity or physical conditioning to maintain the ability of man to return to earth. Such a sudden return could be necessitated by an accident which would require evacuation of STARLAB.

The effect of gravity on the human physiological system can be simulated in two ways; (1) developing an artificial gravity or (2) using devices and exercises which simulate the effect of gravity on the body.

Artificial gravity by rotation of a vehicle is a concept that has been studied by many investigators. Continuous artificial gravity has the potential of eliminating the adverse effects of weightlessness on the crew. Personal functions such as washing and waste elimination are more conveniently carried out in a gravity environment. Also, physical conditioning would be automatic because of the gravity force field and would not require the additional crew time that the simulating systems need.

There are two systems to provide artificial gravity. Both systems use centrifugal force produced by spinning a vehicle as an analog for true gravity. One way is to spin the counterweight against the ship. It is often proposed to use a spent booster stage as the counterweight. Another method is to use a wheel and hub system so that the spinning wheel has artificial gravity while the central hub remains in zero g. The relationship between rotational rate, length of moment arm, and the gravity level is shown in Figure V-17.

Developing artificial gravity by rotation of a vehicle is unattractive from several standpoints. The first and most obvious are the engineering problems associated with spinning, stabilizing, and pointing on a rotating laboratory. Docking to a spinning ship could be difficult and would require that the ship be despun if it were configured using a spent-state counterweight. This would require large recurring expenditures of fuel to spin and despin the ship for each docking. It is anticipated that dockings would be frequent on STARLAB, therefore, the counterweight method did not receive any further consideration.

The artificial gravity system using a wheel turning on a hub requires large diameter seals and bearings to maintain the air-tight integrity of the vehicle. Balancing problems of the turning portion could be difficult to solve. There would be a weight penalty of at least 25 000 lbm for the spinning wheel (Table V-8).

A second limitation is the effect of rotation on the crew. A man moving on a rotating vehicle experiences forces and moments, the magnitude of which depends on his linear and angular velocity, rate of rotation, and the radius of rotation to his location. These forces which are developed as by-products of the artificial gravity will influence a man's performance in proportion to their magnitude.

An illustration of one type of effect can be shown by considering the effect on a man walking along the plane of rotation of the spinning ship. If he walks in the same direction as the spinning ship, he will feel heavier. If his walking

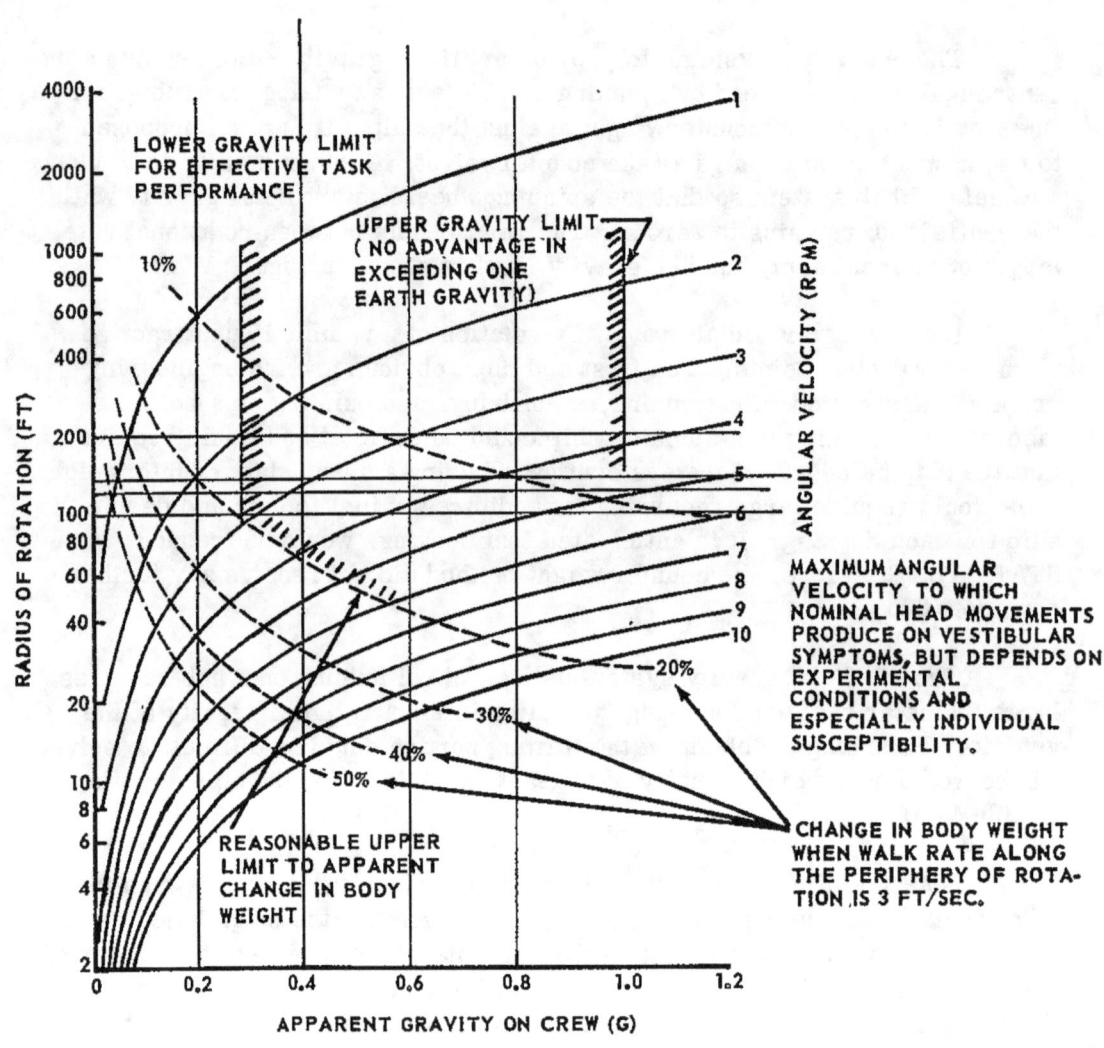

FIGURE V-17. PRELIMINARY TOLERABLE LIMITS FOR ACCEPTABLE HUMAN PERFORMANCE IN ROTATING SPACE SYSTEMS

TABLE V-8. WEIGHT PENALTY FOR ARTIFICIAL GRAVITY
OR DEVICES TO PREVENT PHYSIOLOGICAL
DEGRADATION IN ZERO-g

Device	Weight	Time	Remarks
Pressure Cuffs	20 lb/man	Worn Continually Distractive	Available
LBNP	20 lb/man	2-4 hr/day/man	May be Part of Biomedical Lab on Board
Centrifuge	12 lb/man	2-3 hr/day/man	2 Required (300 lb Each), One Will be Present for Biomedical Lab
Anti-g Suit	15 lb/man	None Donned Prior to Re-entry	Must Come Up With Each Crew Member
Artificial Gravity Wheel-Hub	25 000 lb	No Time	Only Living, Eating, and Waste Disposal

counteracts the velocity of the spinning section he will become lighter. Also, his effective weight will vary as he moves the spokes of the wheel toward the compartments on the end.

In addition to the physical influence of these forces, there are unaccustomed and often disquieting effects on the vestibular system. A complete analysis of these effects has been carefully documented (Ref. V-26). This analysis has shown that man can withstand a rotational rate of up to 4 revolutions per minute without any upsetting effects. At this speed, Figure V-17 shows that the radius of rotation must be about 110 feet to provide 0.6 g which is an acceptable level of gravity.

There are various exercises that can be carried out and devices that can be used to load the circulatory system to cause the heart to beat faster. Thus, the cardiovascular system can be kept in at a higher level of conditioning than would otherwise be possible. The weights and necessary times for some of these are summarized in Table V-8.

If the factors cited above are coupled to the fact that the bulk of the experimental program must be performed in zero g (the main reason for STARLAB), the decision can be reached to have a zero-g station. A zero-g station exposes some further constraints on the system that will be discussed in other parts of this report. However, crew stay-times will be dictated by the ability of man to live and work in space before he is deconditioned to the point where he cannot safely return to earth. Since the space shuttle will be available by the time STARLAB is launched, crew rotation on a regular schedule will be relatively easy. Therefore, for this early station, it was decided that STARLAB will be a zero-g vehicle and that the crew will be rotated often enough to prevent any long-term debilitating effects of weightlessness.

Extravehicular Activity (EVA)

Experience to date shows that EVA is very taxing on man. To have a successful EVA, extensive simulation and careful equipment development is required.

Since the experiment program is varied on STARLAB, the opportunity for detailed practice and simulation is limited. Accordingly, it is recommended that EVA be kept to a minimum. Small vehicles such as the space "taxi" are needed for handling packages and for transport around the exterior of the ship.

For EVA, an improved space suit is needed which will allow greater freedom and dexterity of movement with smaller exertion by the man.

Living Quarters

Highly motivated individuals such as the astronauts can live, work, sleep, and eat for days without leaving their seats. However, for more normal operations or for more extended trips, the various areas should be separated. In particular, the living and sleeping quarters should be separated from the work areas.

The general layout for the ship is shown in Figure V-18. The layout for the living quarters is shown in Figure V-19. Each floor has quarters for six crewmen and has a lounge and a bathroom facility. The floor plan shows that each man has a private room with 90 square feet of floor space. A 6-1/2-foot ceiling gives 586 cubic feet of volume that agrees closely with the

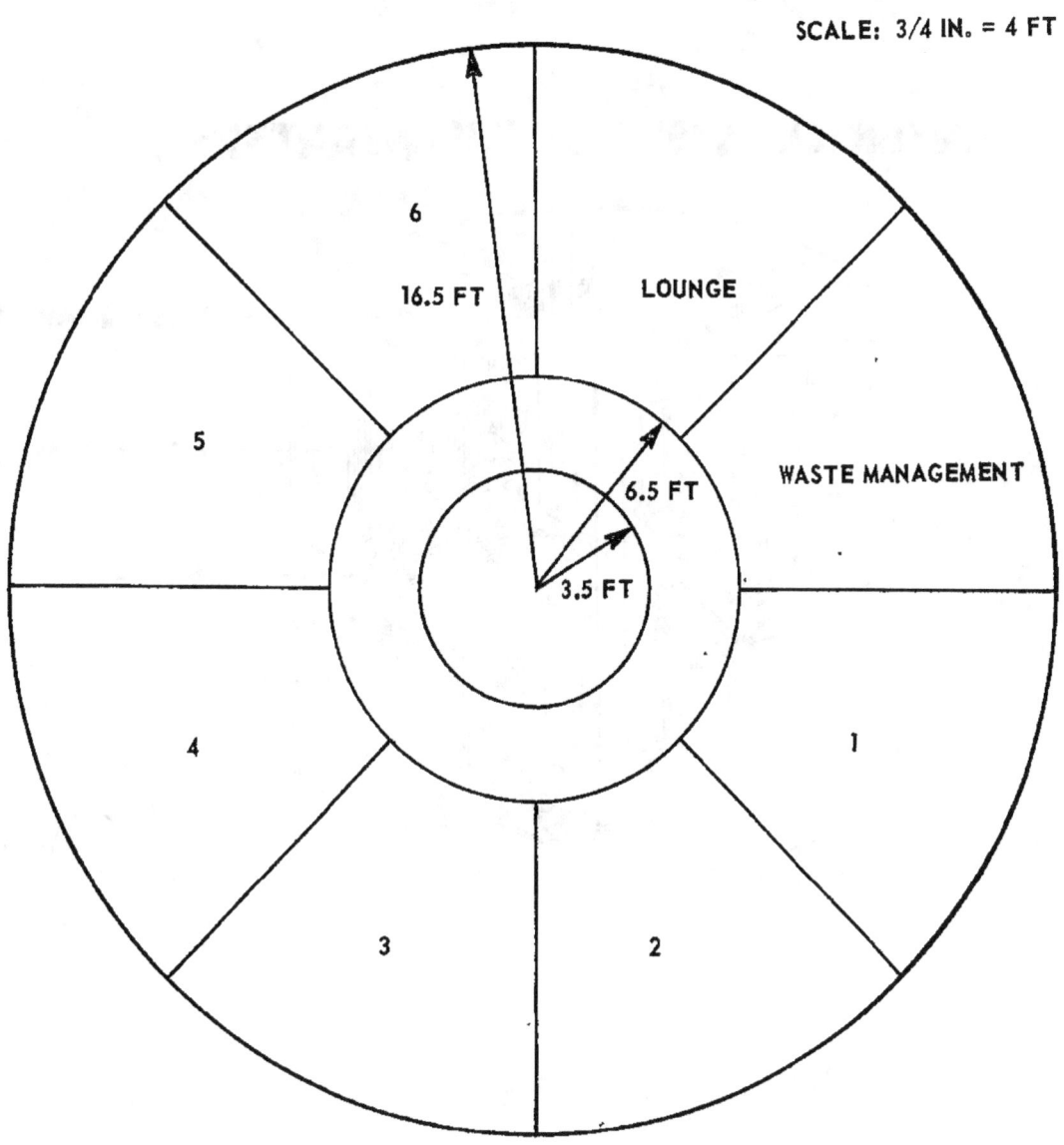

FIGURE V-18. FLOOR PLAN OF LIVING AREAS

5-71

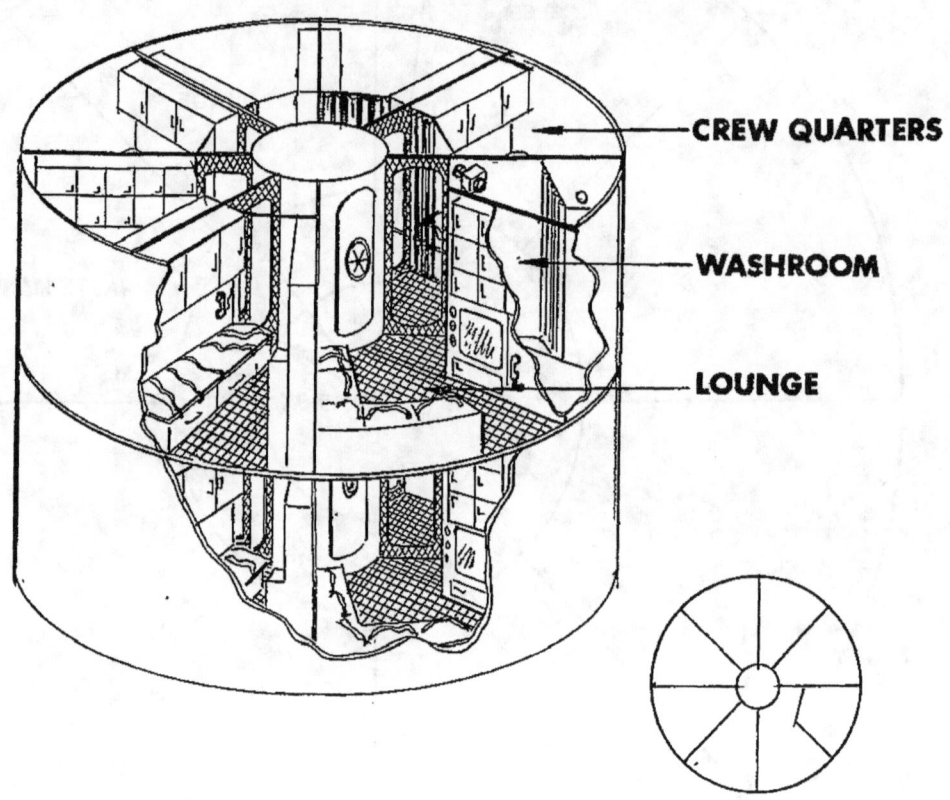

FIGURE V-19. INTERIOR VIEW — CREW QUARTERS

recommendation of 600 cubic feet per man (Ref. V-21). One floor is for the ship officers, the captain and the chief scientist, and any guests who may be on board.

The accommodations require a total of 8 floors to provide quarters for 45 people. It should be emphasized that the required volume for good habitability is reduced for short stay-times. Therefore, if the crew rotation has to be at frequent intervals, the living space could be compressed. More room is then made available for experiments. The living space is easily convertible because of a minimum amount of permanent equipment. Each room contains a personal communication center for the crewman. The crewman is able to talk directly to earth and use his TV for visual communication.

Each of the lounge areas will have view ports. These are arranged so that viewing of earth and stars is possible. It is believed that windows are necessary to prevent claustrophobic feelings and also to admit light.

Waste Management and Personal Hygiene

All work areas generate waste particles which are removed in the areas where produced. All areas of STARLAB should be provided with small hand-operated, battery powered, vacuum cleaners that are used to pick up small particles. Wherever particles are continuously produced, the process must also provide for their immediate removal.

Organic wastes accumulate from food management areas and from toilets. In each case, the organic material is stabilized and not allowed to produce bacteria. The most desirable means of waste collection from toilets is by an air-flow collection system. Wastes should then be dried in the collection bags and the water is not to be reused (Ref. V-24). The system provides for centrifugal separation of urine and feces which are collected by airflow through a permeable bag. A backup system of solid waste storage must be provided. The stored wastes from STARLAB will be returned to earth by the shuttle.

For optimal conditions, provision must be made for personal hygiene in various areas, namely:

1. Dental and oral.

2. Hand and face washing.

3. Body bathing.

4. Hair and nail trimming.

5. Shaving.

6. Clothing and laundry.

The methods for carrying out the above functions have not been well defined at present. It is anticipated that further developments will be available by the time a space station is ready to operate.

Movement and Restraints

Since STARLAB operates in zero-g condition continuously, restraints must be provided to enable man to remain at his station and work effectively (Ref. V-21). The wide diversity of functions and work areas on the ship makes provision of a common restraint system difficult. Each operation and area will be analyzed individually to determine the restraints best suited.

The central trunk will be the passageway for movement from one section of the ship to another. Adequate handholds must be available to make movement along the corridor fast and easy. The "fireman's pole" type of restraint may be suitable for that area.

Since the functions of an area may change, the ability to alter or redesign the restraint system is essential. However, it is not necessary that STARLAB have the manufacturing capability to make the new units. These can better be brought up from the ground.

Safety

On any space vehicle, provision must be made for the safety of the crew in the event of collision, meteorite puncture, high radiation event, or contamination. The following sections outline the provisions that have been made for STARLAB to cope with these emergencies.

In case of collision, the procedure is to evacuate the crew to the central trunk area. This is a "hard" area and has emergency life-support equipment. This area can be sealed off into compartments. From the trunk, the crew could move to a section of the ship which was undamaged. Then, the damage could be repaired, or the area closed off and abandoned.

A meteorite puncture has a low probability of occurring; however, the station must be capable of handling a puncture. A large puncture would be treated like a collision. If a small puncture occurred, the air-tight doors would be left open until the crew members in the damaged section could be removed, or until the pressure in the ship had dropped to the danger level. This time is on the order of several minutes for holes to 6 inches in diameter When the crew members are out of the damaged area, the doors to that area will be sealed. Then, the remaining crewmen return to routine operation and the damaged area could be repaired by a man in a space suit.

The radiation of most concern is from solar flares or proton events. The procedure in both cases is to turn the ship so that it is oriented with the reactor pointing toward the sun (Fig. V-20). The crew would then be shielded behind the reactor shielding. This maneuver can be accomplished in about 1 minute. After the high radiation event is over, the STARLAB returns to the X-POP orientation and resumes normal operations.

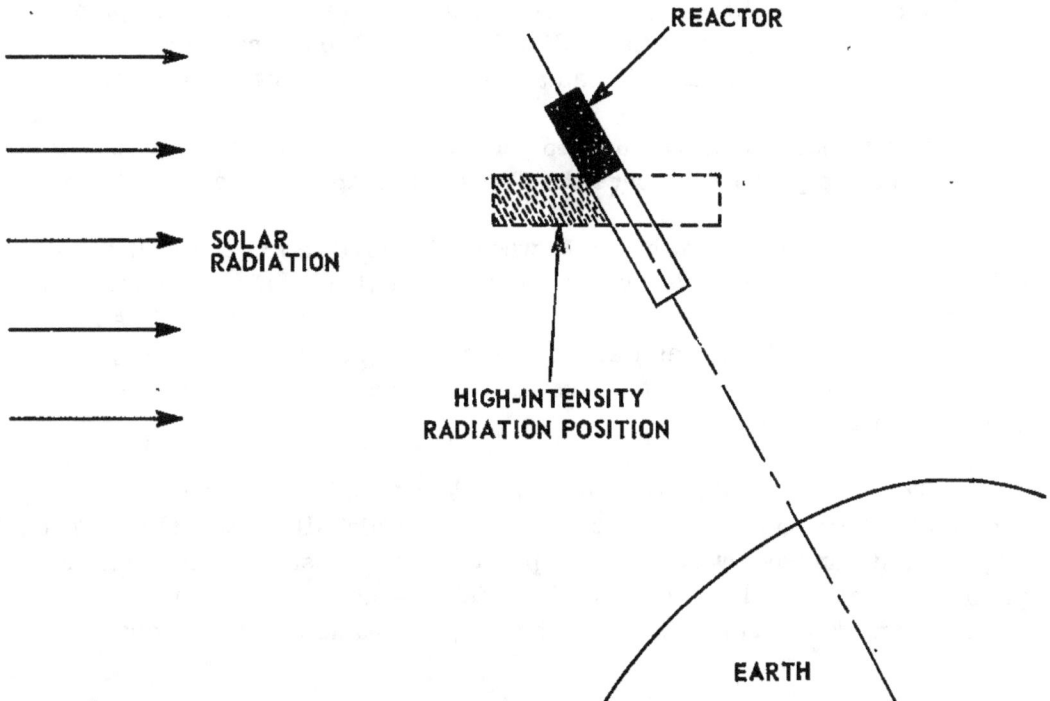

FIGURE V-20. LABORATORY VEHICLE ATTITUDE DURING PERIODS OF HIGH-INTENSITY SOLAR RADIATION

The procedure for chemical or bacterial contamination is to isolate the contaminated area by use of pressure doors. Then after the contaminant is removed, the area could be reopened and returned to use.

If escape from the ship is necessary, the crew uses the RFM's as lifeboats to stay in orbit until they are rescued.

Crew

Living quarters are provided for a maximum of 45 people on STARLAB. The number of people on board ranges from 35 to 45. Of these, there will be eight whose full-time function is running the ship. The breakdown is as follows:

1. Captain.

2. Cook — Stewards (2).

3. Communications technician (3).

4. Life support technicians (3).

In addition, there will be a medical officer aboard who will function primarily as a scientist. Everyone on the station is required to be familiar with emergency procedures and they will function as a part of the operating crew if needed.

The remainder of the 35 to 45 people on board will be the scientific crew and maybe some passengers once the STARLAB is in operation.

It is anticipated that by 1977, when STARLAB is to be launched, research will have been completed for qualifying man for extended stay times in space. The chief limitation on ability of the crew to remain will be the habitability of the station. Accordingly, as was outlined in previous sections, every effort has been, and will be, made to make the space station as habitable and as comfortable as possible.

The training of the crew, technicians, and scientists will start with ground simulation and continue at STARLAB to make all personnel at STARLAB fully effective and as aware of safety procedures as possible. Part of the ground simulation will be used to eliminate personnel who are subject to motion sickness or could not handle the long stay-time necessary for a

STARLAB mission. A ground simulation buttoned-up time period up to a 1-week duration may be necessary to eliminate people who couldn't take a 60- to 90-day mission.

A continuing program of physical training at STARLAB will be under the supervision of the medical officer. This study will be part of the determination on how space stay-time will affect man. Types of athletic equipment could include a "double-ended trampoline," bicycles, and spring-loaded muscle tone devices. All physical exercises should be able to be medically monitored and recorded.

Logistics

The resupply and man loading of STARLAB is totally dependent on the proposed space shuttle and frequency of the S-V launch schedule. General capabilities for the shuttle are 40 000 lbm payload plus 10 people for the STARLAB operation in addition to the shuttle crew and supplies to hold the shuttle for the entire trip and not be dependent on STARLAB for support even if docked. If STARLAB is put into an orbit that reduces the shuttle's payload capability, the frequency of the resupply trips may be increased. Also, if man's stay-time in space is less than 60 days, the more frequent trips could allow a smaller size space shuttle. Thus, how many is affected while on the AAP mission could determine the final design of the shuttle.

The effect of the frequency of shuttle launch (Fig. V-21) will determine how rapid the manpower buildup can be accomplished once the STARLAB construction is completed. When the launch frequency time period is every 8 weeks, 6 weeks will be needed for a 35- to 40-man buildup. If the launch period is restricted to once in 4 weeks, a 12-week time period will be necessary to fill a 35- to 40-man status and man will have to be able to stay 120 days at STARLAB. A more probable 3-week launch cycle that will more nearly use the full launch shuttle capacity and with an approximate 90-day man stay-time, will take 9 weeks to fill a 35- to 40-man crew. The 2-week launch schedule is necessary if a longer stay-time of 90 days or more is obtained; therefore, the problem of crew rotation is reduced to an as-desired bus run.

The crew life-support logistics (Table V-9) show that the daily requirements for various methods of handling the food and water requirements for STARLAB. When no provision is built in for recycling waste products, approximately 25 lbm per person per day of food, clothing, O_2, and N_2 are

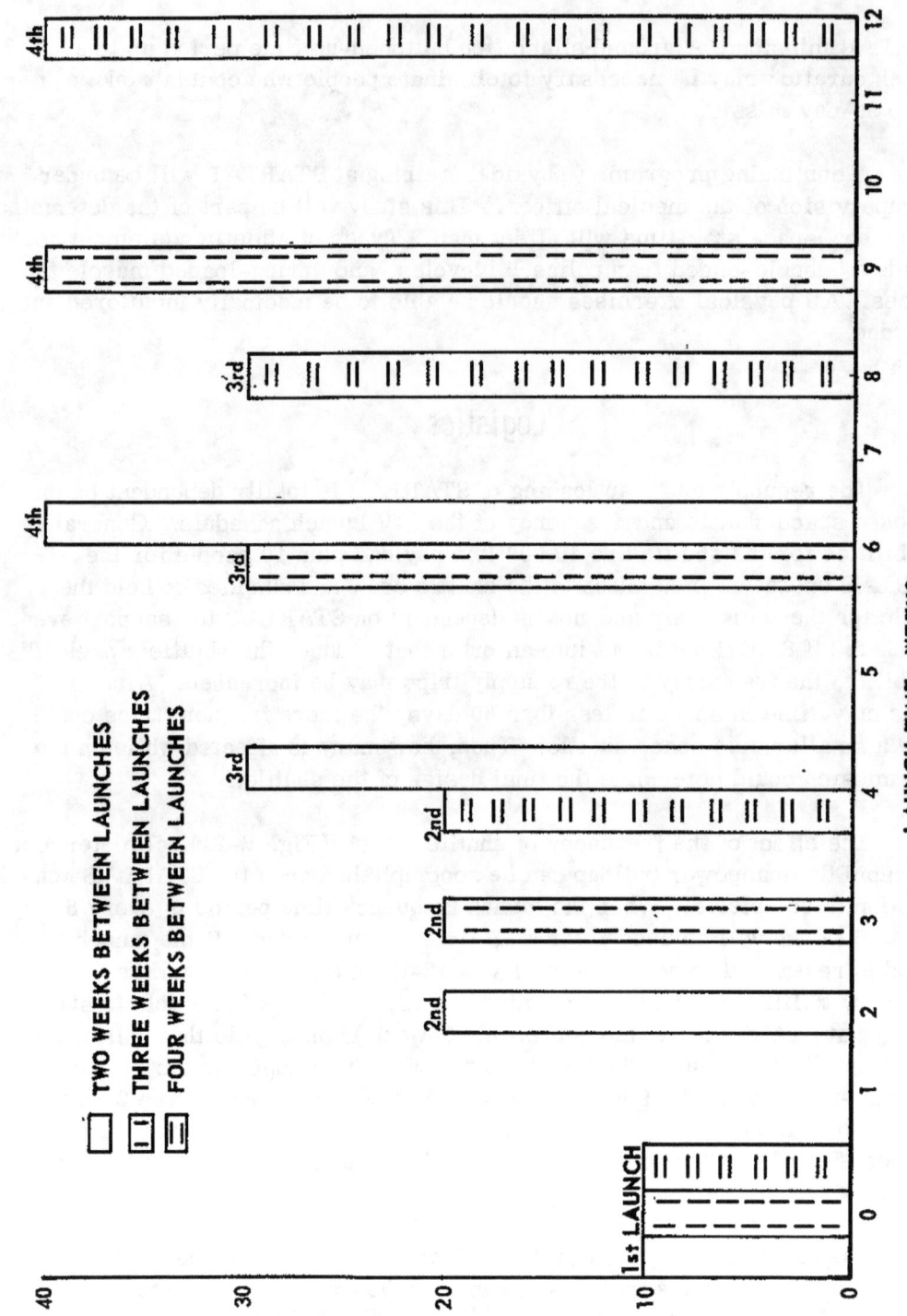

FIGURE V-21. STARLAB MANPOWER BUILDUP VERSUS SHUTTLE LAUNCH PERIOD

TABLE V-9. LOGISTICS ON LIFE SUPPORT RESUPPLY BY SPACE
SHUTTLE TO 35-MAN STARLAB

	lbm/man/day	2-Week Shuttle (lbm)	3-Week Shuttle (lbm)	4-Week Shuttle (lbm)
Totally Expended	25	12 250	18 375	25 500
Closed H$_2$O Loop	15	7350	11 025	14 700
Normal Expected (Closed H$_2$O Loop)	12.5	6125	9187.5	12 250
Apollo Food (Closed H$_2$O Loop)	8.9	4361	6541.5	8722
Absolute Minimums (Near Survival)	6.0	2940	4140	5880

needed. Once the water cycle is a closed-loop process, the requirements should drop to between 12.5 and 15 lbm per person per day. If Apollo-type food packaging is used along with a closed H_2O loop, approximately 8.9 lbm per person per day to an approximate minimum of 6.0 lbm per person per day for survival is required. The possibility that once the closed water loop is in operation, no additional water will be needed to be shuttled to STARLAB if prepackaged frozen food is used insteady of dehydrated food. Water for the "normal expected" is still carried in case of contamination, and the water must be dumped and replaced in one of the four independent systems.

Because of continuous gas leakage from STARLAB, an allowance for $N_2 + O_2$ has been made. The status of the actual leakage rate will probably be more readily calculated from the AAP determinations.

If the space shuttle resupply is restricted to once every 28 days, the proposed logistics will be possible if a closed water loop is in operation (Table V-10). Where the water loop is not closed, a 3-week (21-day) resupply shuttle of 30 000 lbm would be permissible.

TABLE V-10. LOGISTICS FOR THE 35-MAN STARLAB — RESUPPLY TIME

	Closed Water Loop (lbm)		
	14-Day	21-Day	28-Day
Food	2500	3750	5000
Water	500	750	1000
O_2	1500	2250	3000
N_2	500	750	1000
Clothing	1000	1500	2000
Propellants	4000	6000	8000
Experiments	3000	4500	6000
FFM's	2000	3000	4000
Totals	15 000	22 500	30 000

	Open Water Loop (lbm)		
	14-Day	21-Day	28-Day
Totals	20 000	30 000	40 000

REFERENCES

V- 1. Anon.: Saturn-V Workshop Study. Task Team Analysis, OMSF, Vol. II, p. 228.

V- 2. Anon.: Saturn-V Workshop Study. Task Team Analysis, OSMF, Vol. II, p. 139.

V- 3. D5-17525 Study of Structural-Thermal Insulation-Meteoroid Protection Integration. NAS8-21430.

V- 4. Bauer, P.: Batteries for Space Power Systems. NASA SP172, 1968.

V- 5. Anon.: Application of the Saturn/Apollo Hardware to Unmanned Scientific Exploration of the Solar System. Report No. TM-292/3-6-003, Northrop Space Laboratories, Huntsville, Alabama, April 1966.

V- 6. Acker, R.: Electrical Power Conversions System for Manned Earth Orbital Vehicle. Research Achievements Reviews, NASA TM X-53793, Vol. II, Series 9.

V- 7. Scull, J. R.: Spacecraft Guidance and Control. Space Technology, Vol. IV, NASA SP-63.

V- 8. Treble, F. C.: Large Solar Cell Arrays. Royal Aircraft Establishment, Technical Report 68019, January 1968.

V- 9. Apperson, J. L., et al.: Large Area Solar Array, Phase II. NASA CR-97868, October 1968.

V-10. Acker, R.: Electrical Power Conversions System for Manned Earth Orbital Vehicle. Research Achievements Reviews, NASA TM X-53793, Vol. II, Series 9.

V-11. Roberts, J. J., et al.: A Heat-Pipe-Cooled Fast-Reactor Space Power Supply. ANL-7422, Argonne National Laboratory, Reactor Engineering Division, June 1968.

V-12. Anon.: A Study of Jupiter Flyby Missions. Report No. FZM-4627, General Dynamics Co., Ft. Worth, 17 May 1967.

REFERENCES (Continued)

V-13. Anon.: Electric Power Generation in Space. NASA Facts, NF-38, December 1967.

V-14. Macosko, R. P., et al.: Performance of an Experimental SNAP-8 Power Conversion System. NASA TM X-1732, March 1969.

V-15. Gylfe, J. D., and Johnson, F. A.: Reactor-Thermoelectrics System Development Phase. Report No. AI-AEC-MEMO-12840, Atomics International Division, North American Rockwell, 1 July 1969.

V-16. Gylfe, J. D., and Johnson, R. A.: Reactor-Thermoelectric System for NASA Space Stations. Report No. AI-AEC-12839, Atomics International Division, North American Rockwell, 1 July 1969.

V-17. Tonelli, A. D., et al.: Parametric Study of Space Power Systems. Vol. III, Power Subsystem Parametric Data, NASA CR-73281, November 1968.

V-18. Heathcoat, J. R.: Techniques for Maintenance, Repair, or Replacement of Space Station Attitude Thruster Equipment in Orbit Without EVA. R-P&VE-AV-68-184, 18 December 1968.

V-19. Kramer, R. D.: Earth Orbital Workshop Fuel Weight Computer Program. R-P&VE-AV-69-3, Marshall Space Flight Center, 2 January 1969.

V-20. Davis, B. G.: A Tutorial Discussion of Orbital Workshop Orientation and Gravitational Effects. NASA TM X-53829, Marshall Space Flight Center.

V-21. Fraser, T. M.: The Intangibles of Habitability During Long Duration Space Missions. NASA CR-1084, June 1968.

V-22. Armstrong, R. C.: Life Support System for Space Flights of Extended Time Periods. NASA CR-614, November 1966.

V-23. Anon.: Single Wall Cryogen Storage Systems Report. McDonnell-Douglas Astronautics Co.

REFERENCES (Concluded)

V-24. Stone, R. W., and Piland, W. M.: Potential Problems Related to Weightlessness and Artificial Gravity. NASA TN D-4980, January 1969.

V-25. Roth, E. M. (Ed.): Compendium of Human Responses to the Aerospace Environment. NASA CR-1205 (1,2,3), June 1968.

V-26. Livshits, N. N. (Ed.): Effects of Ionizing Radiation and of Dynamic Factors on the Functions of the Central Nervous System — Problems of Space Physiology. NASA TTF-354, August 1965.

V-27. Berry, C. A., and Coons, D. O., et al.: Man's Response to Long Duration Flight in the Gemini Spacecraft. Gemini Midprogram Conference, Including Experiment Results, NASA SP-121, February 1966, pp. 235-261.

V-28. Anon.: Statement of Work — Space Station Program Definition (Phase B). April 1969.

V-29. Mendelsoham, B. T.: Report on the Optimization of the Manned Orbital Research Laboratory (MORL) System Concept. Volum XIII, Laboratory Mechanical Systems — Artificial Gravity Systems, Report SM-46084.

BIBLIOGRAPHY

Austin, L. G.: Fuel Cells. NASA SP-120, 1967.

Bioregenerative Systems. NASA SP-165, November 1966.

Life Support System for Space Flight of Extended Time Periods. NASA-CR-614, General Dynamics, November 1966.

Manning, H. S.: MSFC Memorandum, Annual Radiation Dose for 1975-1980 Space Base Study. 13 March 1969.

NASA Facts. NF-41/12/67.

Saturn V Workshop Study. Vol. II, Task Team Analysis, OMSF Planning Group, 1 April 1968, p. 228.

Standard Human Engineering Design Criteria. MSFC-STD-267A, 23 September 1966.

The Closed Life Support System. Ames Research Center, NASA, April 1966.

CHAPTER VI

ECONOMICS AND COST OF STARLAB

CHAPTER VI. ECONOMICS AND COST OF STARLAB

INTRODUCTION

Considering the time frame of the construction and utilization of STARLAB, one must carefully approach the problem of cost of such a large project. On the other hand, the cost of such a conceptual design, such as presented herein, requires a great amount of latitude since the design is not yet extensively detailed. Nevertheless, cost should be done at this level of design to size the cost of STARLAB as well as attempting to get the economic aspects into the situation of a systems approach.

APPROACH

Initially, attempts were made at employing already developed costing modes. These were discarded because either STARLAB was too large and required invalid extrapolation (sometimes as far as five times the region of definition of several cost-estimating relationship curves), or the model required detailed design parameter data.

The method employed was an approximate "brute-force" method based on appropriate old costings. The nearest study to STARLAB was a space-base cost estimate for a 35- to 50-man base. The space base was smaller than STARLAB but served as a good cost comparison.

After much exchange of information and comprehension discussions, costs were estimated. Members of the design group and the costs group also arrived at a set of assumptions as follows:

1. The five RFM's of STARLAB have the same basic structure.

2. The five modules of STARLAB have the same basic structure.

3. Judgement was the mode of determining costs.

4. Costs were done by modules.

5. Recurring and nonrecurring costs were estimated.

6. Costing was done for each subsystem in a module.

7. The costs of a system of three geosynchronous satellites that would function as laser repeaters to STARBASE on earth were included in the nonrecurring costs of module 1, the critical services module under the communications subsystem.

8. All units of money herein are reported in terms of billions (B) of dollars ($).

9. Crew training costs were included under crew systems subsystem.

10. An inflation factor was not considered herein.

Several design alternatives were as follows:

    a. Remote flying versus onboard nuclear power supply.

    b. S-band versus laser communications.

    c. Open versus closed atmosphere control.

    d. Onearth versus onboard support laboratories.

    e. Onboard versus free-flying materials and manufacturing program.

These design alternatives were considered for a trade-off with costs; however, the method of costing was not precise enough to consider these alternatives.

HOW MUCH WOULD STARLAB COST?

This question (how much would STARLAB cost?) can be most easily answered by discussing costs in a sequence beginning with the grand total, then progressing to more and more detailed statements of the costs.

The entire estimated cost for STARLAB from instigation of the project to its tenth, and assumed final, year of operation is $23.831 B. This amount of money is estimated to be spent over a 17-year period. During the 1970 to

1977 period, an estimated $12.631 B would be spent for the construction of STARLAB. This represents an average of $1.804 B per year for the first 7 years. STARLAB would be launched in 1977 and during the next 10 years the remaining $11.200 B would be spent for support and updating of STARLAB. This represents a $1.20 B spending per year. Of the $11.200 B spent for support, $5.000 B would be spent for mission and support operations and program management ($0.5 B per year). Another $2.000 B would be spent for subsystems maintainability, repairability, long-life capability, and redundancy. The applications and experimental programs would have alloted $0.300 B per year for a total of $3.000 B. Five Saturn V launches at $0.100 B each are necessary to put STARLAB into orbit ($0.5 B).

The logistics costs for the shuttle were computed assuming 14 shuttles per year for 10 years at $0.050 B per shuttle which yields $0.700 B. Using a model which was defined for the Mercury, Gemini, and Apollo data, a ratio of the dollars required for ground test, ground support engineering, and integration for each dollar of total subsystems costs on each module of STARLAB was established. After the detailed subsystems costs of a module in terms of nonrecurring and recurring costs were developed, then total ground test, ground support engineering, test articles, facilities, systems engineering, and integration costs were determined using this ratio.

The modules were numbered in the order in which they are to be deployed. Module 1 refers to the critical services module, module 2 refers to the power module, and modules 3, 4, and 5 make up the remainder of STARLAB proper. These 5 modules will be launched by low cost ($0.100 B each) Saturn V's.

Modules 1 through 5 have a common structure subsystem which will be assembled and checked out on the ground. Hence, module number 1 was chosen to contain the bulk of the nonrecurring costs for developing these 5 common modules.

Also, a part of the STARLAB are the RFM's numbered 6 through 10. These 5 modules have a great deal of commonality in their structure subsystems requiring only minor modifications for adaption to their missions. Modules 6 and 8 must contain life support and environment control systems, hence, module 8 was costed with the nonrecurring costs in modules 9 and 10.

Module Number	Application
6	Life-Behavioral Sciences
7	Solar Astronomy
8	Materials and Manufacturing
9	Stellar Astronomy
10	Physics

A weight statement for the research and support laboratories from the design of STARLAB was helpful in estimating the cost of the experimental programs in the various modules. These weights by laboratories are presented in Tables II-1 through II-4. Cost of experiments in various areas per pound were used in estimating these experimental costs.

Table VI-1 presents a synopsis of the costs of STARLAB. Table VI-2 contains a cost statement by subsystems for typical modules 1, 2, and 3 (modules of STARLAB proper). The subsystem costs for typical RFM's 6, 8, and 9 are also given (Table VI-3), while costs for modules 4, 5, and 7 are omitted. A subdivision of nonrecurring and recurring costs for each of the subsystems and their totals for the entire STARLAB are given in Table VI-4.

Table VI-3 contains a subsystem cost statement for the life-behavioral sciences, materials and manufacturing, and stellar astronomy RFM's. The first two contain a life-support system and the last one does not.

WHAT PART OF NASA'S BUDGET WILL STARLAB CONSUME?

If it is assumed that NASA's budget in 1969 was approximately $4.0 B and that it will grow at the same rate as the Gross National Product (GNP) (Fig. VI-1), then by 1970 (GNP will have gone from $880 B in 1969 to $950 B in 1970) NASA's budget could be estimated as $4.3 B. Assuming that the GNP grows this approximate steady rate (quite a conservative prediction!) of 7 percent per year, then we arrive at the percentage costs of STARLAB in Table VI-5. Note the marked drop of percentage program costs after the launch date of 1977. It is also notable that the most expensive time period is the early 1970's.

TABLE VI-1. COST OF STARLAB (OPTIMUM BUDGET)

	COST (B$)	
STARLAB		
Subsystem Structure, Communications, etc.	5.504	
Ground Support Ground Test, Facilities, etc.	5.675	7 Years Develop
Experiments	1.452	
SUPPORT		
Logistics 5 Saturn V's and 140 Shuttle Launches	1.200	
Mission Support, Operations, and Program Management	5.000	10 Years Maintain
Maintainability, Repairability, Long Life, Capability, and Redundancy	2.000	
Experimental Up-Date	3.000	
	23.831	

COST RATE (B$)

| 1.80/Year | 1970 – 1977 |
| 1.12/Year | 1977 – 1987 |

Note: Ten modules were used.

TABLE VI-2. TYPICAL STARLAB (PROPER) MODULAR SUBSYSTEM COSTS

Subsystem	MODULE 1			MODULE 2			MODULE 3		
	N-R	R	TOT	N-R	R	TOT	N-R	R	TOT
Environment Control	0.315	0.020	0.335	0.010	0.001	0.011	–	0.020	0.020
Structure	0.750	0.300	1.050	0.050	0.300	0.350	0.030	0.300	0.300
Stabilization and Control	0.075	0.010	0.085	–	0.010	0.010	–	0.010	0.010
Reaction Control	0.021	0.007	0.028	–	0.007	0.007	–	0.007	0.007
Guidance and Navigation	0.035	0.011	0.046	–	0.011	0.011	–	0.011	0.011
Electrical Power	0.025	0.003	0.028	0.165	0.040	0.205	–	0.003	0.003
Communications	0.500	0.100	0.600	–	–	–	–	0.100	0.100
Instrumentation	0.010	0.005	0.015	–	0.005	0.005	0.009	0.005	0.014
Crew	0.075	0.015	0.090	–	–	–	0.020	0.015	0.035
Totals	1.806	0.471	2.277	0.225	0.374	0.599	0.059	0.471	0.530

Notes: All costs in B$
N-R = Nonrecurring
R = Recurring
TOT = Total

TABLE VI-3. RFM'S SUBSYSTEM COSTS

Subsystem	MODULE 6			MODULE 8			MODULE 9		
	N-R	R	TOT	N-R	R	TOT	N-R	R	TOT
Environment Control	0.050	0.020	0.070	0.077	0.006	0.083	–	–	–
Structure	0.065	0.030	0.095	0.250	0.030	0.280	0.065	0.030	0.095
Stabilization and Control	–	0.010	0.010	0.023	0.007	0.030	–	0.007	0.007
Reaction Control	–	0.007	0.007	0.019	0.006	0.025	–	0.006	0.006
Guidance and Navigation	–	0.011	0.011	0.015	0.011	0.026	–	0.011	0.011
Electrical Power	–	0.003	0.003	0.005	0.003	0.008	–	0.003	0.003
Communications	–	0.005	0.005	0.015	0.005	0.020	–	0.005	0.005
Instrumentation	0.005	0.005	0.010	0.003	0.001	0.004	0.005	0.001	0.006
Crew	–	–	–	0.010	0.005	0.015	–	–	–
Totals	0.120	0.091	0.211	0.417	0.074	0.491	0.070	0.063	0.133

With Life Support: Modules 6 and 8
Without Life Support: Module 9

Note: All costs in B$, N-R = Nonrecurring, R = Recurring, TOT = Total.

6-7

TABLE VI-4. SUBDIVISION OF RECURRING AND NONRECURRING COSTS FOR THE ENTIRE STARLAB AND RFM'S (10 MODULES TOTAL)

Subsystem	N-R	R	TOT
Environment Control	0.452	0.107	0.509
Structure	1.400	1.650	3.050
Stabilization Control	0.098	0.088	0.186
Reaction Control	0.040	0.066	0.106
Guidance and Navigation	0.050	0.110	0.160
Electrical Power	0.195	0.067	0.262
Communications	0.515	0.425	0.940
Instrumentation	0.060	0.011	0.071
Crew	0.145	0.025	0.170
Totals	2.955	2.549	5.504

Note: All costs in B$
     N-R = Nonrecurring
       R = Recurring
    TOT = Total

TABLE VI-5. PERCENT PROGRAM COST OF STARLAB

Year	Postulated NASA Budget (B$)	% STARLAB	Phase
1969	4.0	–	DEVELOPMENT
1970	4.3	42	
1971	4.7	39	
1972	5.0	36	
1973	5.4	33	
1974	5.8	31	
1975	6.2	28	
1976	6.6	26	
1977	7.1	16	LAUNCH
1978	7.6	15	MAINTAIN
1979	8.1	14	
1980	8.7	13	
1981	9.3	12	
1982	9.9	11	
1983	10.6	10	
1984	11.4	9.8	
1985	12.2	9.2	
1986	13.1	8.5	
1987	14.0	8	

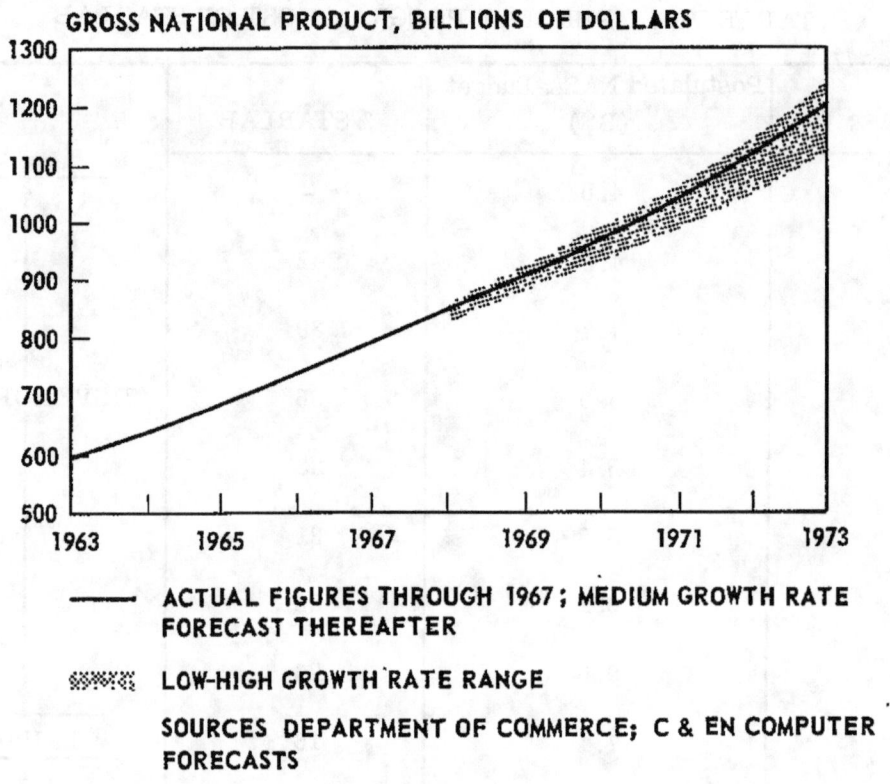

FIGURE VI-1. PROBABLE TRILLION DOLLAR ECONOMY BY 1971

ALTERNATIVE COSTS OF STARLAB

Three interesting alternative costing models were developed. The first model involves immediate deployment of the five large modules of STARLAB proper and the five RFM's. This cost synopsis is presented in Table VI-1. Upon considering Table VI-5, however, we note that STARLAB would have very high initial cost (note 42 percent of NASA's projected 1970 budget). Two other alternatives were considered. Alternate 1 was defined to have a delayed FFM development, essentially pushing some of the development costs into the 10-year operational period of STARLAB. This would impair somewhat STARLAB's ability to procure space environment data, but all areas of study having their FFM's delayed have lab space in STARLAB already except astronomy. Astronomy's program would be retarded if STARLAB were the only platform for astronomy experiments. However, it is anticipated that astronomy will have several platforms in other programs. Most importantly, the earth resources and the life-behavioral sciences program would be unaltered. Table VI-6 is a synopsis of the costs of alternative 1. It achieves the objective of a lower initial cost rate in 1970-1971 ($1.365 versus $1.800 B per year).

The final alternative, alternative 2 (Table VI-7), deletes the FFM's. The effects of this deletion would have more serious effects on the ability of STARLAB to gather data on the space environment and man. This alternative could be useful only if a severe funding perturbation occurred and is not recommended because of the lower data returns. However, note that even under these modes of operation, STARLAB could still be made completely productive by the later addition of FFM's. The estimated effects of delayed or deleted FFM deployment upon STARLAB's effectiveness are given for each research area in Table VI-8.

These last two alternatives are examples of so-called "reversible" losses in a program. These types of losses can be added later when funding is better and one has somewhat reduced losses in the effectiveness of a program.

The recommended mode of budget is the earliest deployment scheme (Table VI-1). It was felt that this method optimizes the information and data acquisition of STARLAB.

RECOMMENDATIONS

The costing models that were considered for costing STARLAB were found to be inadequate. There seemed to be a lack of costing data available for model building despite the fact that much is available in terms of manned space flight systems.

It is further suggested that the present ideas of costs and economics be defined to include the assignment of value to the production of ideas, processes, and materials from STARLAB. This aspect of accounting which includes the assignment of dollar values to intangibles could be handled through STARAD. It is felt that a complete total must be kept of the ideas, processes, or materials that have been produced on STARLAB including the spin-offs. A 10-year laboratory such as STARLAB is a golden opportunity to illustrate that the space environment can be utilized profitably by man.

TABLE VI-6. ALTERNATE DEPLOYMENT NO. 1, LATE DEPLOYED
RFM'S (LOW INITIAL COST BUDGET)

STARLAB Proper (5 Modules)	Cost (B$)	
Subsystem (Structure, Communications, etc.)	4.456	} 7-Year Develop
Ground Support (Ground Test, Facilities, etc.)	4.549	
Experiments	0.552	
Modules Later (5 FFM's)		Launch
Subsystem	1.052	} 10-Year Maintain and Develop
Ground Support	1.126	
Experiments	0.900	
Support	11.200	
Cost Rate		
$1.365 B/year 1970-1977		
$1.427 B/year 1977-1987		

TABLE VI-7. ALTERNATE DEPLOYMENT NO. 2 COSTING, NO
REMOTE FLYING MODULES (DISASTER BUDGET)

STARLAB Proper Only (5 Modules)	Cost (B$)	
Subsystem (Structure, Communications, etc.)	4.456	} 7-Year Develop
Ground Support (Ground Test, Facilities, etc.)	4.549	
Experiments	0.552	
		Launch
Support	11.200	} 10-Year Maintain
	20.757	
Cost Rate $1.365 B/year 1970-1977 $1.120 B/year 1977-1987		

TABLE VI-8. ESTIMATED EFFECTS OF DELAYED OR DELETED FFM DEPLOYMENT UPON STARLAB EFFECTIVENESS

	Using Optimal Cost Rate at Base (%)	Alternate No. 1, Delayed FFM's (%)	Alternate No. 2, Deleted FFM's (%)
Chemistry	100	100	100
Earth Resources	100	100	100
Life-Behavioral Sciences	100	90	60
Materials and Manufacturing	100	90	80
Physics	100	85	50
Astronomy	100	50	0
R & D Engineering	100	95	95
Communication/Navigation and Traffic Control	100	100	100
Totals	800	710	585
% Goal Completion	100	89	73

The final recommendation is that a study be made of configurations and concepts that could be classified as "reversible" expendables. It seems that any large-scale programs should have many of these expendables such that a reversible flexibility is designed into the program. This increases the viability of such a program by enabling it to survive budgetary squeezes and self-heal itself later.

CAN WE AFFORD STARLAB?

If man is to continue to gain more and more control over his environment, and as he learns more about managing himself he slowly realizes that he must work to improve present conditions on earth or possibly face extinction.

STARLAB and the space program emphasize the need for the management of earth's resources. STARLAB has been designed to optimize the benefits of utilizing the space environment as well as easily and handily carrying out earthward-looking surveillance and monitoring.

As a species, striving to perpetuate itself, man must learn to control and harness all that he is capable of perceiving. STARLAB will help harness our newest natural resource — the space environment.

In terms of more practical matters, the emphasis of the scientific mission is man-focused and to have immediate benefits. As mentioned earlier in this report, the direct and indirect benefits of STARLAB could easily pay for itself as well as provide a handsome profit in arresting human misery and suffering.

APPENDIX A

INVENTION DISCLOSURES

APPENDIX A. INVENTION DISCLOSURES

INTRODUCTION

This appendix is composed of the actual invention disclosure submittal forms[1] and accompanying descriptions and sketches which resulted from the summer program; these are as follows:

1. Fabrication method of hollow spheres.

2. High density packaging of space modules — Dixie Kup concept.

3. High density packaging of space modules — drinking barrel concept.

4. Rhombic units for constructing a variety of structures.

5. Space shoe locking device.

6. Flexible coupling for docking two vehicles in space.

---

1. Parts of the forms are omitted to reduce their size to standard report size.

Branch No. _____

NASA Case No. _____

NATIONAL AERONAUTICS AND SPACE ADMINISTRATION

Office of Assistant General Counsel for Patents

DISCLOSURE OF INVENTION

This is an important legal document. It should be carefully completed by the inventor(s) and forwarded to the Patent Representative. Two copies of each document are desired.

1. Descriptive Title of Invention

   FABRICATION METHOD OF HOLLOW SPHERES

2. Name(s), Title(s), and Home Address(es) of Inventor(s)

   Dimitrios Constantinos Agouridis, Associate Professor of Electrical Engineering
   3186 Dickman Ave., Memphis, Tennessee 38111

3. Name and Address of Employer

   Auburn University, NASA ASEE, 1969, MSFC.

4. Stage of Development	Date Month/Yr.	Location	Identify persons or records supporting facts stated in 4a-4e
a. First disclosure to others	7/8/69	MSFC	Meeting with H. F. Wuenscher and Other ASEE Fellows
b. First sketch or drawing			
c. First written description	7/10/69	MSFC	E. R. Lawson — M. L. Forthun
d. Completion of first model or full size device			
e. First successful operational test			

5. List other pertinent notebook entries, photographs, reports, drawings.

6. If the invention was disclosed outside of NASA, identify the individuals, the companies or activities they represent, and the date of disclosure.

7. List any known or contemplated public use, publication, or oral presentation of the invention.

8. Indicate any past, present, or contrmplated Government use of the invention.

9.

FABRICATION METHOD OF HOLLOW SPHERES

Description

The purpose of this disclosure is to introduce a unique procedure for fabrication of hollow spheres. Such spheres are of considerable technological importance in applications such as ball bearings. An analysis of the capabilities of such spheres is given in Reference A-1. The uniqueness of this invention lies on the way by which the hollow region can be made. Figure A-1 represents the desired product. The construction procedure is described by the following steps.

Fabrication Procedure

One fabrication procedure is to construct a spherical thin shell or balloon (Fig. A-1). This can be done by welding together two premachined semispherical sections or blown out of a soft material melt by a procedure similar to glass blowing. The thickness of the spherical shell should be no more than that needed to withstand the heat and the mechanical strain introduced by the deposition of the material of the hollow sphere (see below). The material of which the spherical shell is made can be either the same as that of the desired sphere or some different kind of material, preferably of lower density to keep the overall weight low. The surface of the outer shell should be treated so that the deposited material on it dampens (sticks on) the surface. The radius of this spherical shell will be equal to the radius of the hollow region (inner radius) of the desired hollow sphere and should be designed according to this specification.

The deposition of the material of the desired hollow sphere is discussed in this paragraph. With the spherical shell described above properly prepared, the deposition of the material of the hollow sphere can be done by two methods as follows:

1. Premelt the material to be deposited. Immerse the spherical shell into the material so that the melted material will flow over the spherical shell. Let enough of the melted material flow over the spherical shell to obtain the desired thickness. Cool to solidify.

The spherical perfection of the sphere so formed will depend on the environmental condition:

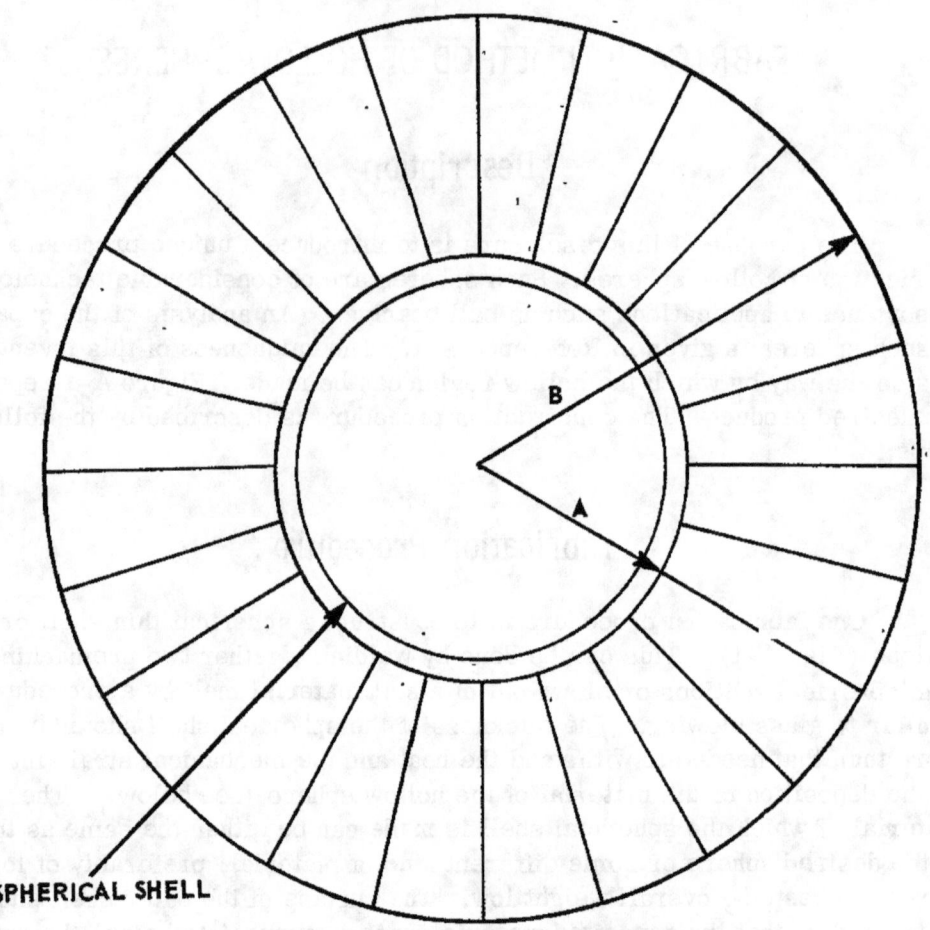

FIGURE A-1. HOLLOW SPHERE

    a. If the fabrication occurs in outer space where the influence of the earth's gravity is nearly zero (zero g), the fabricated sphere will be perfect because the spherical configuration is most natural and the total energy is minimum.

    b. If the fabrication occurs on the surface of the earth, the outer shape of the solidified product will normally have the shape of the container in which the solidification occurred. However, one can obtain an approximate spherical shape by spinning the sphere around while melting or by free-falling. Under these circumstances, the final shape of the sphere and its smoothness are obtained by proper machining.

It is obvious that the zero-g environment is ideal for such fabrication. On the other hand, such fabrication can occur on the suface of the earth where with additional refinement of the fabrication one can hope to obtain a practical product with existing technology.

2. Evaporate (spray on) the material on the spherical shell uniformly until the desired thickness is achieved. Again, the zero-g environment will be advantageous since the deposition can be done with the spherical shell floating in space (no support is needed). On the other hand, a very good product can be fabricated on earth by properly rotating the spherical shell during deposition.

Obviously, the deposition by evaporation is not very practical when the thickness of the sphere is large.

Branch No. _____
NASA Case No. _____

NATIONAL AERONAUTICS AND SPACE ADMINISTRATION

Office of Assistant General Counsel for Patents

DISCLOSURE OF INVENTION

This is an important legal document. It should be carefully completed by the inventor(s) and forwarded to the Patent Representative. Two copies of each document are desired.

1. Descriptive Title of Invention

   STACKED "DIXIE KUPS" HIGH DENSITY PACKAGING OF SPACE MODULES

2. Name(s), Title(s), and Home Address(es) of Inventor(s)

   Melvin L. Forthun, Associate Professor, Mechanical Engineering, North Dakota State University

   John R. Luoma, Chemistry Department, North Dakota State University, Fargo, North Dakota

3. Name and Address of Employer

   NASA — Auburn Design Group

4. Stage of Development	Date Month/Yr.	Location	Identify persons or records supporting facts stated in 4a-4e
a. First disclosure to others	July/1969	Huntsville	Auburn Design Interim Report
b. First sketch or drawing	"	"	" " " "
c. First written description	"	"	" " " "
d. Completion of first model or full size device	"	"	" " " "
e. First successful operational test	—	—	— — — —

5. List other pertinent notebook entries, photographs, reports, drawings.

   See "4-b"

6. If the invention was disclosed outside of NASA, identify the individuals, the companies or activities they represent, and the date of disclosure.

   None

7. List any known or contemplated public use, publication, or oral presentation of the invention.

   Design Review Presentation in August 1969.

8. Indicate any past, present, or contrmplated Government use of the invention.

   None at present.

9.

A-6

STACKED "DIXIE KUPS" HIGH DENSITY PACKAGING OF SPACE MODULES

The purpose of this invention is to provide a method of efficient packaging of space modules that are carried by a booster rocket into orbit or to a planetary surface. The method permits the packing of many modules that could provide a large amount of volume while still maintaining a short-stack height that is mounted on top of a booster. These frustra of cones, called "Dixie Kups," can be used to construct large-volume shelter in space (Fig. A-2).

Previously, space housing for workshops, space stations, barges, space bases, and planetary bases considered using the right cylinders available from the spent S-IVB or an S-II stage of a Saturn V.

Previously, suggested right cylinders do not pack efficiently, requiring an unmanageably tall rocket or a large number of launches to place (e.g., six) the modules somewhere in space. This new stacking of Dixie Kups could significantly reduce the number of launches needed to put so many cubic feet of living space at the disposal of a crew of astronauts.

The stack of Dixie Kups can be packed (for example) six high, one inside another, with the large end of the cup sitting on top of the S-II stage (Fig. A-2). The Kup could be 30-feet-diameter base, 20-feet-diameter top, and 40 feet high. The Kups could be separated and rotated 180 degrees and docked to form a double-Kup assembly section (Fig. A-3).

A modification of the frustrum shown in Figure A-3 can be used to construct a wheel space station.

Stacked figures are similar to a truncated cone or a frustrum of a cone (e.g., truncated pyramid, truncated tapered hexagon, and truncated trigonal pyramid), and the results are as follows:

1. Fewer launches to obtain large-volume shelter.

2. Lower stack height for rocket.

3. Efficient packing for large-volume shelter.

This is a new method of packing a modified module to facilitate launching of large-volume shelter with low-volume launch configuration.

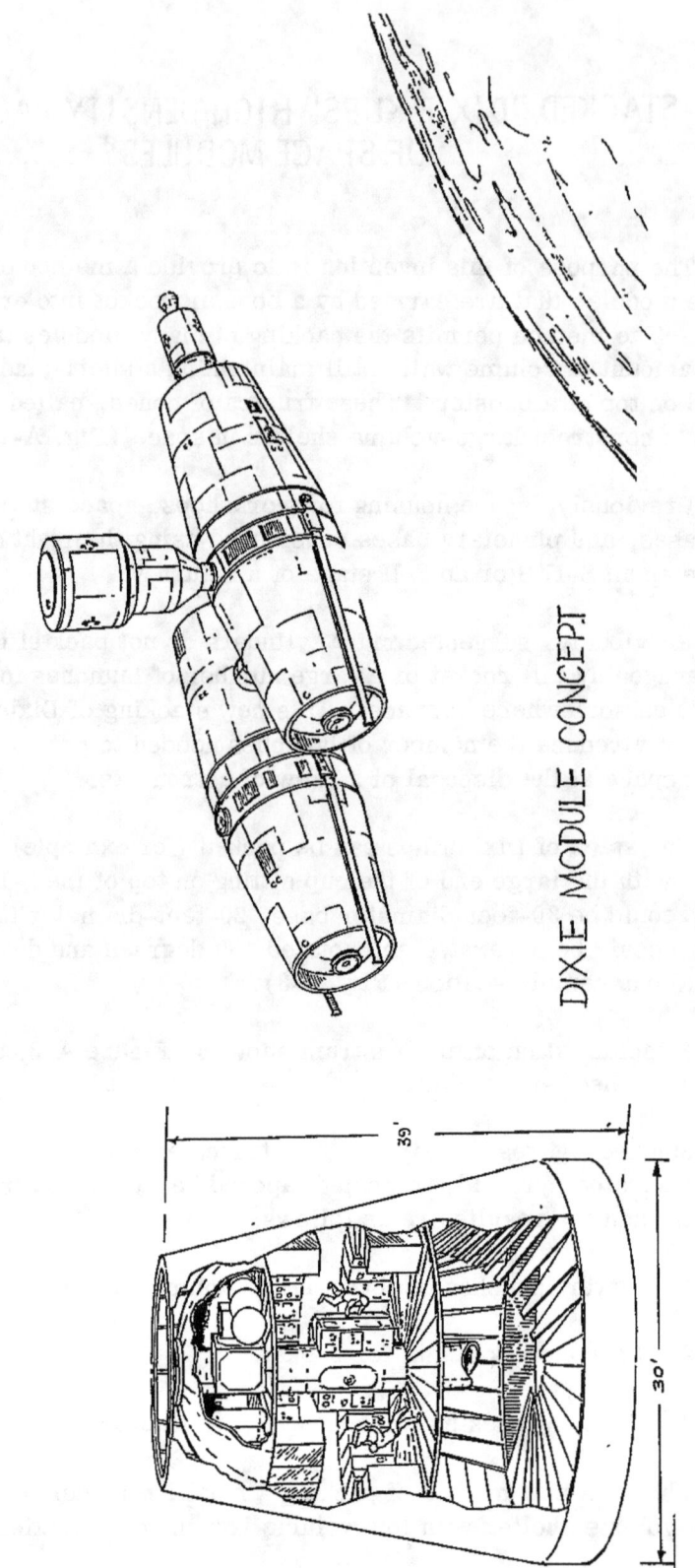

FIGURE A-2. NESTING CONCEPTS "DIXIE MODULES"

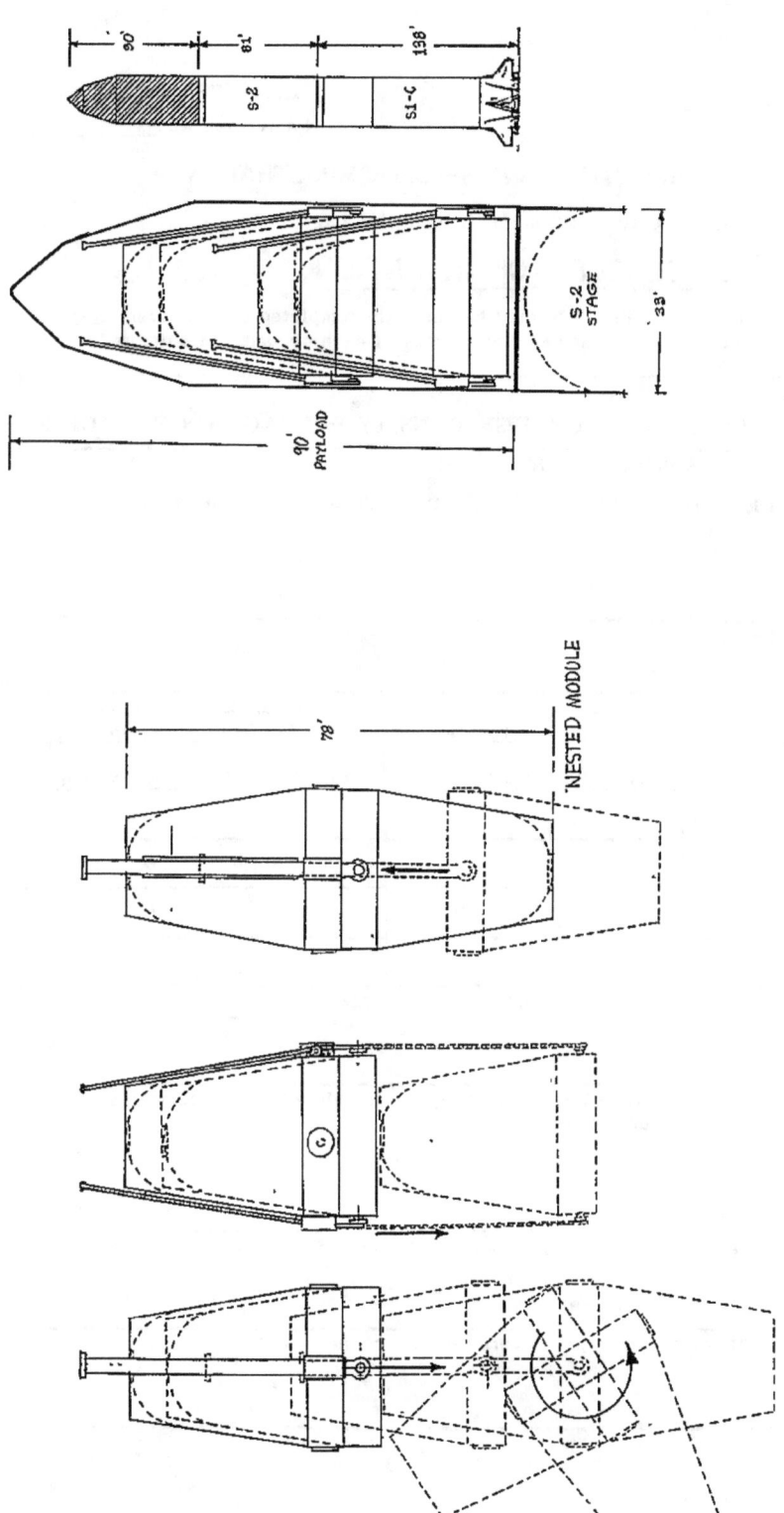

FIGURE A-3. NESTED MODULE ASSEMBLY

A-9

Branch No _____

NASA Case No. _____

NATIONAL AERONAUTICS AND SPACE ADMINISTRATION

Office of Assistant General Counsel for Patents

DISCLOSURE OF INVENTION

This is an important legal document. It should be carefully completed by the inventor(s) and forwarded to the Patent Representative. Two copies of each document are desired.

1. Descriptive Title of Invention

   "DRINKING BARREL" CONCEPT FOR HIGH-DENSITY PACKAGING OF SPACE MODULES

2. Name(s), Title(s), and Home Address(es) of Inventor(s)

   Donald C. Raney, Associate Professor, Mechanical Systems Engineering, University of Alabama, Tuscaloosa, Alabama

3. Name and Address of Employer

   NASA — Auburn Design Group

4. Stage of Development	Date Month/Yr.	Location	Identify persons or records supporting facts stated in 4a-4e
a. First disclosure to others	July 1969	Huntsville	Auburn Design Interim Report
b. First sketch or drawing	"	"	" " " "
c. First written description	"	"	" " " "
d. Completion of first model or full size device	"	"	" " " "
e. First successful operational test	—	—	— — — —

5. List other pertinent notebook entries, photographs, reports, drawings.

   See attached sketch

6. If the invention was disclosed outside of NASA, identify the individuals, the companies or activities they represent, and the date of disclosure.

   None.

7. List any known or contemplated public use, publication, or oral presentation of the invention.

   Design Review Presentation in August 1969

8. Indicate any past, present, or contemplated Government use of the invention.

   None at present

9.

A-10

DRINKING BARREL CONCEPT FOR HIGH-DENSITY PACKAGING OF SPACE MODULES

The purpose of this invention is to provide a method for high-density packing of space modules into a given launch configuration. The method permits the packaging of many modules which provides a large volume when deployed while still maintaining stack-height limitations imposed on the launch vehicle. By proper design, the internal floor and much equipment can be mounted so that little effort is required to place the module in operation other than the basic deployment of the structure (Fig. A-4).

As previously discussed, right circular cylinders as basic building blocks for space vehicles do not pack efficiently for launch. This leads to prohibitively tall launch vehicles or to multiple launches to have a desired working volume in space. This new stacking concept can significantly reduce the number of launches required to assemble a large-volume space vehicle in earth orbit for use as an earth satellite or for planetary missions.

This stacking concept can be used for various size structures. Typical module sizes for possible use with existing launch systems can be 30 feet in diameter base with 20-foot-diameter top. The compressed configuration height can be about 20 feet, with a deployed height in space of about 45 feet. These modules can be used as building blocks to construct various larger configurations in space.

The structure can be deployed by simply pressurizing the interior of the modules and providing guides on the moving sections to prevent jamming.

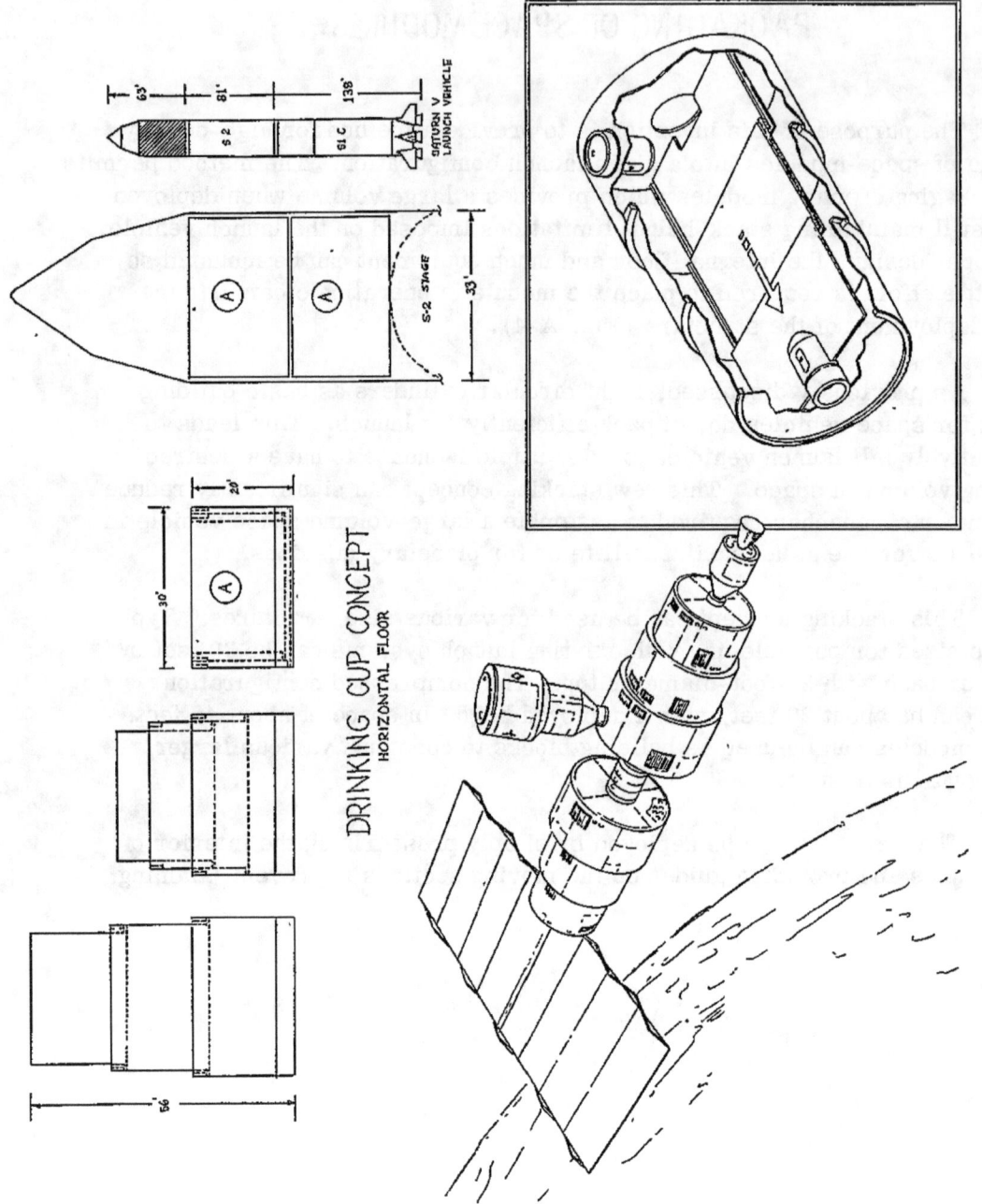

FIGURE A-4. "DRINKING BARREL" CONCEPT FOR SPACE STRUCTURE

_____Branch No._____

NASA Case No._____

NATIONAL AERONAUTICS AND SPACE ADMINISTRATION

Office of Assistant General Counsel for Patents

DISCLOSURE OF INVENTION

This is an important legal document. It should be carefully completed by the inventor(s) and forwarded to the Patent Representative. Two copies of each document are desired.

1. Descriptive Title of Invention

   RHOMBIC UNITS FOR CONSTRUCTING A VARIETY OF STRUCTURES

2. Name(s), Title(s), and Home Address(es) of Inventor(s)

   John R. Luoma, Chemistry Department, North Dakota State University, Fargo, North Dakota

3. Name and Address of Employer

   NASA — Auburn Summer Design Program

4. Stage of Development	Date Month/Yr.	Location	Identify persons or records supporting facts stated in 4a-4e
a. First disclosure to others	July/69	Huntsville	Auburn Design Interim Report
b. First sketch or drawing	"	"	" " " "
c. First written description	"	"	" " " "
d. Completion of first model or full size device	"	"	" " " "
e. First successful operational test	—	—	— — — —

5. List other pertinent notebook entries, photographs, reports, drawings.

   See "4b"

6. If the invention was disclosed outside of NASA, identify the individuals, the companies or activities they represent, and the date of disclosure.

   None

7. List any known or contemplated public use, publication, or oral presentation of the invention.

   Design Review Presentation in August 1969

8. Indicate any past, present, or contemplated Government use of the invention.

   None

9.

A-13

RHOMBIC UNITS FOR CONSTRUCTING A
VARIETY OF STRUCTURES

The general purpose of this invention is to provide a compact, sturdy, versatile, reusable, easily manufactured structural unit that can be used to produce in a space environment (or on planets) a variety of structural shapes and sizes for space workshops, space stations, space bases, planetary bases, interplanetary construction, and repair hangers, etc.

The previous method for providing structural units was the employment of spent or previously outfitted S-IVB or S-II Saturn V stages.

Disadvantages of the S-IVB stage are as follows:

1. More difficult to manufacture cylinders.

2. Little structural versatility (all same size).

3. At least twofold design of unit partially designed for living and workshop and partially designed as a structural unit of a launch vehicle (or fuel tank).

4. Cannot be packaged together efficiently.

A unit is defined as two equilateral triangles hinged together with a two-way hinge. Using these units, one can construct the following structures:

1. One unit — Space shelter, radiation shield solar cell panel frame [Fig. A-5(a)].

2. Two units — Tetrahedron, domicile with optional door, power-communications-life support module to power other structures below [Fig. A-5(b)].

3. Three units — Trigonal bipyramid, gravity-gradient orienting, separately flying module [Fig. A-5(c)].

4. Four units - Octahedron, experiment containing module. Easily controlled dynamical character. Useful as a structural unit for a large structure [Fig. A-5(d)].

5. Five units — See Figure A-5(e).

6. Ten units — Icosihedron, large structural unit, crew quarters, large experimental housing, assembly hanger for deep space probes [Fig. A-5(f)].

7. Twelve units — Pseudocylinder [Fig. A-5(g)]. Variable cylinder height.

8. Fifteen units — Pentadome to enclose a planetary base [Fig. A-5(h)].

9. Thirty units — Pentasphere, large construction hanger, can be peeled away when work is completed [Fig. A-5(i)].

10. Structural doubling, each of the above structures can be doubled in size by making a double unit by packing four units together [Fig. A-5(j)].

Larger structural units can be composed of combinations of several of the nine described above; e.g., tetrahedrons and octahedrons can be used to make a thick plane platform. A large space-filling structure can be made by packing octahedrons together and using trigonal bipyramids to fill the interstices.

The unit [Fig. A-5(h)] consists of two equilateral triangular plates joined by a two-way hinge. The triangle is made from a triangular pipe frame. This frame is covered with an easily removable metal skin which encloses a heat and meteorite protection shield. This triangle is easier to manufacture than a cylinder. The two triangles are joined together by a two-way removable hinge, and they can be made into a rigid rhombus by addition of suitable stiffening rods. Combinations of many rhombi can easily be made into a floor (platform).

To make a larger structure (e.g., icosihedron) one can, by joining four rhombi together [Fig. A-5(i)], a rhombi unit can be made that will be twice the size of the original one.

The sealing of these structures can be accomplished by a heavy-duty tape that is followed by a sprayed chemical that will seal the enclosure. The tape could then be stripped for reusing the units.

Geometrical figures other than equilateral triangles can be employed to construct structures; e.g., hexagons, pentagons, squares, etc. The advantages are as follows:

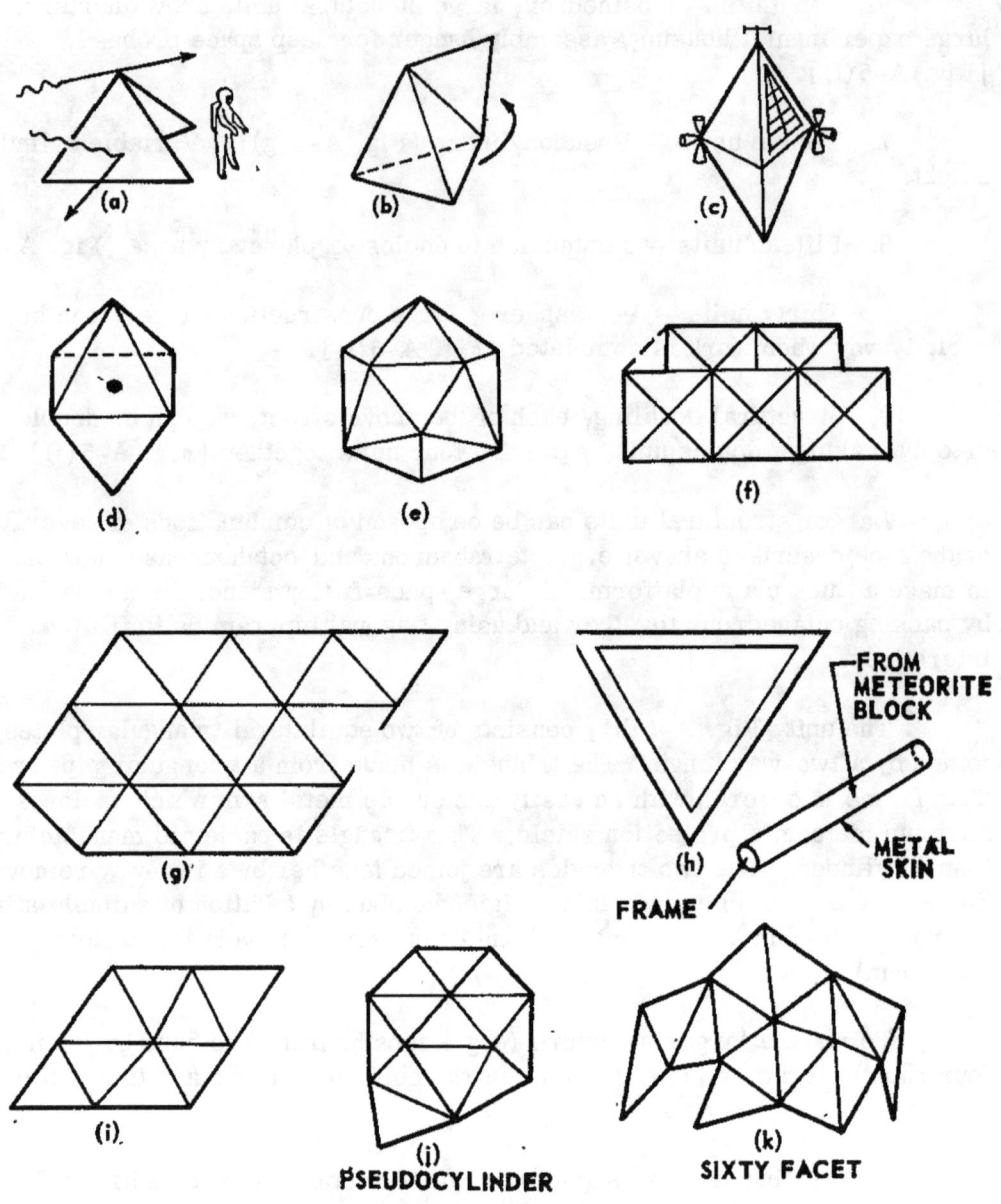

FIGURE A-5. RHOMBIC UNITS FOR CONSTRUCTING A VARIETY OF STRUCTURES

1. Easily manufactured.

2. Reusable.

3. Easily packed in folded form; stackable.

4. A variety of structures with several shapes and sizes can be made.

5. Strong structural unit (triangular brace).

6. Common construction unit; all structures made employing same unit.

7. Interchangeability of units.

This invention contains the application of a new structural unit (a rhombus with a hinged middle joint) to construct a variety of structures of varying shapes and sizes.

Branch No. _____
NASA Case No. _____

NATIONAL AERONAUTICS AND SPACE ADMINISTRATION

Office of Assistant General Counsel for Patents

DISCLOSURE OF INVENTION

This is an important legal document. It should be carefully completed by the inventor(s) and forwarded to the Patent Representative. Two copies of each document are desired.

1. Descriptive Title of Invention

   SPACE SHOE-LOCKING DEVICE

2. Name(s), Title(s), and Home Address(es) of Inventor(s)

   Melvin L. Forthun, Associate Professor, Mechanical Engineering, North Dakota State University, Fargo, North Dakota

3. Name and Address of Employer

   NASA — Auburn Design Group

4. Stage of Development	Date Month/Yr.	Location	Identify persons or records supporting facts stated in 4a-4e
a. First disclosure to others	July 1969	Huntsville	Auburn Design Interim Report
b. First sketch or drawing	"	"	" " " "
c. First written description	"	"	" " " "
d. Completion of first model or full size device	—	—	— — — —
e. First successful operational test	—	—	— — — —

5. List other pertinent notebook entries, photographs, reports, drawings.

   NASA — Auburn Design Program Laboratory Book

6. If the invention was disclosed outside of NASA, identify the individuals, the companies or activities they represent, and the date of disclosure.

7. List any known or contemplated public use, publication, or oral presentation of the invention.

8. Indicate any past, present, or contemplated Government use of the invention.

   For use in AAP program

9.

SPACE SHOE-LOCKING DEVICE

The necessity of using a method to hold a person at a given location within a space vehicle is of primary importance. Locomotion within a zero-g environment is by pulling (using the hands) or by pushing off using the hands, feet, or body. If a person needs to use both hands for an operation, then the rest of the body must be restrained by some locking device; i.e., safety belt or the body wedged into some fixed projection.

Present shoe-locking devices have devices attached to the shoe that require the person to twist his feet; thus, his body position will lock into position. The present method requires a grate-type floor material with a large grate opening. The proposed device will be a smaller-type hook with two hooks placed on each shoe. One hook will be placed on the toe of one shoe and one hook on the heel of the other shoe, then one hook will be placed on the outside of each shoe. The new feature of this invention will allow leverage to be applied through the legs where the legs force outward against each other, fixing the body so that independent motion of the arms and upper trunk of the body is possible. The proposed method will also utilize a floor grate and the length of the shoehook will again depend on the floor-grating design as shown in Figure A-6.

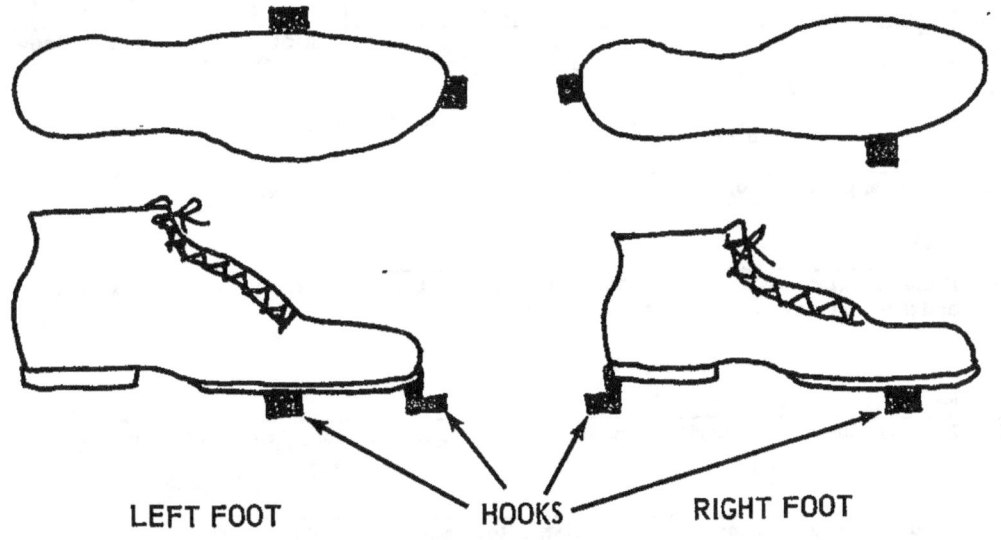

FIGURE A-6. SKETCH OF HOOK LOCATION ON SPACE SHOES, TOP VIEW AND SIDE VIEW

Branch No. _____

NASA Case No. _____

NATIONAL AERONAUTICS AND SPACE ADMINISTRATION

Office of Assistant General Counsel for Patents

DISCLOSURE OF INVENTION

This is an important legal document. It should be carefully completed by the inventor(s) and forwarded to the Patent Representative. Two copies of each document are desired.

1. Descriptive Title of Invention

   FLEXIBLE COUPLING FOR DOCKING TWO VEHICLES IN SPACE

2. Name(s), Title(s), and Home Address(es) of Inventor(s)

   L. Robert Beardsley, Associate Professor of Mechanical Engineering
   Carrier Hall
   University of Mississippi, Oxford, Mississippi 386‾‾

3. Name and Address of Employer

4. Stage of Development	Date Month/Yr.	Location	Identify persons or records supporting facts stated in 4a-4e
a. First disclosure to others	8 Aug 69	Huntsville	Major Kelley and Dr. Brewer Written Disclosure
b. First sketch or drawing			
c. First written description	8 Aug 69	Huntsville	Major Kelley and Dr. Brewer Written Disclosure
d. Completion of first model or full size device			
e. First successful operational test			

5. List other pertinent notebook entries, photographs, reports, drawings.

   None

6. If the invention was disclosed outside of NASA, identify the individuals, the companies or activities they represent, and the date of disclosure.

   .None

7. List any known or contemplated public use, publication, or oral presentation of the invention.

   None

8. Indicate any past, present, or contemplated Government use of the invention.

   None yet

9.

A-20

FLEXIBLE COUPLING FOR DOCKING OF TWO VEHICLES IN SPACE

The purpose of this invention (Fig. A-7) is to provide a means of bringing a docking module or vehicle to a space station and dock it without providing the harmful effects of the inertia and mass of the approaching module to upset the attitude control of the station. Methods currently used and proposed require that the module be firmly and rigidly attached to the station at the point of docking. The disadvantage of using this method is that the mass and the resulting moment of inertia of the module affects the stability of the entire station and also displaces the center of mass of the assembly. This condition makes attitude control more difficult and requires more power to provide a stable attitude of the station.

To avoid the effects of a mass of an attached or docked module, it is proposed that the module have the flexible docking device which will permit engagement with the station but not provide a rigid attachment. This would require a bellows or flexible tube through which the people and equipment would be transferred from the module to the station. This bellows and the attaching devices would remain on the vehicle or module and would engage with the station only when the module or vehicle is docked to the station. With this flexible connection between the station and the smaller mass, the effect of this mass on the mass of the station would be negligible because, instead of forces and moments being transferred between the masses, the forces and moments are absent and the motion of one relative to the other is permitted. This relative motion must be limited to keep the relative position within prescribed limits. The attitude control system of the module or vehicle will be used to maintain a satisfactory relative position or orientation. It is also proposed that the sensors governing the attitude control during the docked period will be connected to both the docked vehicle as well as the station. This relative motion will be sensed by displacement sensors which will keep the relative position of the vehicle and the station within the prescribed working limits. The displacement sensors have the advantage that with an array of sensors all degrees of freedom can be accounted for and these can be used to control the attitude of the smaller vehicle. Also, in general, the attitude of the logistics vehicle or module will not be as stringent as the limits imposed on the station. Thus, less effort is required and less fuel expended in attitude control of the smaller vehicle when it is allowed to deviate from the optimum orientation and position, a condition which could not exist with rigid attachment. By having the sensors on the small vehicle, the attitude control system and its sensors

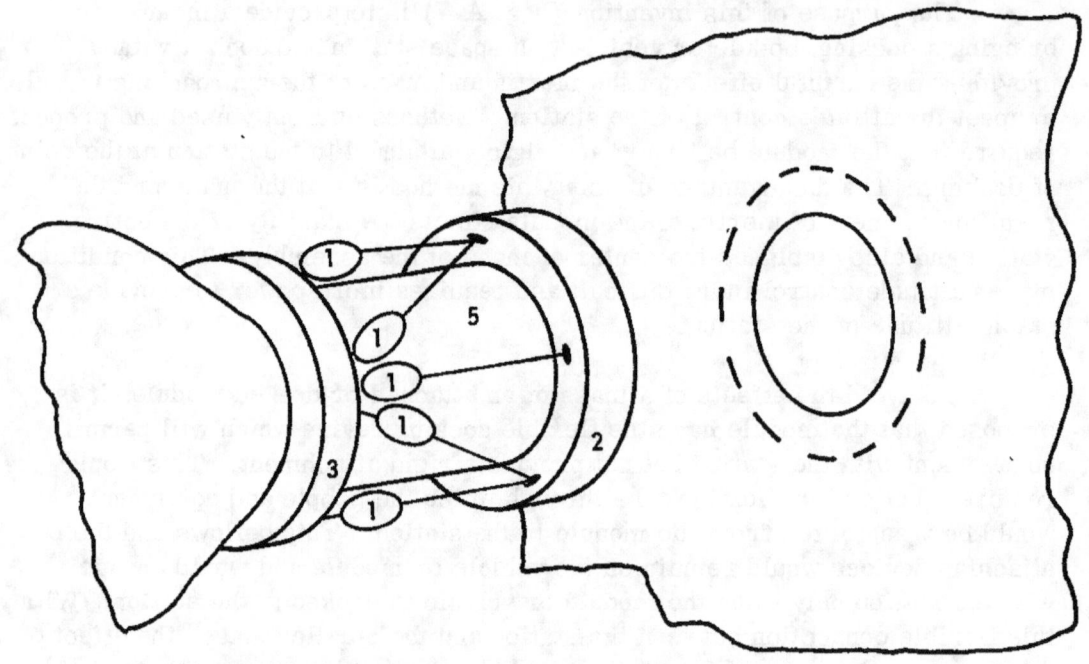

ITEM DESCRIPTION

1. DISPLACEMENT SENSORS
2. CENTERING AND LATCHING ASSEMBLY
3. REFERENCE RING
4. SHUTTLE, MODULE, OR OTHER DOCKING VEHICLE
5. FLEXIBLE BELLOWS OR SIMILAR PRESSURE TIGHT ENCLOSURE THROUGH WHICH MEN AND EQUIPMENT WILL BE TRANSFERRED TO OR FROM VEHICLE
6. DOCKING ADAPTER ON STATION OR LARGER VEHICLE

FIGURE A-7. FLEXIBLE COUPLING FOR DOCKING TWO VEHICLES IN SPACE

are on the same vehicle and no electrical connection is necessary. If the bellows and sensors are applied to the station or the larger mass, the electrical connection would be necessary at the time of docking. This creates a greater difficulty in aligning the module or vehicle with the docking facility on the station. The advantages of the system are:

1. By using the attitude control of the smaller mass to keep the distances and orientations within prescribed limits, no forces are transmitted to the main station and the effect of the small mass on the large one does not exist, because they are not rigidly coupled.

2. By not having the masses coupled, it is easier to keep the attitude of the station within limits and minimize the effective fuel requirements for attitude control.

3. By not actually docking mass to mass, the impact forces between the masses are eliminated; this prevents excessive stresses on the materials of the structure both of the station and of the vehicle. The force involved in coupling the bellows between the masses is reduced by more than an order of magnitude from those values obtainable with rigid coupling.

4. By having the flexible coupling or bellows and the sensors on the smaller vehicle, it has the advantage that if this vehicle is a logistics vehicle or shuttle, these components can be serviced and renewed while the vehicle is on the ground.

The features of this invention believed to be new are:

1. The use of a flexible coupling system.

2. The use of displacement sensors in place of radar or electronics to provide a distance between the vehicles.

3. Not only distance but rotational displacement components can be sensed with a sufficient number of these sensors.

4. Servicing and maintenance of these coupling and sensing devices can be done on the ground for the shuttle vehicle which will dock with this space station.

REFERENCE

A-1. Wuenscher, H. F.: Manufacturing in Space Low and Zero Gravity Manufacturing in Orbit. Manufacturing Technology Unique to Zero-Gravity Environment. NASA 454, Revised October 1967.

APPENDIX B

COMMUNICATION SUBSYSTEM AND CIRCUIT STUDIES

APPENDIX B. COMMUNICATION SUBSYSTEM AND CIRCUIT STUDIES

Backup and Emergency Subsystems

The S-band subsystem operating at 2.2 GHz provides the main backup communications link to earth. This is done in either of two ways: (1) Table B-1 shows that an S-band link can be provided directly to the ground using omni antennas; this could be used with the present network of Apollo-type communications facilities; and (2) Tables B-2, B-3, and B-4 show the various combinations of links using synchronous satellites. All will function to provide continuous communications with CIMC and provide initial aperture alignment for the laser communications link.

The VHF emergency voice link operates using 3 w of transmitted power and essentially onmi antennas. Table B-5 shows a 20-decibel circuit quality margin for this link.

A summary of the frequency allocations utilized external to the STARLAB operation is shown in Table B-6.

Circuit Studies

The basic concept of the circuit quality chart is concerned with estimating radio transmission loss. Since the communications circuits considered are line of sight, the radio transmission loss for each circuit is essentially the inverse square law in optics applied to radio transmissions. Thus, for a one-wavelength separation between nondirective (isotropic) antennas, the free-space loss is 22 decibels, and it increases 6 decibels each time the distance is doubled (Ref. B-1). The free-space transmission ratio at a distance d is given by

$$\frac{P_r}{P_t} = \left(\frac{\lambda}{4\pi d}\right) G_t^2 G_r \qquad (B-1)$$

TABLE B-1. CIRCUIT QUALITY CHART NO. 2, S-BAND
CRITICAL DIRECTLY TO GROUND EMERGENCY

Type of Modulation — PCM/PM
Frequency — 2200 MHz
Communication Range — 900 n. mi. (1)
Transmitter Power — 100 w (12)

Parameter	Nominal Value
Computation of Received Carrier Power	
Transmitter Carrier Power	20 dBw (12)
Transmitter Circuit Losses	-5.5 dB (12)
Transmitting Antenna Gain	-3.0 dB (12)
Polarization Loss	-3.0 dB (12)
Free Space Loss	-163.9 dB (12)
Receiver Antenna Gain	48.4 dB (14)
Receiver Circuit Loss	-0.5 dB (12)
Received Carrier Power	-106.6 dBw
Computation of Desired Carrier Power	
Equivalent Receiver Noise Density	-200.0 dBw/Hz (15)
Receiver Noise Bandwidth	73.0 dB (16)
Equivalent Receiver Noise Power	-127.0 dB
Desired Signal-to-Noise Ratio	10.4 dB (16)
Desired Carrier Power	-116.6 dBw
Circuit Quality Margin	9.9 dB

TABLE B-2. S-BAND CRITICAL CIRCUIT QUALITY CHART NO. 3, SATELLITE TO CIMC

Type of Modulation — PCM/PM	
Frequency — 2.2 GHz	
Communication Range — 24 000 miles (17)	
Transmitter Power — 100 w (12)	

Parameter	Nominal Value
Computation of Received Carrier Power	
Transmitter Carrier Power	20 dBw (12)
Transmitter Circuit Losses	-5.5 (12)
Transmitting Antenna Gain	35.9 (18)
Polarization Loss	-3.0 (12)
Free Space Loss	-191.0 (19)
Receiver Antenna Gain	48.4 (14)
Receiver Circuit Loss	-0.5
Received Carrier Power	-95.7 dBw
Computation of Desired Carrier Power	
Equivalent Receiver Noise Density	-200.0 dBw/Hz (15)
Receiver Noise Bandwidth	73.0 dB (16)
Equivalent Receiver Noise Power	-127.0 dBw
Desired Signal-to-Noise Ratio	10.4 dB (16)
Desired Carrier Power	-116.6 dBw
Circuit Quality Margin	20.9

TABLE B-3. CIRCUIT QUALITY CHART NO. 4, S-BAND CRITICAL, STARLAB TO SATELLITE

Type of Modulation — PCM/PM Frequency — 2.2 GHz Communication Range — 24 000 miles (17) Transmitter Power — 100 w (12)	
Parameter	Nominal Value
<u>Computation of Received Carrier Power</u>	
Transmitter Carrier Power	20 dBw
Transmitter Circuit Losses	-5.5
Transmitting Antenna Gain	35.9
Polarization Loss	-3.0
Free Space Loss	91.0
Receiver Antenna Gain	35.9
Receiver Circuit Loss	-0.5
Received Carrier Power	-108.2
<u>Computation of Desired Carrier Power</u>	
Equivalent Receiver Noise Density	-200.0 dBw/Hz (15)
Receiver Noise Bandwidth	73.0 dB (16)
Equivalent Receiver Noise Power	-127.0 dBw
Desired Signal-to-Noise Ratio	10.4 dB (16)
Desired Carrier Power	-116.6 dBw
Circuit Quality Margin	8.4 dB

TABLE B-4. CIRCUIT QUALITY CHART NO. 5, S-BAND CRITICAL, SATELLITE TO SATELLITE

\multicolumn{2}{c}{Type of Modulation — PCM/PM}
\multicolumn{2}{c}{Frequency — 2.2 GHz}

Parameter	Nominal Value
Type of Modulation — PCM/PM	
Frequency — 2.2 GHz	
Communication Range — 34 600 miles (20)	
Transmitter Power — 100 w (12)	
Computation of Received Carrier Power	
Transmitter Carrier Power	20 dBw (12)
Transmitter Circuit Losses	-5.5 (12)
Transmitting Antenna Gain	35.9 (18)
Polarization Loss	-3.0 (12)
Free Space Loss	-194.0 (21)
Receiver Antenna Gain	35.9 (18)
Receiver Circuit Loss	-0.5 (12)
Received Carrier Power	-111.2
Computation of Desired Carrier Power	
Equivalent Receiver Noise Density	-200.0 dBw/Hz (15)
Receiver Noise Bandwidth	73.0 dB (16)
Equivalent Receiver Noise Power	-127.0 dBw
Desired Signal-to-Noise Ratio	10.4 dB (16)
Desired Carrier Power	-116.6 dBw
Circuit Quality Margin	5.4

TABLE B-5. CIRCUIT QUALITY CHART NO. 1, VHF VOICE,
STARLAB TO GROUND (EMERGENCY)

Type of Modulation — AM Frequency — 296.8 MHz Communication Range — 900 n. mi. (1) Transmitter Power — 3 w (2)	
Parameter	Nominal Value
Computation of Received Carrier Power	
Transmitter Carrier Power	4.8 dBw (2)
Transmitter Circuit Losses	-2.3 dB (3)
Transmitting Antenna Gain	-4.0 dB (4)
Polarization Loss	-3.0 dB (5)
Free Space Loss	-146.5 dB (6)
Receiver Antenna Gain	18.0 dB (7)
Receiver Circuit Loss	-0.2 dB (8)
Received Carrier Power	-132.2 dBw
Equivalent Receiver Noise Density	-199.4 dBw/Hz (9)
Desired Peak Signal-to-rms-Noise Power Density	44.5 dB (10)
Modulation Characteristic	1.4 dB (11)
Desired Carrier Power	-109.0 dBw (12)
Circuit Quality Margin	20.3 dB

TABLE B-6. CHANNEL FREQUENCY ALLOCATIONS

Lasers for Primary STARLAB — CIMC Communications

S-Band for Secondary STARLAB — CIMC Communications

C-Band for STARLAB — Ancillary Modules Communications

X-Band Radar for Tracing and Navigation of Ancillary Modules from STARLAB

VHF+HF Omni for Emergency Voice

---

where P_r, P_t = received power and radiated power, respectively, measured in the same units. λ = wavelength, in the same units as d, G_t (or G_r) = power gain of transmitting (or receiving antenna).

By taking the log of equation (B-1) and introducing several significant factors which are real losses in the communications circuit, we obtain equation (B-2) which calculates the received carrier power P_r,

$$P_r \text{(dBw)} = P_t \text{(dBw)} - L_t \text{(dB)} + G_t \text{(dB)}$$

$$- L_p \text{(dB)} - L_{fs} \text{(dB)} + G_r \text{(dB)} - L_r \text{(dB)} \qquad \text{(B-2)}$$

where

P_t(dBw) = the transmitted carrier, equation (B-2), power in decibels above a watt

L_t(dB) = the losses in the transmitting circuit

G_t(dB) = the power gain of the transmitting antenna

L_p(dB) = the polarization loss (a mismatch of transmitting and receiving antenna polarization)

L_{fs} (dB) = the free space loss is $37.8 + 20 \log d + 20 \log f$ where f is in MHz and d is in n. mi.

G_r (dB) = the power gain of the receiving antenna.

L_r (dB) = the losses in the receiving circuit.

For the computation of desired carrier power, the receiver equivalent noise spectral density is computed from

$$N_o = 10 \log k + 10 \log T + F,$$

where

k = Boltzmanns constant of 1.38×10^{-23} per °K

T = system temperature in °K

F = preamplifier noise figure.

The receiver noise bandwidth is determined by the bandwidth of the signal-to-noise ratio and by the demodulation technique and the allowable error rates, fidelity required, etc.

The circuit quality margin is then given by the difference between the received carrier power and the required carrier power. Where the circuit quality margin is positive (the received carrier power is greater than the desired carrier power), the circuit will provide an adequate communications link. If the circuit quality margin is negative, corrective steps must be taken in the communications circuit design, such as increasing the antenna gains, or increasing the power transmitted, or decreasing the receiver (preamplifier) noise figure, etc. In this sense, the circuit quality chart can be used as a design tool even though it is an analysis technique. The primary purposes of the circuit quality charts in this report are to show that the proposed communications systems are feasible and that the state-of-the-art will provide the communications links discussed earlier in this appendix. The following notes are provided to completely describe the source of information and computations.

1. Communications range of 900 n. mi. is near the maximum range.

2. Assumed value but equivalent to VHF-AM Gemini transmitter-receiver equipment.

3. Transmitter circuit loss includes 0.5-decibel multiplexer loss, 20 feet of 8.3 decibel per 100 feet of cable loss, 0.05 decibel per connector loss.

4. Assumed value — essentially a practical isotropic antenna.

5. Assume circular polarization on ground, linear on STARLAB.

6. Free space loss is $37.8 + 20 \log d + 20 \log f$ where f is in MHz (296.8) and d is in n. mi. (900).

7. A quadhelix antenna is assumed on ground. The gain for a typical tacco quadhelix antenna in the frequency range 225 MHz to 300 MHz.

8. Typical Gemini ground equipment, 0.2-decibel receiver line loss, noise figure 4.5 decibel, transmitter power 100 w, transmitter line loss 2.5 decibels.

9. Receiver noise density is $-228.6 + 10 \log Ts$ where

$$Ts = \frac{Ta}{L} + \left(1 - \frac{1}{L}\right) 290 + (N - 1) 290$$

and $Ta = 300 \lambda^2$, L is receiver circuit loss ratio, N is receiver noise figure. (See 8. above for pertinent numbers.)

10. The threshold of perceptibility for a rather difficult-connected discourse and which corresponds to approximately 75 percent monosyllabic phonetically balanced work intelligibility can be determined from

$$\frac{S_p}{dN/df} = 50 + \Sigma C_n$$

where S_p is the peak instantaneous audio signal power in w, dN/df is the noise power spectral density in w/Hz and C_n are correction factors for speech processing. The correction factor for -12 decibels of peak audio clipping in AM system is 9 decibels. The audio bandwidth of the spacecraft transmitter is from 300 to 3000 Hz (assumed value but equivalent to Gemini and Apollo equipment). The 300-Hz cutoff requires a correction factor of 3.5 decibels while the upper cutoff has little effect. Thus, the required peak signal-to-noise power spectral density requirement becomes 44.5 decibels (Ref. B-2).

11. For amplitude modulation, the required rms carrier power to prediction noise spectral density is $50 + 20 \log 1/M + \Sigma C_m$, M is assumed to be 0.85 (Ref. B-2).

12. Assumed value.

13. Distance is 900 n.mi., frequency 2200 MHz. (Note 6. above gives the formula for free space loss.)

14. Antenna gain is given by $20 \log f + 20 \log D - 52.6$ where f is in MHz and D is diameter in feet. f is 2200 MHz, D is a 60-foot reflector, with an 0.54 illumination factor. Beam angle is $7 \times 10^4/(fD)$.

15. The equivalent noise spectral density as seen by the receiver is computed (in decibels) from $N_o = 10 \log K = 10 \log T + F$, where K is Boltzmann's constant (1.38×10^{-23} joules per °K, T is system temperature (290°K), and F is preamplifier noise figure (assumed 4 decibels).

16. Bandwidth = $10 \log \Delta w$ where Δw = 20 MHz. For coherent PSK, 10.4 decibels is the required signal-to-noise ratio in the 20 megabit rate bandwidth for 10^{-6} bit error rate (Ref. B-3).

17. 24 000 miles was used as the approximate maximum range for computation.

18. D is 12 feet. (See note 14. above for formula.)

19. d = 24 000 miles = 20 800 n.mi. (See note 6. above for formula.)

20. d = 34 000 miles = 29 600 n.mi. is used as the distance between satellites.

21. d = 29 600 from note 20. above. (See note 6. above for formula.)

REFERENCES

B-1. Jasik, Henry: Antenna Engineering Handbook. McGraw-Hill Book Co., Section 33, by Kenneth Bullington, 1961.

B-2. Craiglow, R. L.: Power Requirements for Speech Communications Systems. IRE Transactions on Audio, November-December 1961, pp. 186-190.

B-3. Lawton, John G.: Comparison of Binary Data Transmission Systems. IRE 2nd National Conference Proceedings on Military Electronics, 16-18 June 1958, pp. 54-61.

APPENDIX C

ALTERNATIVE CONSIDERATIONS

STARLAB: AN ORBITING SPACE TECHNOLOGY APPLICATIONS AND
RESEARCH LABORATORY

Auburn University
Auburn, Alabama

August 1969

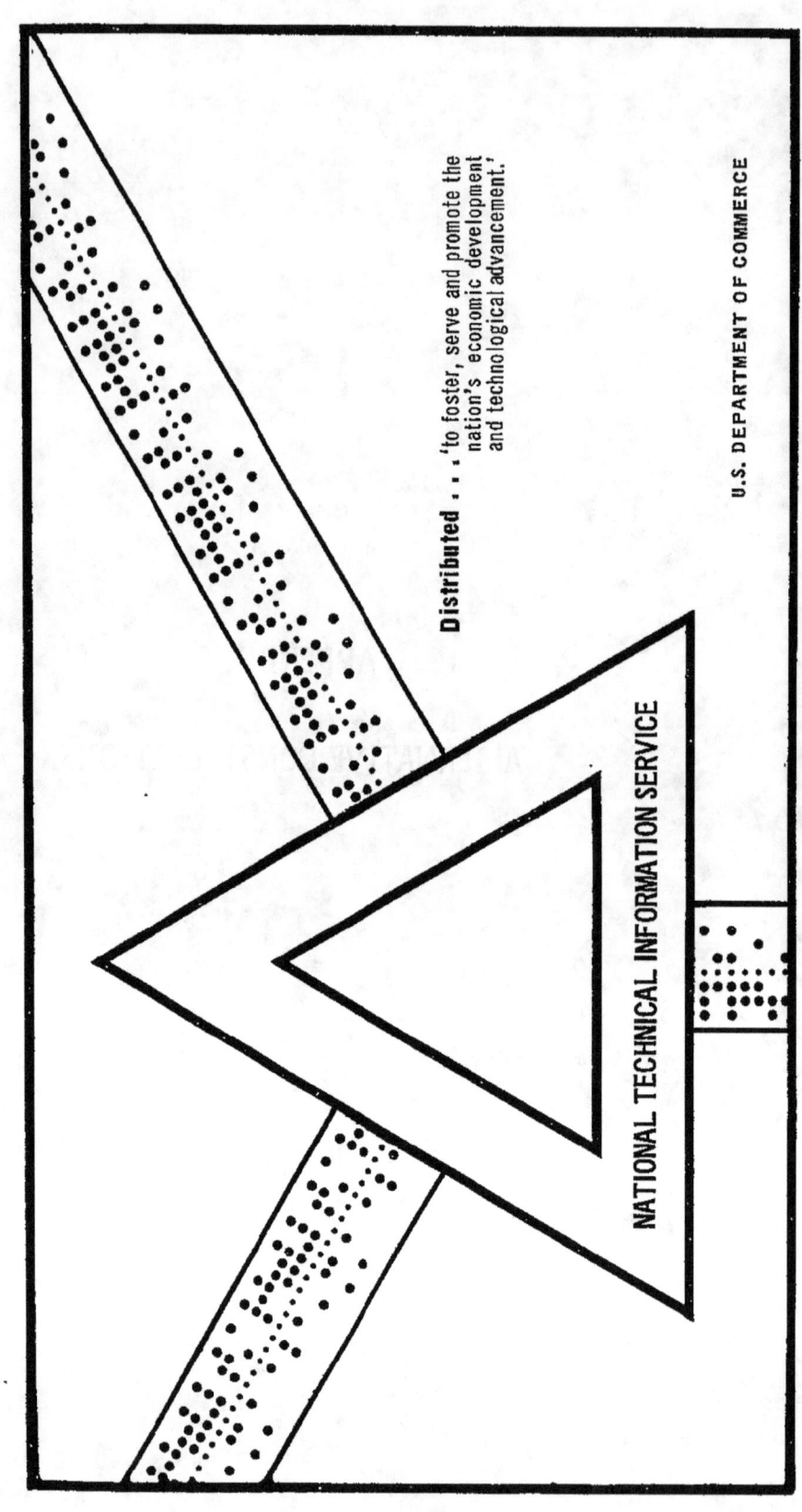

APPENDIX C. ALTERNATIVE CONSIDERATIONS

CONFIGURATION ALTERNATIVES

Nuclear power systems have a severe impact on configuration alternatives. Early in the investigations, it was thought that the power system weight could be kept well under 50 000 pounds. Two alternatives were considered as discussed in the paragraphs below.

Free-Flying Microwave Power Station

Safety and ease of replacement were the attractive features of microwave power transmission from a free-flying power station; it would fly at least a mile. This station could be launched initially and replaced using shuttle transportation. Power would be transmitted between two 30-foot antennas. This system simplified the external configuration of STARLAB (Fig. C-1). One less S-V launch is required.

Integral Power Source

This is the usual method of deploying a nuclear power plant. Isolation from radiation is achieved by placing the power source a distance from the center of operations so that a partial or spot shield may be employed to create a conical zone of safety. While total reactor weight remained well under 50 000 pounds, this kind of unit could be deployed from the critical services module by using a series of telescoping structural elements (Fig. C-1) deployed and (Fig. C-2) folded in launch configuration. Guy-wire tension members are an optional alternative to increasing material in the central beam assembly. Separations achieved using this configuration are about three times those proposed for current programs that are not expected to function as long as the 10-year STARLAB lifetime. These separations would be achieved at the expense of design simplicity. This system is still a state-of-the-art approach whereas the microwave power station concept requires further development of transmission efficiencies to be a really attractive alternative.

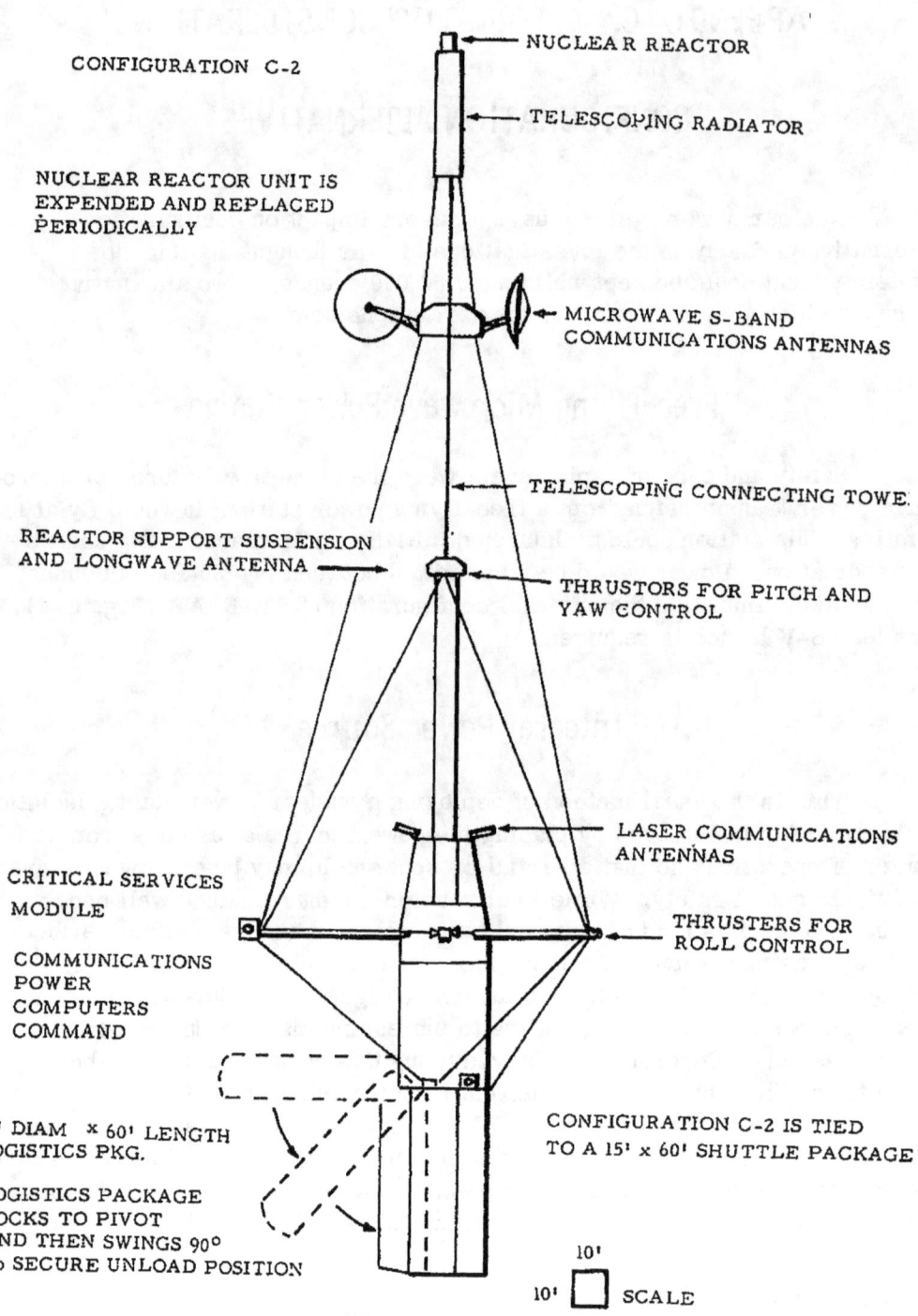

FIGURE C-1. ALTERNATIVE CONSIDERATIONS — DEPLOYMENT

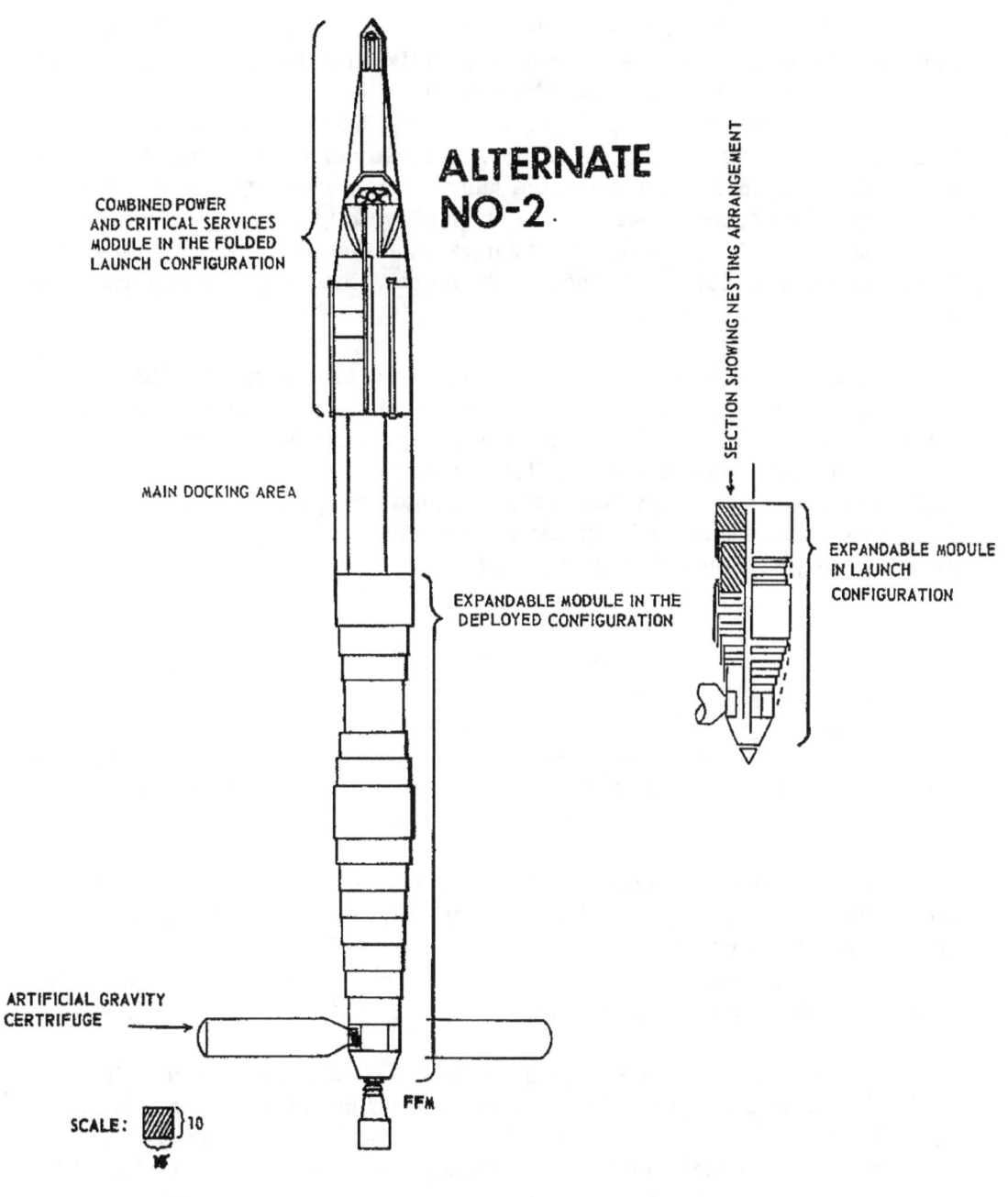

FIGURE C-2. ALTERNATIVE CONSIDERATIONS — LAUNCH

Final Choice of Power Unit

Both of the above concepts were abandoned when the power demands increased to the point where there was little likelihood of having a power unit weighing under 50 000 pounds developed in time to be used. The power unit weight was approaching 70.000 pounds and would necessitate an additional S-V launch. This ILM unit would have to be replaced periodically and filled with expendable equipment. Considerable launch weight capacity would still be available for additional shielding so design simplicity could be achieved by eliminating complicated telescoping deployment systems. A separation of 80 feet or more would fit into the S-V envelope without any deployment extensions.

Additional usable volume could be made available to STARLAB as power units are replaced. This would be accomplished by designing separately the expended portions of the power module to jettison, leaving the bulk of the structure (used to separate the radiation source) behind. This portion could be designed for easy integration into the usable occupied portion of STARLAB. The new structure launched with the replacement power system would dock to the remaining portion of the former unit.

Configuration C-2

Figures C-1 and C-2 are a composite of a number of alternatives considered early in the design trade-off phase. These features have advantages that were sacrificed, usually to achieve design simplicity and functional reliability.

Placing reaction control units at a distance from the major inertial axes about which they must control motions would substantially reduce propellant requirements. Contamination of the local atmosphere by propellants would be not only reduced, but the concentrations would be moved farther from the STARLAB vehicle equipment sites that might suffer from contamination.

The logistics and docking facility in this configuration differs from the final proposal in that the features depend on the dimensions of the shuttle payload. This design is for a small diameter 15- by 60-foot logistics package. As in the final proposal, the logistics package (not the shuttle vehicle) will dock to STARLAB. Docking area structure is made up of three specially designed shuttle payloads in a close-pack parallel installation (Fig. C-3). Each would be a separate pressure hull. Logistics packages would dock to

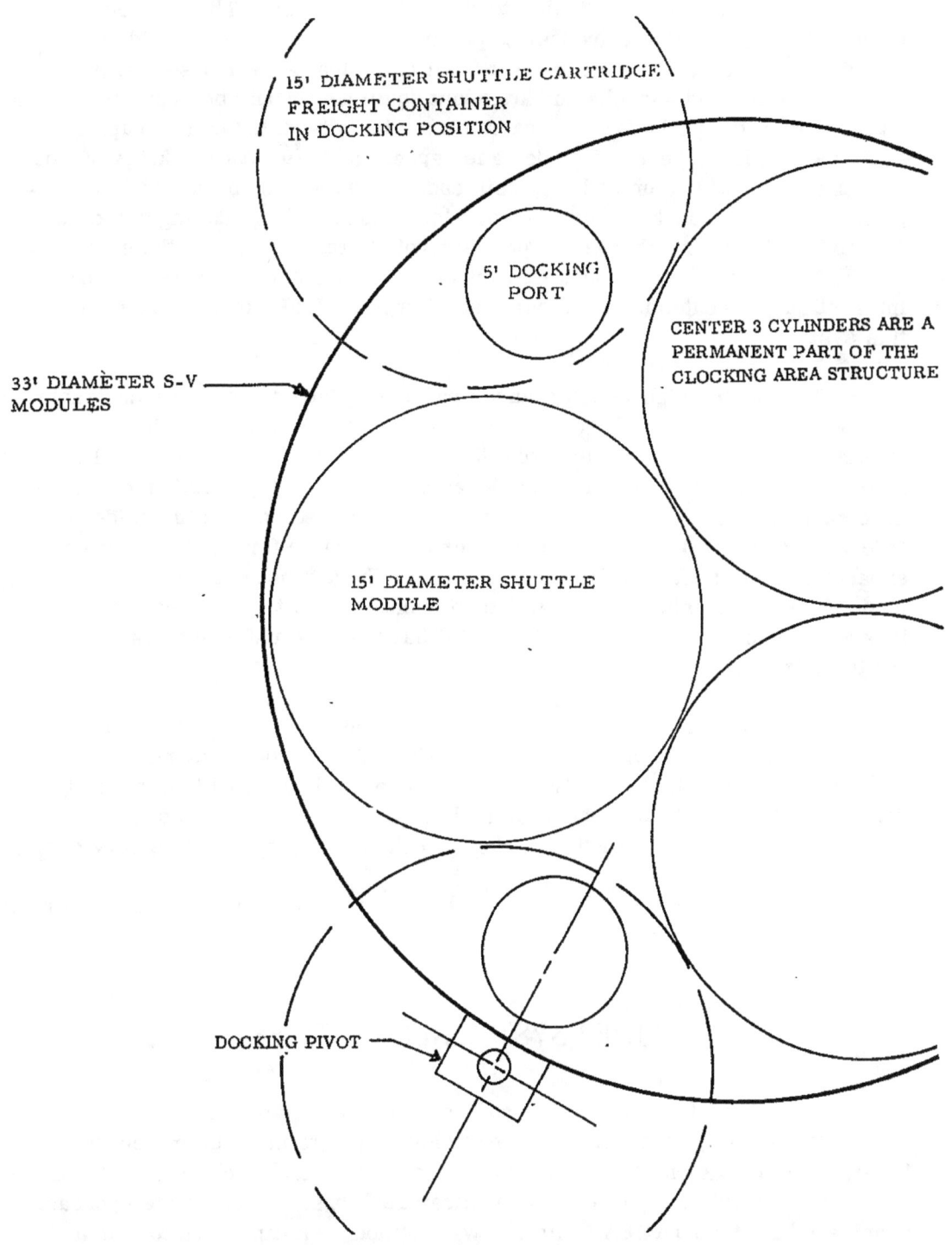

FIGURE C-3. CLOSE-PACK PARALLEL INSTALLATION OF SHUTTLE PAYLOADS

a docking pivot axis perpendicular to the STARLAB axis. This is a comparatively easy docking maneuver to perform. Then the pivot would swing the logistics package to a close-pack position so that it is in a secure parallel attitude. The docking probe and attendant equipment would be centrally located on the axis of the package and separate from the off-center hatch coupling equipment, allowing each to be designed specifically for one task. A 60-foot-long package could be unloaded at both ends and at points on the side. Close-packed paralleled docking positions are less susceptible to damage by collision, easier to unload, and change attitude-control characteristics of the composite STARLAB vehicle less by docking very close to the center of mass. Three internal supply routes near the outer shell of STARLAB could be used with this system.

Expandable high-density launch packages place a large pressurizable volume in orbit at a small transportation cost. The lower half, shown in Figure C-2, has one such unit providing 2-1/2 times the volume of an ILM module whereas both require one S-V launch. This collapsible "drinking cup" concept (Figs. A-2, A-3, and C-4) would be deployed on suitable guides or tracks (Fig. C-5) using internal pressure as a motive force. The ratio of expanded-to-collapsed volume is a function of the amount of integral equipment launched with the unit. Nonintegral equipment may be transported using inexpensive shuttle payloads, but in-orbit hardware transfer and installation would be required.

Artificial gravity for experimentation is available in much greater quantities with this configuration. A centrifuge hub would be launched integral with the expandable unit. The centrifuge spoke gondolas would be specially designed shuttle payloads docked to the hub (Fig. C-2). Continuous access to the centrifuge is not provided, and the wheel must be stopped to be serviced.

Artificial gravity (discussed earlier) for a human habitat was eliminated early in the investigations.

POWER SYSTEM ALTERNATIVES

Current state-of-the-art of nuclear reactors for orbiting laboratory power systems requires that consideration be given to methods of separating the power source from the vehicle even through advanced technology may prove the reliability and safety of onboard nuclear reactors. An alternate approach considered for the STARLAB mission was to mount multiple reactors in a

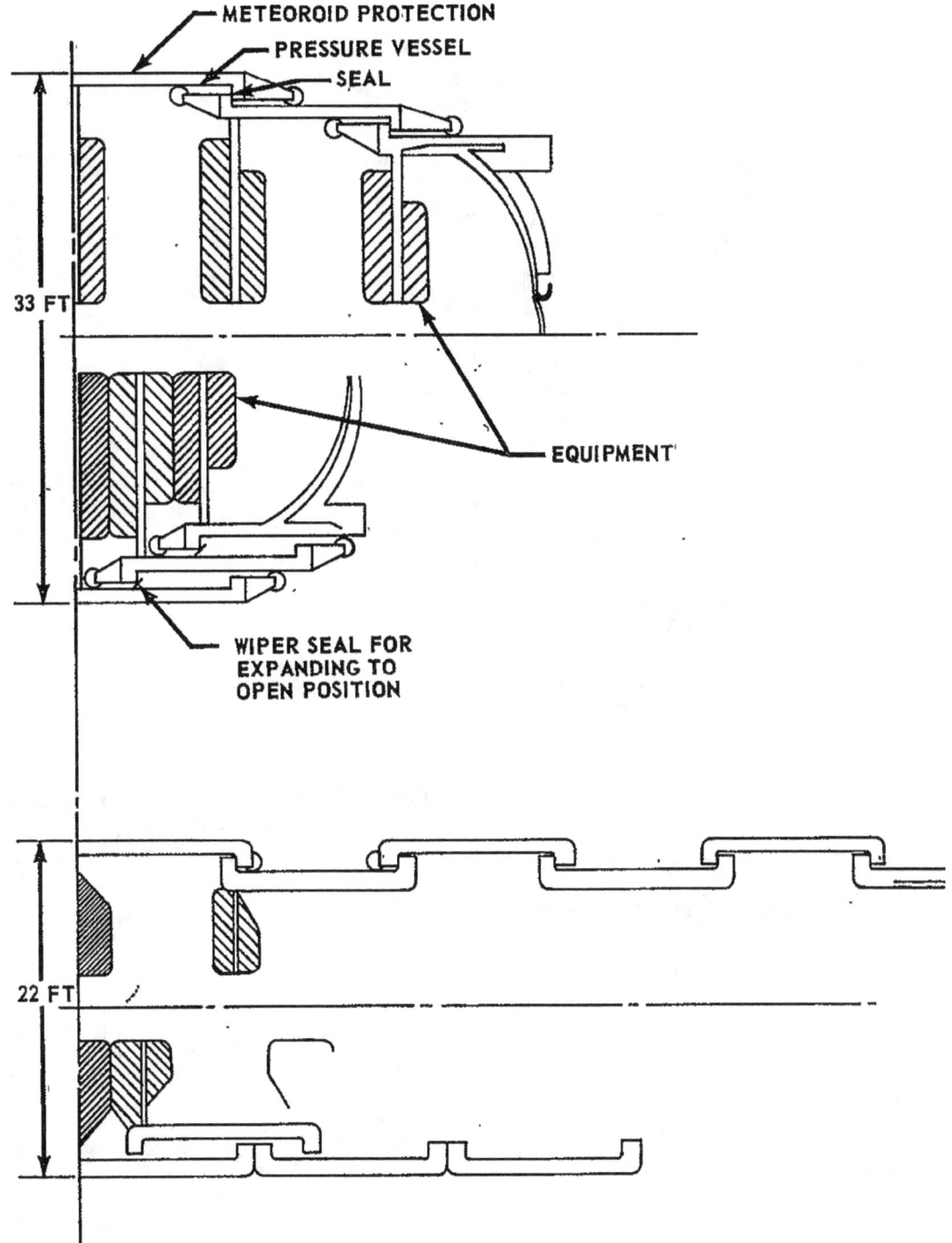

FIGURE C-4. DEPLOYMENT SCHEME — "DRINKING CUP" CONCEPT

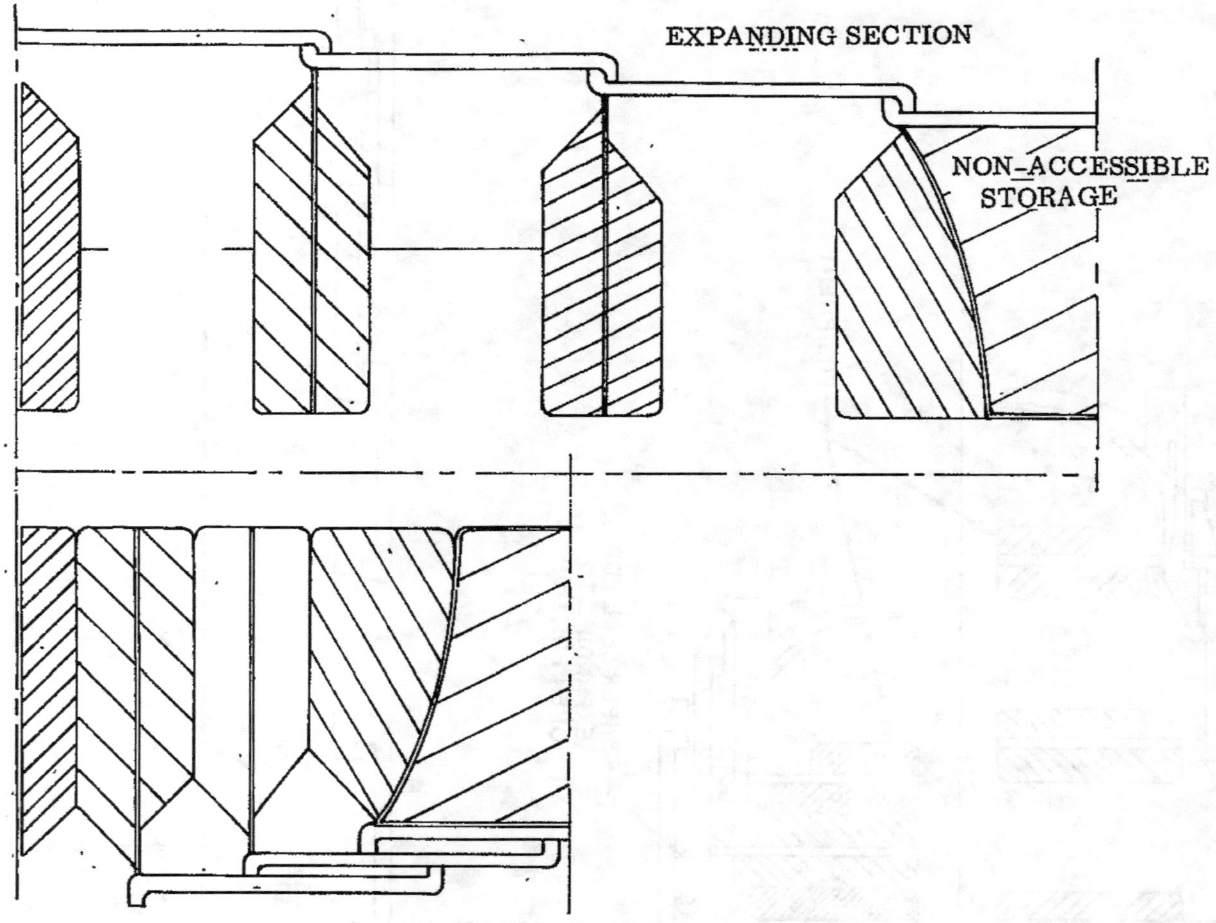

FIGURE C-5. DETAILS OF DEPLOYMENT SCHEME FOR "DRINKING CUP" CONCEPT

free-flying module (Fig. C-6) and transient power by microwave transmission. This could be a serious alternative if large power system converters fail to materialize in the near future. It would become necessary to use numerous smaller systems to supply the large amount of power required in an orbiting laboratory. The use of modular power relieves the laboratory vehicle of the problem of providing radiator surface area for the power system. This could be a formidable area for large systems. For instance, a 100-kW$_e$ power system requires approximately 5000 feet. With a remote free-flying module, the power system would not be an integral part of the laboratory vehicle; hence, extravehicular activity would be possible in the vicinity of the space vehicle. This is not possible when using an integral configuration. Another advantage is that radiation levels inside the spacecraft, as well as in adjoining vehicles and modules, would be minimized.

For the proposed launch date, electrical power could be transmitted dc to dc by microwave techniques with an efficiency of 47 percent. With advanced technology, future efficiencies of 70 percent are possible. When space travel is not power limited, this method of power transmission may become important. The power module could be inserted into orbit and stationed in front of or behind the orbiting laboratory. Power transmission at 1 km using a 3-cm wavelength signal requires an antenna diameter of 10 meters. The antenna could be electrically aimed and interlocked for maximum power transmission.

Two methods of supplying power were investigated; direct and indirect. The simplest in terms of orbital flight is direct. In this configuration, the main nuclear power source could be physically attached to the space vehicle by means of a long structural boom with necessary reinforcement structures. Conductors would then be routed directly to the main power distribution system. The inherent dangers of such a system are obvious. First, with the extension of a boom to 200 feet, it is possible that some EVA will be required; this is not a desirable condition. Another problem arises in attitude control. If the space vehicle executes a violent maneuver because of a malfunction in a control system or natural phenomena, it is unlikely that a mass of 40 000 pounds, concentrated at the end of a 200-foot support member, could be safely retained. It is more likely that the supporting structure would fail with a resulting form of major damage inflicted on the space vehicle. Another objectionable feature is presented by the consideration of maintenance. It should be expected that continuous operation of the power source will lead to a major breakdown within 2 years. The replacement of the faulty system could require EVA with possible exposure to hazardous radiation.

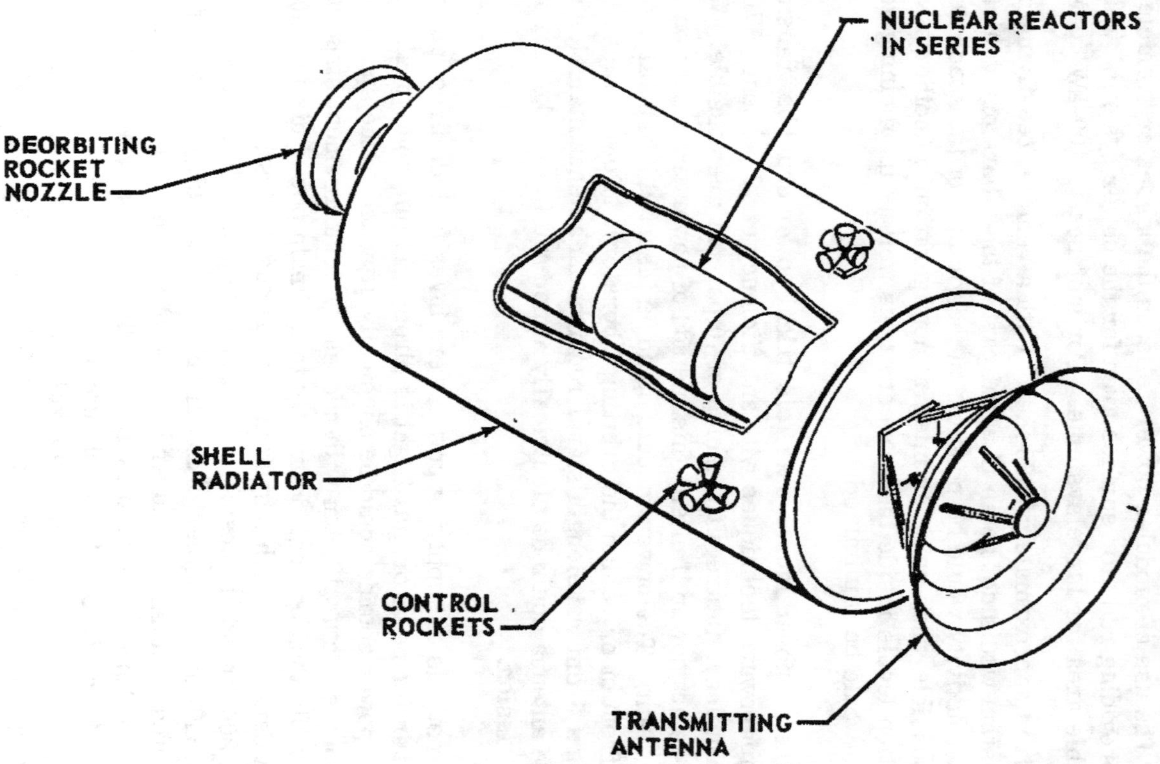

FIGURE C-6. POWER MODULE

The alternative approach is to orbit a self-contained nuclear-powered system on the same orbital path as the manned laboratory. Power would then be transmitted from the supply system to the space station by means of microwave techniques. The presently demonstrated overall efficiency for dc input/output is 26.2 percent. It is expected that with additional development, this efficiency could be increased.

The benefits of such a system are worth serious consideration. For instance, the space station could be placed in orbit, solar panels deployed for power, and the system made ready for habitation. In the future, the nuclear power source could be orbited into position or transported by means of the predicted space shuttle to a holding position up to 1 mile from the space station. The module would have to be fitted with a propulsion and guidance system for station keeping. If a dangerous condition should occur, the vehicle could be moved rapidly away from the space station for safety reasons or it could be sent into a storage orbit. Replacement of this type system is relatively simple. A burned-out power system could be propelled away from the space station and replaced by a new unit without the necessity of EVA.

REFERENCE

C-1. Robinson, W. J.: The Feasibility of Wireless Power Transmission for an Orbiting Astronomical Station. NASA TM X-53806, May 1969.

www.ingramcontent.com/pod-product-compliance
Lightning Source LLC
Chambersburg PA
CBHW081716170526
45167CB00009B/3601